THE
BUFFALO
BOOK

THE BUFFALO

SWALLOW PRESS
OHIO UNIVERSITY PRESS

BOOK
The Full Saga of the American Animal

DAVID A. DARY

97

5 7 9 10 8 6

Swallow Press/Ohio University Press books are printed on acid-free paper. ∞

Library of Congress Cataloging-in-Publication Data

Dary, David.
The buffalo book : the full saga of the American animal / David A. Dary.
p.   cm.   "New edition"—Acknowledgements.   "Sage books."
Bibliography: p.   Includes index.
ISBN 0-8040-0653-9 cloth   ISBN 0-8040-0931-7 pbk.
1. Bison, American—North America—History.   2. Bison, American—
North America—Folklore.   3. West (U.S.)—History.   I. Title.
QL737.U53D37   1989   599.73′58—dc20   89-35104

Endpaper photos: Dot Island, buffalo bull, considered largest living buffalo when picture made in 1905 at Yellowstone National Park. Courtesy Colorado State Historical Society. Mounted buffalo skeleton. Courtesy Smithsonian Institution.

Title page photo: part of Pawnee Bill's buffalo herd early in this century. Courtesy Oklahoma Historical Society. (Original photo was retouched for clarity, which accounts for its somewhat "painted" appearance.)

Cartoon on page 297 courtesy Mike Peters of the *Dayton Daily News*.

To

the memory of

TED YATES

who loved the West and the
buffalo. He died too young.

Contents

Illustrations

Tables

Introduction

PERHAPS no animal in the history of any nation has ever played a more important role than the American buffalo.

A little more than a hundred years ago, within a few miles of where I was born, near the junction of the Big Blue and the Kansas, thousands of wild buffalo grazed in the river valleys and upon the rich bluestem prairies. These prairies—the Flint Hills—stretch for nearly 300 miles from north of Manhattan, Kansas, south into Oklahoma, where they become the Osage Hills. The late Dan D. Casement, Kansas rancher, once described these prairies as "Mother Nature's round, undulating breasts, soft and warm in the sunshine, restfully inviting and rich in the promise of nurture."[1]

Today the Flint Hills are as beautiful as ever. They are still much as they must have been 100 or even 1,000 years ago. Most of the land is still virgin, having never been tilled. The jackrabbit, bobcat, coyote, and rattlesnake still reside on the land as do many different kinds of birds and increasing numbers of deer. But cattle have replaced the buffalo. The white man's buffalo, as the Indian called cattle, graze on the lush bluestem grasses, sometimes knee deep or higher, which were once claimed by the shaggy monarch of the plains.

The wild buffalo is gone.

The American buffalo *was* the most important animal on the western plains and prairies. He even surpassed the mustang and the longhorn in importance. As a social factor, the buffalo's influence on Indian and white man alike was tremendous. In America, the economic and social impact of the buffalo exceeded that of the beaver and the whale.

Yet no complete scientific study of the buffalo was made during those days when the animal roamed the plains in vast numbers. Few

1

book-educated men ventured westward then. Most of those who did were interested only in killing the buffalo for sport. As the late J. Frank Dobie once observed,

> Nearly all men who got out on the plains were "wrathy to kill" buffaloes above all else. Mountain men, emigrants crossing the plains, Santa Fe traders, railroad builders, Indian fighters, settlers on the edge of the plains, European sportsmen, all slaughtered and slew. Some observed, but the average American hunter's observations on game animals are about as illuminating as the trophy-stuffed den of a rich oilman or the lockers of a packing house.[2]

The educated remained chair-bound in the East, content to accept as final the stories, legends, and tales of the frontiersmen. Unfortunately most frontiersmen lacked book learning and scientific training. Often their stories, sometimes referred to as "oldtimer history," contained more fiction than fact. Their history usually asserts that something either *always* happens or *never* happens. There was seldom a *sometimes*. Frank Gilbert Roe, the buffalo historian, called them the "always-and-never" witnesses.

Actually the observations of early fur traders, those before 1840, are, on the whole, more factual concerning the buffalo than are those of the immigrant settlers who came later. Those European immigrants who came to America and settled on the plains experienced tremendous change; the New World was in fact a world full of marvels. What seemed like a miracle one day was surpassed the next by things soon found to be everyday occurrences.

Had the more educated arrived on the plains and prairies with the early settlers, we probably would have a more accurate picture of the wild plains buffalo of a century or more ago. But it is doubtful whether they could have saved the animal from its eventual fate. The killing of thousands upon thousands of buffalo was inevitable.

The Indian blocked the way west for the white man. The countless herds of buffalo roaming the western plains and prairies provided food, clothing, and other necessities for the Plains Indians. So long as there were buffalo, the Indians maintained their strength, courage, and freedom. But without their buffalo, the Plains Indians were conquered and the land taken by ranchers and settlers alike.

Even if there had been no Indians on the land, the buffalo would still have been in the way of the white men. Buffalo ate the free grass that the ranchers wanted for their cattle, and the shaggy animals roamed the land the settlers wanted to till to raise crops. The buffalo had to be cleared from the land, and they were.

If it had not been for a few *thinking* North Americans late in the 19th century and early in this century, the buffalo long ago would have gone the way of the passenger pigeon. But because a handful of men captured a few wild buffalo and raised them in captivity during the 1880s and 1890s, the buffalo was saved.

The governments in Washington and Ottawa and quite a few private citizens in both countries preserve the buffalo today. Yet the vast majority of people in both countries look on the buffalo as an oddity. Even the scientific world gives more attention to grizzly bears or jackrabbits or blackbirds than it does to buffalo.

The purpose in writing this book was to gather between two covers the best information relating to the American buffalo from the time the animal first arrived in North America until today. This account is factual, and where possible conclusions have been drawn. But much lore and legend about the buffalo have also been included; these are part of the animal's story, and without them the story is not complete.

What we call the American buffalo is, of course, not a buffalo at all. It is a *Bison,* which is related to the European *Wisent.* The scientific world insists that the word "buffalo" should be used only to describe the African buffalo or the water buffalo of Asia. But for more than 150 years this animal, the bison, has been called a "buffalo." To millions of persons he is a buffalo, and on these pages this is what I call him.

DAVID A. DARY

Along the Kaw
Lawrence, Kansas

I

The First Buffalo

The history of any land begins with nature, and all histories must end with nature.

J. Frank Dobie 1943

It was September 1925. The morning was hot and muggy. Jim O'Connel hated to leave the coolness of his ranch house, but he had work to do. As he stepped outside, the bright sunlight almost blinded him. He squinted and stopped until his eyes adjusted to the brightness. Then he crossed the yard to the corral where the horses were kept. He saddled one of the animals, mounted, and began to ride out over his ranch.

The first thing O'Connel wanted to do was check his cattle. A bad storm with heavy rain and lightning had swept across his land during the night. Maybe some of his cattle had been injured or drowned in the creek or struck by lightning. As he headed toward Cottonwood Creek, O'Connel hoped none of these things had happened. The small stream was not far ahead; it twisted and curved through much of the ranch, about twenty-five miles southeast of Coldwater, Kansas.

As he reached the stream, he could tell that the water had been high. Brush, debris, and matted grass lined the creek banks, sure signs of high water. But the stream had receded since the storm and was only a few feet deep, almost normal.

As O'Connel followed a bend in the stream, he spotted something sticking out of the sandy creek bank. He had seen the object before and thought it was a dead root from an old cottonwood tree nearby. There were many big and old cottonwoods along the creek, but somehow the object looked different now. The ground had been washed out around it. Something told him it was not a root. Getting off his horse, O'Connel knelt down and grasped the object. It felt bony, not

4

like a root at all. It would not budge when he tugged at it. So with his hands, he dug into the damp earth and freed the object. It was part of a buffalo skull.

At first O'Connel was disappointed. He had found many buffalo skulls on his ranch, remnants of those days sixty years earlier when thousands, perhaps millions, of buffalo had roamed the Kansas prairies and plains. But as he studied the skull, he knew it was different. It was larger than any he had found before, and it looked a lot older. The horn cores—broken into several pieces—were long. When he gathered up all the pieces he could find and laid them out on the damp ground, end to end, they stretched more than six feet from tip to tip. "What a horn spread!" he thought to himself.

It was obvious to O'Connel that the remains were not those of a modern day buffalo. The skull was the biggest he had ever seen and the horn spread—why, it was longer than the spread of most Texas longhorn cattle. It did not take O'Connel long to conclude that he had made an unusual discovery. Very carefully he carried his find back to the ranch house. Later he saw to it that the remains were taken to the University of Kansas at Lawrence, where people smarter than he about such things might be able to tell him more. At the university, Handel T. Martin, a paleontologist, was excited by the find. It was not long before Professor Martin had an answer for rancher O'Connel. The remains were those of a huge buffalo several hundred thousand years old. Such animals, said Martin, were classified as *Bison latifrons*.[1]

O'Connel's find was not the first. In 1925 there were three or four known specimens of this species, the first race of buffalo in North America. They came over from Asia sometime between 200,000 and 800,000 years ago during the Illinoisan glacial age, the third major glaciation of the four Pleistocene ice advances. From Asia, the center of mammal development, these shaggy animals crossed over to North America on the land mass that geologists tell us once connected Asia to what is today Alaska. These first buffalo were stout animals compared to the buffalo that roamed the western plains of North America in the 1800s. And they had very large horns which were thick and curved. The horns came out from the skull at right angles to the midline. Some fossil horns measure more than six feet across. Because of the horn size, some scientists believe the first buffalo were "solitary" in their habits and did not live in large herds. Apparently, Nature gave the animals their large horns as a means of self-protection.

For thousands of years these first buffalo spread slowly southward and eastward across North America. They ranged as far west as California, as far east as Florida, and perhaps as far south as Honduras and Nicaragua in Central America. But slowly the cold ice age climate in which these animals lived began to change. It was about 120,000 years ago, toward the end of the last interglacial (Sangamon) period, that the climate became warmer and the first buffalo died out. In their place appeared two new forms. One of these, *Bison antiquus*, may have originated from the earlier buffalo; it is a little smaller than the first buffalo. Fossil remains of this animal have been found in Texas, in Mexico, in the tar pits of the Rancho la Brea in California, and elsewhere. But like the earlier buffalo, this form also died out, becoming extinct during the last glaciation (Wisconsin) period, about 9,000 to 11,000 years ago. This form was associated with early man in North America.

However, the other new form, *Bison occidentalis,* did not die out. It prospered. Unlike the earlier forms, it had horns that angled backward from the midline of the skull, like those of the modern day buffalo. Fossil remains of this form were first discovered in the early 1800s by Sir John Richardson at Fort Yukon, Alaska, near where the Porcupine River empties into the Yukon. Since then other fossils of this form have been found, including a complete skeleton near Russell Springs in Logan County, Kansas.

From this species—*Bison occidentalis*—two races of buffalo developed. The plains buffalo (*Bison bison*) and the mountain or wood buffalo (*Bison athabascae*). These are our modern day buffalo.

A noted buffalo authority, W. A. Fuller of the University of Alberta at Edmonton, wrote in 1968:

> The Wood Bison and the Plains Bison are clearly related as both are clearly related to the European Wisent. I think that the most likely explanation of this relationship is that during the Ice Age there were bison in the general region of the Bering Strait. As these animals were enabled to spread out, some went west into Eurasia and some moved southeast into North America. In North America some remained in wooded regions—the so-called Wood

Buffalo—and some stayed along the Rocky Mountain chain—the so-called Mountain Buffalo. Comparison of skeletal materials of these two races indicates that they were really the same animal. Other bison moved out onto the plains and became modified in minor ways.[2]

By A.D. 1000 the two species of buffalo could be found over a wide area of the North American continent. The mountain or wood buffalo ranged from near the Great Slave Lake southward through the Rocky Mountains into what is today New Mexico. It is possible that some of these animals also lived in the Appalachian Mountains where they became known as Pennsylvania Bison. But the largest number of buffalo were those on the plains and prairie lands of North America. And it was there that the animals prospered and multiplied to a far greater extent than those in the mountains and woodlands.

There is no record that Columbus, an Italian, ever saw a buffalo in the New World. The Spanish explorer Hernando Cortez was the first white man we know of to have seen the animal. Cortez saw the buffalo in 1521 in a menagerie kept by Montezuma near where Mexico City stands today. Of the discovery, the Spanish historian De Solis later wrote:

> In the second Square of the same House were the Wild Beasts, which were either presents to Montezuma, or taken by his Hunters, in strong Cages of Timber, ranged in good Order, and under Cover: Lions, Tygers, Bears, and all others of the savage Kind, which *New-Spain* produced; among which, the greatest Rarity was the *Mexican* Bull; a wonderful composition of divers Animals: It has a Bunch on its Back like a Camel: its Flanks dry, its Tail large, and its Neck covered with Hair like a Lion: It is cloven-footed, its Head armed like that of a Bull, which it resembles in Fierceness, with no less strength and Agility.[3]

Unfortunately Cortez and other Spanish explorers of the 16th and 17th centuries made little more than passing mention of the animal. To these adventuresome Spaniards the buffalo apparently was just another of the new marvels, new discoveries, made in what they called New Spain. When the buffalo were mentioned, the early Spanish descriptions were not always complete, and they were often misleading.

Alvar Muñez Cabeza de Vaca was the second Spanish explorer we know of to see a buffalo. About 1530 he was shipwrecked on the Gulf Coast, probably not far from where Houston, Texas stands today. As he wandered westward across Texas, de Vaca presumably became the first white man to see buffalo in the wild. At least three times he saw vast herds of the shaggy animals grazing peacefully on the rich grasslands, and he ate buffalo steak several times. He compared the buffalo to Spanish cattle:

> I think they are about the size of those in Spain. They have small horns like those of Morocco, and the hair long and flocky, like that of the merino. Some are light brown and others black. To my judgement the flesh is finer and sweeter than that of this country [Spain]. The Indians make blankets of those that are not full grown, and of the larger they make shoes and bucklers. They come as far as the sea-coast of Florida [now Texas], and in a direction from the north, and range over a district of more than 400 leagues. In the whole extent of plain over which they roam, the people who live bordering upon it descend and kill them for food, and thus a great many skins are scattered throughout the country.[4]

After de Vaca it was ten years before the Spaniards again had contact with buffalo. Then, about 1540, Coronado's expedition saw buffalo for the first time somewhere on the southern plains, perhaps in the area of the Texas panhandle. Pedro de Castañeda, the expedition's historian, recorded that the first time they saw buffalo their horses "took to flight." Castañeda said the animals were horrible to look at. He described them as having a "broad and short face, eyes two palms from each other," and he noted that their eyes projected sideways from their head, making it possible for them to "see a pursuer." He compared a buffalo's beard to a goat's and said the buffalo's beard was so long that it "drags the ground when they lower their head." The buffalo's horns, Castañeda observed, were very short and thick and sometimes could scarcely be seen through their hair.[5]

During the 16th century, other Spaniards crisscrossing the American southwest encountered buffalo. They frequently killed the animals for food. Sometimes they would take the robe or hide. And by the beginning of the 17th century the Spanish knew more about the

Pictograph from Utah canyon wall, probably Fremont culture, c. A.D. 600-1250. *Courtesy Miles- Renzetti.*

The first known drawing of an American buffalo published in Europe (1552-53), from Francisco Lopez de Gomara's *Historia de las Indias. Courtesy Western History Department, Denver Public Library.*

Another early illustration of an American buffalo, from André Thevet's *Les Singularitez de la France Antarctique,* Antwerp 1558.

Father Louis Hennepin's depiction of an American buffalo, first published in 1683.

cibola, as they called the buffalo, than anyone else except the Indians. But then other Europeans had not yet seen the animal.

There is no record that the English, along the Atlantic Coast, had contact with the buffalo during the 16th century. Nearly all of their settlements were well to the east of the buffalo's range. It was not until the early 17th century, as the English began exploring inland, that they discovered the buffalo.

In 1612 Captain Samuell Argoll came upon the animal near where Washington, D.C. now stands. Argoll is the first Englishman known to have seen buffalo. But if he had not told of the event in a letter to a young friend back in England the fact might have been lost in time. On March 18, 1612, according to the letter, Argoll—somewhere along the coast of Virginia—unloaded corn from his ship and ordered his men to cut trees for wood to build a frigate. Then he went up the Pembrook River—the Potomac—as far as he could, tied up his ship, and went ashore. From there, guided by Indians, Argoll moved inland. Just how far he went or in which direction, no one knows. It is known that he soon came upon buffalo. The Indian guides killed two of the animals for food. Argoll wrote his young friend that the meat tasted "very good and wholesome" and that the buffalo were "heavy, slow, and not so wild" as other animals in the wilderness.[6]

Unfortunately, Argoll, like the Spanish explorers, made only casual reference to the buffalo. Very little was recorded during the 16th and 17th centuries about the buffalo's habits, range, or number. And it appears that many of the early reports of sighting buffalo may have been false. What some observers reported as buffalo were probably moose, musk oxen, or elk. This was because many explorers had only a vague idea of what a buffalo looked like. Then, too, the buffalo was not the only new animal to be found in the New World. The large wapiti, for one, was also unknown to most early explorers.

Anthonie Parkhurst, in a letter to Richard Hakluyt in 1578 about Parkhurst's exploration of Newfoundland, wrote:

> Nowe again, for Venison plentie, especially to the North about the grand baie, and in the South neere Cap Race and Plesance: there are many other kinds of beasts, as Luzarnes, and other mighty beastes like to camels in greatnesse, and their feete cloven, I did see them farre off not able to discerne them per-

fectly, but their steps shewed that their feete were cloven, and bigger than the feete of Camels, I supposed them to bee a kind of Buffes which I read to be in the countreyes adjacent, and very many in the firme lande.[7]

There is little evidence that Parkhurst actually saw buffalo. What he probably saw were moose. So far as is known, the modern day buffalo never inhabited Newfoundland. Moose, however, were plentiful there.

Names given to new animals found in North America often added confusion to their identification years later. Many names have been misinterpreted. Early French Jesuits and English explorers usually used the term "wild cows" to describe elk and the words *buffu* or *buffle* to identify moose, not buffalo. And some French Jesuit travelers, writing of their adventures in the new land, referred to vast herds of *vaches sauvages*. Some of these animals may have been buffalo, but when the Jesuits wrote complete descriptions of what they saw, there is little doubt that the animals were elk or wapiti and not buffalo.[8]

One of the first persons to study the buffalo in North America with more than a passing interest was Father Louis Hennepin, probably the first Frenchman to see a buffalo. In 1669 Sieur de La Salle sent Father Hennepin down the Saint Lawrence River, across the Great Lakes, to explore what is now Illinois and Indiana. On the Illinois River, near modern Peoria, Hennepin saw buffalo grazing on the prairie. That was in December 1679. He took much time to observe the animals, especially an old bull, the first buffalo he saw, which was mired near the Illinois River. After killing the bull, Hennepin noted, "We . . . had much ado to get him out of the Mud."[9]

After nearly two years observing buffalo, Hennepin recorded that the bulls had a very fine coat or robe which seemed to be more like wool than hair. He noticed that during the winter the cows had longer coats than did the bulls. Their horns, Hennepin said, were almost black and much thicker but shorter than those of the buffalo's European cousin, the Wisent.

Just how the word "buffalo" came into use is uncertain. To the Spanish explorers the animal was frequently called *cibola*. Some Spanish writers spoke of it as *bisonte*. Others called it *armenta*. Early French colonists usually called buffalo *Bison d' Amerique*. Canadian voyageurs used the term *boeuf*—"ox" or "bullock." Later French explorers called the animal the *bufflo* and later *buffelo*. It may have been from these terms, adopted by the English colonists, that the word *buffalo* originated. Although the word was used by the colonists beginning around 1710, it first appeared in print in 1754 in Mark Catesby's *A Natural History of Carolina*. Since then the word *buffalo* has been used by nearly everyone, save a few exacting scientists and writers, to describe the American bison.

There were about 260,000 persons living in the thirteen American colonies as the 18th century began. Most of them lived along the Atlantic Coast and had little, if any, contact with buffalo. The animal's range was well to the west of the English settlements. However, as men pushed westward to open new areas for settlement, these pathfinders began to see the animal.

One early and well documented account is dated October 1729. It is found in the journals kept by Colonel William Byrd. A survey party under Colonel Byrd's command were determining the boundary line between North Carolina and Virginia when they met three buffalo on Sugar-Tree Creek, about 155 miles from the coast. Byrd's party were very curious about the animals. They probably had never seen buffalo before. Since the party had a good supply of meat, none of the animals was killed. Instead, Byrd and his men stopped and observed them. This is perhaps the first instance of white men's not trying to kill wild buffalo when seeing them for the first time. However, several weeks later, as Byrd and his men were returning to the coast, they found a young buffalo bull drinking from Sugar-Tree Creek. Needing meat, they killed the animal. As Byrd later wrote in his journal, "It was found all alone, tho' Buffaloes Seldom are."

Four years later, in October 1733, Colonel Byrd led another survey party through the same general area. Again they came upon buffalo, and another bull was killed for meat. "It was welcome, too, for providence threw the animal in our way, just as our provisions began to fail us," wrote Byrd.[10]

That same year, a few hundred miles to the north, Sir William Berkeley sent Captain Henry Batt to explore the interior of Virginia. Seven days after leaving Appomattox, Batt and his party reached the heart of the Appalachian Mountains. There they found much wild game, including buffalo. Batt later recalled that the buffalo were "so gentle and undisturbed" that they had no fear of man. He reported that his men could almost pet them.[11]

History records several other such encounters by Englishmen exploring the eastern slopes of the Appalachians. From what is today southern New York State southward to southeastern Georgia, small herds of buffalo were found and usually killed by white men. Few buffalo were left alone. And in a matter of years they began to disappear. Those that were not killed by man or by natural causes apparently wandered westward over the mountains. By 1730 the white man had killed most of the wild buffalo in Virginia. None existed in the Carolinas or northern or eastern Georgia by 1760. The wild buffalo east of the Appalachians had been destroyed.[12]

During the 1750s, Daniel Boone had hunted buffalo near his home on the Yadkin River in North Carolina. By the late 1760s the animals became scarce. The eastern boundary of the buffalo range was then along the western slopes of the Appalachians. It was there in June 1769 that Daniel Boone again found the animal in abundance. Boone had left his Yadkin home and crossed the Appalachians to explore the country of Kentucky. He and his five companions frequently found buffalo "numbering up in the thousands—at one lick a hundred acres were densely massed with these bulky animals, who exhibited no fear until the wind blew from the hunters toward them, and then they would 'dash wildly away in large droves and disappear.' "[13]

Although a handful of frontiersmen like Boone saw and killed buffalo along the western slopes of the Appalachians, the animals still prospered because the land was still a wilderness. For the most part, the animals were not bothered by man. They enjoyed the lush grasses that grew in the meadows or on the tree-covered slopes, and water was plentiful. But the peace and tranquility that existed were soon to vanish.

Until the War for Independence, the British had forbidden whites to settle west of the crest of the Appalachians. The "Proclamation Line," as the British called the settlement boundary, had been established in 1763. But when the War for Independence ended in January 1783, thousands of soldiers were paid in land scrip and encouraged to migrate westward across the Appalachians. The government described the new land as very attractive. It was said to be good for raising crops such as corn and potatoes, and livestock such as hogs, chickens, and cattle. Then, too, said the government, there were the natural benefits: plenty of wild game to see a man and his family through until the land would pay. It sounded good, and by the late 1700s the once virgin lands of western Pennsylvania and the Ohio Valley were being settled.

In those areas buffalo were rather plentiful. For many of the settlers, the buffalo, of all wild game, must have appeared, at times, to be a gift from God. A single animal provided a good supply of meat for a week or more; and for many settlers, the buffalo meant the difference between survival and starvation. But as more and more people crossed the Appalachians to make their homes on the new land, greater numbers of buffalo were killed. In many areas the animals became scarce.

In 1790 there were few, if any, buffalo along what is today the West Virginia and Ohio border on the Ohio River. In that year a group of Frenchmen arrived there and settled the community of Gallipolis (Ohio). There is no record of a single buffalo having been seen in the area until one summer day in 1795 when Charles F. Duteil went out hunting for deer. About two miles west of Gallipolis he saw a herd of buffalo. In a letter to J. A. Allen, written in April 1876, George Graham of Cincinnati described what happened:

> Duteil fired without aiming at any particular one, and luckily a large one fell. He was so elated with this feat that without stopping to examine the animal he ran as fast as he could to the town, and, having announced his luck, came back, followed by the entire body of colonists, men, women, and children. They quickly formed a procession with musicians playing violins, flutes, and hautboys in front, the fortunate hunter proudly marching with his gun on his shoulder, and the animal swinging from poles thrust through between its tied feet, followed by the crowd, singing and rejoicing at the prospect of good and hearty fare. The animal was quickly skinned and dressed on its arrival at the town, and for several days there was feasting, as the first and last buffalo of Gallipolis was served up in such a variety of ways and means as none but the French could devise; Charles Francis Duteil remaining until his death the renowned marksman

who killed the first and last buffalo of all the emigrants from France who settled the town of Gallipolis.[14]

But not all buffalo killed east of the Mississippi were slaughtered for food. A handful of men made a business of killing the animal for its robe. They perhaps have the dubious honor of being the first white men to slaughter buffalo en masse for profit and not for necessity's sake. Thomas Ashe tells the story in *Travels in America,* published in 1808, a book detailing his experiences while exploring the Allegheny, Monongahela, Ohio, and Mississippi Rivers. The setting for the story is the area that today is Union County in central Pennsylvania.

An old man, one of the first settlers in this country, built his log house on the immediate borders of a salt spring. He informed me that for the first several seasons the buffaloes paid him their visits with the utmost regularity; they travelled in single files, always following each other at equal distances, forming droves, on their arrival, of about three hundred each. The first and second years, so unacquainted were these poor brutes with the use of this man's house, or with his nature, that in a few hours they rubbed the house completely down; taking delight in turning the logs off with their horns, while he had some difficulty to escape from being trampled under their feet, or crushed to death in his own ruins.

At that period he supposed there could not have been less than two thousand in the neighborhood of the spring. They sought for no manner of food, but only bathed and drank three or four times a day, and rolled in the earth, or reposed, with their flanks distended, in the adjacent shades; and on the fifth and sixth days separated into distinct droves, bathed, drank, and departed in single files, according to the exact order of their arrival. They all rolled successively in the same hole, and each thus carried away a coat of mud to preserve the moisture on their skin, which, when hardened and baked in the sun, would resist stings of millions of insects that otherwise would persecute these peaceful travellers to madness or even death.

In the first and second years this old man, with some companions, killed from six to seven hundred of these noble creatures, merely for the sake of their skins, which to them were worth only two shillings each; and after this "work of death" they were obliged to leave the place till the following season, or till the wolves, bears, panthers, eagles, rooks, ravens, etc., had devoured the carcasses and abandoned the place for other prey. In the following years the same persons killed great numbers out of the first droves that arrived, skinned them, and left their bodies exposed to the sun and air; but they soon had reason to repent of this, for the remaining droves, as they came up in succession, stopped, gazed on the mangled and putrid bodies, sorrowfully moaned or furiously lowed aloud, and returned instantly to the wilderness in an unusual run, without testing their favorite spring, or licking the impregnated earth, which was also once their most agreeable occupation; nor did they, nor any of their race, even revisit the neighborhood. . . .

The simple history of this spring is that of every other in the settled parts of this Western World; the carnage of beasts was everywhere the same; I met with a man who had killed two thousand buffaloes with his own hands; and others, no doubt, have done the same."

By the winter of 1799-1800 there were between 300 and 400 buffalo remaining in Pennsylvania. Most of the animals had taken refuge in the wilds of the Seven Mountains in what is today Snyder County near Middleburg in central Pennsylvania. Below the mountains in Middle Creek Valley the presence of settlers prevented the animals from wintering as they had done in the past. But as the winter became more severe, the animals had to move to the sheltered valley where they could dig out grass from under the snow; otherwise they would have starved.

The story of what happened to these buffalo was related years later by Flavel Bergstresser to Henry W. Shoemaker, who put the story in a little booklet called *A Pennsylvania Bison Hunt,* published in 1915:

Led by a giant coal black bull called "Old Logan," after the Mingo chieftain of that name, the herd started in single file one winter's morning for the clear and comfortable stretches of the Valley of Middle Creek. While passing through the woods at the edge of a clearing belonging to a young man named Samuel

McClellan, they were attacked by that nimrod, who killed four fine cows. Previously, while still on the mountain, a count of the herd had been made, and it numbered three hundred and forty-five animals. Passing from the McClellan property the herd fell afoul of the barnyard and haystack of Martin Bergstresser, a settler who had recently arrived from Berks County. His first season's hay crop, a good-sized pile, stood beside his recently completed log barn. This hay was needed to feed for the winter to a number of cows and sheep, and a team of horses. The cattle and sheep were sidling close to the stack, when they scented the approaching buffaloes. With "Old Logan" at their head, the famished bison herd broke through the stump fence, crushing the helpless domestic animals beneath their mighty rush, and were soon complacently pulling to pieces the hay-pile.

Bergstresser, who was in a nearby field cutting wood, heard the commotion, and rushed to the scene. Aided by his daughter Katie, a girl of eighteen, and Samuel McClellan, who joined the party, four buffaloes were slain. The deaths of their comrades and the attacks of the settlers' dogs terrified the buffaloes and they swept out of the barnyard and up the frozen bed of the creek. When they were gone, awful was the desolation left behind. The barn was still standing, but the fences, spring house, and haystack were gone, as if swept away by a flood. Six cows, four calves, and thirty-five sheep lay crushed and dead among the ruins. The horses which were inside the barn remained unharmed.

McClellan started homeward after the departure of the buffaloes, but when he got within sight of his clearing he uttered a cry of surprise and horror. Three hundred or more bison were snorting and trotting around the lot where his cabin stood, obscuring the structure by their huge dark bodies. The pioneer rushed bravely through the roaring, crazy, surging mass, only to find "Old Logan," his eyes bloodshot and flaming, standing in front of the cabin door. He fired at the monster, wounding him which so further infuriated the giant bull, that he plunged headlong through the door of the cabin. The herd, accustomed at all times to follow their leader, forced their way after him as best they could through the narrow opening. Vainly did McClellan fire his musket, and when the ammunition was exhausted, he drove his bear knife into the beasts' flanks to try and stop them in their mad course. Inside were the pioneer's wife and three little children, the oldest five years, and he dreaded to think of their awful fate. He could not stop the buffaloes, which continued filing through the doorway until they were jammed in the cabin as tightly as wooden animals in a toy Noah's ark.

No sound came from the victims inside; all he could hear was the snorting and bumping of the giant beasts in their cramped quarters. The sound of the crazy stampede brought Martin Bergstresser and three other neighbors to the spot, all carrying guns. It was decided to tear down the cabin, as the only possible means of saving the lives of the McClellan family. When the cabin had been battered down, the bison, headed by "Old Logan," swarmed from the ruins like giant black bees from a hive. McClellan had the pleasure of shooting "Old Logan" as he emerged, but it was small satisfaction. When the men entered the cabin, they were shocked to find the bodies of the pioneer's wife and three children dead and crushed deep into the mud of the earthen floor by the cruel hoofs. Of the furniture, nothing remained of larger size than a handspike. The news of this terrible tragedy spread all over the valley, and it was suggested on all sides that murderous bison be completely exterminated. The idea took concrete form when Bergstresser and McClellan started on horseback, one riding towards the river and the other towards the headwaters of Middle Creek, to invite the settlers to join the hunt.

Meanwhile, there was another blizzard but every man invited accepted with alacrity. About fifty hunters assembled at the Bergstresser home, and marched like an invading army in the direction of the mountains. . . . Many dogs, some partly wolf, accompanied the hunters. They were out two days before discovering their quarry, as the fresh snow had covered all the buffalo paths. The brutes were all huddled together up to their necks in snow in a great hollow space known as the "Sink" formed by Boonestiel's Tongue in the heart of the White Mountains, near the present [1915] town of Weikert, Union County, and the hunters looking down on them from the high plateau

above, now known as the Big Flats, estimated their number at
three hundred.

When they got among the animals they found them numb
from cold and hunger, but had they been physically able they
could not have moved, so deeply were they "crusted" in the
drifts. The work of slaughter quickly began. Some used guns,
but the most killed them by cutting their throats with long bear
knives. The snow was too deep to attempt skinning them, but
many tongues were saved, and these the backwoodsmen shoved
into the huge pockets of their deerskin coats until they could
hold no more. After the last buffalo had been dispatched, the
triumphant hunters climbed back to the summit of Council Kup
where they lit a huge bonfire which was to be a signal to the
women and children in the valleys below that the last herd of
Pennsylvania bison was no more, and that the McClellan family
had been avenged.[15]

What probably was the last wild buffalo killed in Pennsylvania was
shot about a year later, on January 19, 1801, by Colonel John Kelly.
He had been an Indian fighter and a soldier in the Revolutionary War.
One morning Kelly was riding his favorite horse, Brandywine, to a
nearby mill on business. Suddenly he saw a large buffalo bull standing in the middle of the road ahead. Kelly dismounted, took careful
aim, and shot the animal through the heart.

Sometime later Kelly had the buffalo's skull nailed to a pitch-pine
tree at the spot where the animal had been killed. For many years the
skull was a familiar landmark until about 1820 when it blew down
during a storm. One of Kelly's relatives picked up the skull and kept
it in his attic for several years, but sometime after Colonel Kelly died
in 1832, at the age of eighty-eight, the skull was thrown out and
burned with some trash by a careless housecleaner.[16]

The fate met by Colonel Kelly's buffalo was the same that met
hundreds of thousands of buffalo east of the Mississippi during the
1700s and early 1800s. The white man destroyed them.

By 1820 only a handful of wild buffalo remained in the East. A
small number lived in very isolated areas yet untouched by man. Perhaps a few dozen others had been captured, fenced up like ordinary

milk cows, and placed on public display. A few others were placed in captivity for the purpose of breeding with domestic cattle.

Finally, when some Sioux Indians on a river in northern Wisconsin killed two wild buffalo in 1832—probably the last wild buffalo killed east of the Mississippi—the story of the wild buffalo in the East became history. Never again would settlers reach for their rifles upon hearing the news that wild buffalo had been seen in the neighborhood. But the tales of hunting and killing the animal would be recounted time and again in front of a roaring fire on a cold winter's night or around a campfire during a squirrel hunt in the fall.

Such tales about killing the buffalo east of the Mississippi were tame compared to stories yet to be written in the West. What happened to the buffalo in the East was only a prelude to what man would do to the largest game animal in North America during the fifty years that followed.

One of the earliest known photographs of a wild buffalo on the plains. Taken in 1869 during a hunt on the Big Timber in Ellis County, Kansas. The raised tail signifies the buffalo's anger. *Courtesy Kansas State Historical Society.*

II

Millions on the Prairies and Plains

*Of all the quadrupeds that have lived upon the earth, prob-
ably no other species has ever marshaled such innumerable
hosts as those of the American bison.*

William T. Hornaday 1889

When the white man began exploring the plains and prairies from
the east in the early 1800s, he was astonished to find millions of buffalo. He saw more than he had ever dreamed existed and certainly
more than he had ever seen east of the Mississippi. "Teeming myr-
iads," "incredible numbers," "the world looked like one robe"—these
were just a few of the superlatives the white man used to describe the
untold numbers of buffalo he encountered.

The sight of hundreds or thousands of buffalo on the rolling prairie
or on a vast expanse of level plains was something few men could
forget. Imagine yourself standing on a high ridge, perhaps 200 feet
above a grass-covered prairie somewhere in central Kansas one hundred or more years ago. The warm breeze, blowing up from the south,
from Indian Territory and from the Staked Plains of Texas, brushes
past your face. From your vantage point you can see for nearly fifty
miles in any direction. As you turn your head from the southeast to
the south and southwest there are buffalo—hundreds, thousands, perhaps millions of the shaggy animals. Below the ridge, perhaps half a
mile away, you can see the individual animals, their brown, almost
blackish bodies slowly moving as they graze on the pale green buffalo
grass which has almost turned a light brown under the hot August
sun. Here and there you can make out a cow with a calf, some nursing, others playing. But as you look farther away the buffalo blend

20

together. A mile or so away you cannot make out the individual animals nor can you see the prairie under their feet. There is only a blanket of dark brown stretching to the cloudless horizon. You know you are seeing buffalo, yet it is an unbelievable sight, one that must be seen to be believed.

The first explorers saw many such sights as they moved westward first onto the prairie and then the plains. Though these regions are both without trees and generally have a sub-humid climate, they are different. The plains are level. The prairie is more rolling. The prairie begins where the woodlands of the East end. The plains begin where the rolling prairie ends.

Nearly all the early explorers who crossed the prairie and then the plains attempted to estimate how many buffalo they saw. One of the first such accounts during the early 1800s came from Captains Lewis and Clark. They had been sent west by President Thomas Jefferson to explore the northern expanse of the country called Louisiana which had just been purchased by the United States from France. As they traveled up the Missouri River in 1804 in a keelboat, they saw "three thousand" buffalo and were amazed at the number.[1] But as they moved deeper into the wilderness even larger herds were seen. By the time they reached the Pacific Ocean and started back through buffalo country, seeing vast herds of the shaggy animals had become commonplace with them, yet they continued to be amazed at the untold numbers they were seeing. When they reached the White River in August 1806, Lewis and Clark recorded in their journal, "We discovered more than we had ever seen before at one time; and if it be not impossible to calculate the moving multitude, which darkened the whole plains, we are convinced that twenty thousand would be no exaggerated number."[2]

Zebulon Pike, another early explorer on the plains, was also impressed at the vast numbers of buffalo. In 1806, before Lewis and Clark completed their journey, Pike took an expedition across the southern plains. In the rolling country along the Arkansas River near the present-day town of Cimarron, Pike saw many, many buffalo. "I do not think it an exaggeration to say there were 3,000 in one view," he reported.[3]

Nearby, a detail of Pike's men under the command of Lt. James B. Wilkinson also encountered the animals while exploring what later was called the Arkansas River. The young officer saw so many buffalo that he wondered whether he was seeing things. "I do solemnly assert," he later reported, "that, if I saw one, I saw more than 9,000 buffaloes during the day's march."[4]

It is understandable that these early explorers might question their own sanity when their eyes first gazed upon so many buffalo. Certainly nowhere east of the Mississippi had there been so many of these animals years before. But here they were, the huge bulls, the smaller cows with their calves; the light tan, almost yellow, colors of the calves and the darker brown, almost black, tint of the older animals; the buffalo's colors contrasting even more with the browns and soft greens of the plains that stretched as far as the eye could see.

Buffalo had made the plains and prairies their home for centuries. From the Rio Grande northward into Canada and from the Rockies east to the Missouri the plains buffalo ranged, eating the rich grasses, drinking from the numerous rivers and small streams. The animals had room to roam on millions of square miles of open space. There they had freedom. And that was as it should be for the wild American buffalo.

For the early explorers the vast expanse of open country—treeless for the most part—was a spectacle in itself. Most had never seen anything like it. Add the sight of hundreds or thousands of buffalo grazing mile after mile after mile over this land, and the picture was truly unbelievable. How could these men describe the sight of thousands of buffalo or the vast distances they could see or the insignificance they themselves felt? The plains and prairies made men feel uncomfortable, but how could they convey in words these feelings and thoughts? How could they describe the grandeur and magnitude and humility they felt?

For most it was difficult, almost impossible. Many tried, but they failed, perhaps because they were explorers and military men, not poets. And when they returned east and tried to describe the new land, the only ones who understood what they were saying were those few who had already ventured westward from the woodlands to the plains and prairies.

People who had never left the East could get only a vague idea of what the plains and prairies were like. Then, too, some of those early explorers who had crossed them and then returned to the East may

"Buffalo on the March: A Drawing from Eye-Witness Accounts," a crude but dramatic attempt to picture the numberless numbers of buffalo blanketing the land. M. S. Garretson drawing. *Courtesy Kansas State Historical Society*.

William Goetzmann called this John Mix Stanley drawing "one ot the best scenes of a buffalo herd ever done." Reproduced as illustration, "Herd of Bison near Lake Jessie," in the Pacific Railroad Surveys reports. *Courtesy Richard Fitch, Santa Fe.*

Steamer *General Meade* on the Yellowstone River, June 10, 1878. *Courtesy Kansas State Historical Society.*

"A 'Holdup' on the Kansas Pacific Railroad," M. S. Garretson drawing. *Courtesy Kansas State Historical Society.*

have wondered themselves what they had seen. What could the land be used for? Unfortunately, there were many false impressions of the plains and prairies. To most the land appeared to be useless. The vague descriptions of the plains and prairies, accounts of hot summers and cold winters, stories about the wind that "always" swept across the land conjured up a negative impression in the minds of Easterners as to what the vast area west of the Missouri and east of the Rocky Mountains was really like.

And when Major Stephen H. Long returned from his Yellowstone expedition in 1820 and prepared a map labeling the plains area the "Great American Desert," this only added further substance to the belief that the land was of little use to anyone but the Indians and the buffalo.

During the twenty years that followed, 1820-1840, there was little inclination in the East to settle what was considered by most to be a vast wasteland. It was only something that had to be crossed in order to get to Oregon or California. Immigrants and traders followed the Oregon Trail on the north and the Santa Fe Trail on the south; as they pushed westward onto the plains and prairies, nearly every wagon train saw buffalo, and often the animals were killed for food and sport.

One traveler, Thomas J. Farnham, crossing Kansas on the Santa Fe Trail in 1839, spent three days moving through what apparently was one large herd of buffalo. Later Farnham recalled:

> It appeared oftentimes extremely dangerous even for the immense cavalcade of the Santa Fe traders to attempt to break its way through them. We traveled at the rate of 15 miles a day. Fifteen times three equals forty-five. Take forty-five times thirty and you get 1350 square miles of country so thickly covered with these noble animals, that when viewed from a height it scarcely afforded a sight of a square league of its surface.[5]

One of the most graphic descriptions of untold numbers of buffalo was made by George Catlin, who wrote:

> In one instance, near the mouth of the White River, we met the most immense herd crossing the Missouri River, and from an imprudence got our boat into imminent danger amongst them,

from which we were highly delighted to make our escape. It was in the midst of the "running season," and we had heard the "roaring" (as it is called) of the herd when we were several miles from them. When we came in sight, we were actually terrified at the immense numbers that were streaming down the green hills on one side of the river, and galloping up over the bluffs on the other. The river was filled and in parts blackened with their heads and horns, as they were swimming about, following up their objects, and making desperate battle whilst they were swimming.

I deemed it imprudent for our canoe to be dodging amongst them, and we ran it ashore for a few hours, where we laid, waiting for the opportunity of seeing the river clear, but we waited in vain. Their numbers, however, got somewhat diminished at last, and we pushed off, and successfully made our way amongst them. From the immense numbers that had passed the river at that place, they had torn the prairie bank of fifteen feet in height, so as to form a sort of road or landing-place, where they all in succession clambered up. Many in their turmoil had been wafted below this landing, and unable to regain it against the swiftness of the current, had fastened themselves along in crowds, hugging close to the high bank under which they were standing.

As we were drifting by these, and supposing ourselves out of danger, I drew up my rifle and shot one of them in the head, which tumbled into the water, and brought with him a hundred others, which plunged in, and in a moment were swimming about our canoe, and placing it in great danger. No attack was made upon us, and in the confusion the poor beasts knew not, perhaps, the enemy that was amongst them; but we were liable to be sunk by them as they were furiously hooking and climbing onto each other. I rose in my canoe, and by my gestures and hallooing kept them from coming in contact with us until we were out of their reach.[6]

As civilization inched its way onto the prairies and plains during the 1850s and '60s, it was a rare occasion when settlers, traders, hunters, and even soldiers after a day on the trail did not discuss the large numbers of buffalo they had seen. How many did you see? How many do you think were in that herd we saw today? How many miles long was that herd? These were common questions.

William D. Street, a pioneer settler in northwest Kansas, asked such questions after an experience with buffalo in the late 1860s. It was in 1904 that he recalled:

> One night in June, 1869, company D, Second battalion, Kansas state militia, then out on a scouting expedition to protect the frontier settlements, camped on Buffalo Creek, where Jewell City is now located. All night long the guards reported hearing the roar of the buffalo herd, and in the stillness of the bright morning it sounded more like distant thunder than anything else it could be compared with. It was the tramping of the mighty herd and the moaning of the bulls. Just west of Jewell City is a high point of bluff that projects south of the main range of hills between Buffalo and Brown Creeks, now [1904] known, we believe, as Scarborough's Peak. When the camp was broken, the scouts were sent in advance to reconnoiter from the point of the bluff, to ascertain, if possible, whether the column was in the proximity of any prowling Indians. They advanced with great care, scanning the country far and near. After a time they signaled the command to advance by way of the bluff, and awaited our approach. When we reached the top of the bluff what a bewildering scene awaited our anxious gaze.
>
> To the northwest, toward the head of the Limestone, for about twelve or fifteen miles, west across that valley to Oak Creek, about the same distance, away to the southwest to the forks of the Solomon, past where Cawker City is now located, about twenty-five miles south to the Solomon river, and southeast toward where Beloit is now situated, say fifteen or twenty miles, and away across the Solomon river as far as the field-glasses would carry the vision, toward the Blue Hills, there was a moving, black mass of buffalo, all traveling slowly to the northwest at a rate of about one or two miles an hour. The northeast side of the line was about one mile from us; all other sides, beginning and ending, were undefined. They were moving deliberately and undisturbed, which told us that no Indians were in the vicinity. We marched down and into them. A few shots

were fired. The herd opened as we passed through and closed up behind us, while those to the windward ran away.

That night we camped behind a sheltered bend and bluff of one of the branches of the Limestone. The advance had killed several fine animals, which were dressed and loaded into the wagons for our meat rations. All night the buffalo were passing, with a continual roar; guards were doubled and every precaution taken to prevent them from running over the camp. The next morning we turned our course, marching north toward White Rock Creek, and about noon passed out of the herd. Looking back from the high bluffs we gazed long at that black mass still moving northwest.

Many times has the question come to my mind, How many buffaloes were in that herd? And the answer, no one could tell. The herd was not less than twenty miles in width—we never saw the other side—at least sixty miles in length, maybe much longer; two counties of buffaloes! There might have been 100,000, or 1,000,000, or 100,000,000. I don't know. In the cowboy days in western Kansas we saw 7,000 head of cattle in one round-up. After gazing at them a few moments our thoughts turned to that buffalo herd. For a comparison, imagine a large pail of water; take from it or add to it a drop, and there you have it. Seven thousand head of cattle was not a drop in the bucket as compared with that herd of buffalo.[7]

Another soldier, Colonel Richard I. Dodge, liked to talk about buffalo. He was fascinated with the animals. During the late 1860s and '70s, Dodge spent much time on the plains talking with buffalo hunters and frequently recording his own firsthand observations of the animal. On one occasion, in May of 1871, Dodge was driving a wagon from Fort Zarah to Fort Larned, Kansas. The distance between the two military posts was thirty-four miles. "At least twenty-five miles of this distance," wrote Dodge, "was through one immense herd, composed of countless smaller herds of buffalo." But Dodge made no attempt to count or even estimate the number of buffalo. He was too busy trying to get safely through the animals.[8]

In 1867 the U.S. government scheduled a large treaty-making conference along the Medicine River in Kansas. A. A. Taylor, son of the Commissioner of Indian Affairs, accompanied his father on the journey and later recalled: "We stood on a sand hill as the buffalo were gathering toward the stream of water in the evening, and it was estimated by old buffalo hunters in the party that there were a hundred thousand within our view."[9]

One buffalo hunter who often viewed vast numbers of buffalo was Robert M. Wright of Dodge City, who first crossed Kansas in 1859. Forty-two years later, as he looked back on his buffalo days, Wright recalled, "I have indeed traveled through the buffaloes along the Arkansas River for 200 miles, almost one continuous herd, as close together as it is customary to herd cattle. You might go north or south as far as you pleased and there would seem no diminution of their numbers." The old plainsman told about the night General Philip Sheridan and Major Henry Inman asked him to sit in on a discussion about how many buffalo there were between Fort Dodge, Kansas, and Fort Supply, seventy-six miles to the south-southeast, in Indian Territory. Wright at the time, the 1860s, was post trader at Fort Dodge. He recollected the discussion:

> Taking a strip of land fifty miles east and fifty miles west, they had first made it ten billion buffalo. But General Sheridan said, "That won't do." They figured it again and made it one billion. Finally they reached the conclusion that there must be one-hundred million buffalo. But you know, they were afraid to give out these figures. They didn't think anybody would believe them. Nevertheless, *they* believed them.[10]

Of all the men who would try to determine the size of a large buffalo herd, L. C. Fouquet may have been one of the most accurate. Fouquet, a Frenchman, came to America sometime around the middle of the 1860s and finally settled in Kansas on a plot of earth near what became the town of Wichita. In time his neighbors taught him to speak English, and since his house was next to the Chisholm Trail, he learned much from talking to the cowboys who headed Texas cattle up the trail. The cowboys would tell Fouquet how many cattle they had in their herds. Fouquet would then study the size, and through the years he developed the ability to accurately estimate the number of animals in any herd that passed his house.

About Christmas time 1870, a neighbor asked Fouquet to go

buffalo hunting with him. Fouquet agreed. The days were sunny and temperatures were unusually mild for December. Anyway, Fouquet had been getting restless and jumped at the chance to go out on the plains.

The two men left Wichita and traveled southwestward. They had gone about thirty miles when they came to a long ridge. From the top they could see fifteen miles or more in the distance. The view, said Fouquet, reminded him "of the Ocean where water and skies seem to join." From their vantage point they saw buffalo, more than either man had ever seen. As far as they could see, there were buffalo scattered in small herds of various sizes. Fouquet at once began to concentrate on their numbers. He soon concluded that the size of the small inner herds ranged from 500 to 5,000 buffalo. Then he began counting the numbers of herds. "I averaged them at two-thousand per band and counted sixty-three herds in full sight. Then there were a few straggling animals here and there. No one but real buffalo hunters will believe this," recalled Fouquet many years later.[11]

If we take Fouquet's average of 2,000 buffalo per inner herd and multiply it by sixty-three, the number of herds, Fouquet and his friend viewed at least 126,000 buffalo on the plains of Kansas that mild December day in 1870.

Gradually, as men tired of speculating on how many buffalo they might have seen at one time in a given herd, some undertook to "guess" at how many there might be on all the plains and prairies from Texas northward into Canada. Horace Greeley was one of these "speculators." Greeley guessed, to add spice to one of his newspaper columns, that there must have been five million buffalo over all the plains. But when Robert Wright heard Greeley's figure, he disagreed. He thought there were "nearly five times that number." Buffalo Jones got into the act and compiled a table which was later published. He estimated that 15 million buffalo were on the plains in 1865, only 14 million by 1870, and so on, to mention only two of Jones' figures.[12] On what Jones based his yearly totals is unknown, but like other such estimates, then and now, it was probably a calculated guess.

Of all such total buffalo estimates, perhaps the most accurate was made by Ernest Thompson Seton, a naturalist, just after the turn of the century. Seton decided to approach the question with some mathematics, logic, and comparison of domestic animals then living on the plains and prairies. He began by determining the size of the original buffalo range. Using maps and other data, he calculated that the modern day buffalo had once occupied about one-and-a-half million square miles of plains area, about one-half million square miles of rich prairie country, and about one million square miles of forest area. The total range of the buffalo, before the white man arrived, was about three million square miles.

Then the naturalist tackled the problem of how many buffalo might have lived on that range. He felt that the plains would certainly support at least the same number of horses and cattle as buffalo. The 1900 census of cattle, horses, and sheep on the ranges of the Dakotas, Montana, Wyoming, Nebraska, Kansas, Colorado, Texas, and Oklahoma was 24 million horses and cattle and 6 million head of sheep in an area of about 750,000 square miles, only one-half of the total plains area. Even so, the figures were adequate for Seton's purpose. He figured that if one-half the entire plains area could support 24 million horses and cattle, then that same area, before the white man turned some of the land into farms, probably supported at least 40 million buffalo. This figure, he felt, was a moderate estimate.

The six million sheep reported in the 1900 census Seton did not include. He felt they represented the other animals such as mustangs, antelope, deer, and wapiti that once competed with the buffalo on the plains for grass and water.

Seton next considered the half-million square miles of prairie lands east of the plains. Remembering that the prairie lands were richer than the vast plains region and recalling that "stockmen reckon one prairie acre equal to four acres on the plains in productivity," Seton concluded, "The prairies sustained nearly as many head of buffaloes as the plains." He set the buffalo's early prairie population at 30 million.

Finally, Seton looked at the forest areas. He knew they were the "lowest in the rate of population" for the buffalo. After some figuring, Seton said that the one million square miles of woodlands where buffalo had once roamed probably supported no more than 5 million buffalo.

Adding this total figure for forest land to those of the prairie and plains, Ernest Thompson Seton figured there were 75 million buffalo in North America before the white man arrived.[13]

III

The Wild Plains Buffalo

The buffalo and bison, wild and fierce,
Roam the wide plains, exulting in their strength.
John Bigland 1844

When the early Spanish explorers first saw buffalo on the plains and prairies, they called them horrible looking and ugly. Later, when traders, travelers, soldiers, and settlers moved out on the vast expanse of open country for the first time and saw buffalo, they too considered them strange-looking creatures. Even today the same thing holds true when people see live buffalo for the first time. But most persons change their minds after taking time to watch the animal. A full grown buffalo, especially a bull, is a magnificent-looking creature in spite of his odd build.

A buffalo bull's large head and forequarters make him appear somewhat out of proportion to his hind quarters. Perhaps this is why many first impressions of the buffalo are negative. Unlike his head, shoulders, and hump, his hind portion is small, including his tail, which is a mere wisp rarely more than two feet long.

A full grown bull might stand five or six feet tall at the shoulder, or high part of his hump, stretch nine feet or more in length, and weigh anywhere from 1,900 to 2,500 pounds. Some oldtimers on the plains said they saw wild bulls weighing 3,000 pounds or more. They probably did, but such animals were rare. Most plains buffalo bulls reached a weight of about 2,000 pounds, but then there are the larger legendary "buffalo-ox."

Some plainsmen believed that if a young buffalo bull was castrated, he would grow to an immense size. They would call such an animal a "buffalo-ox." There are stories telling of white hunters' castrating

30

young bulls and a few accounts of Indians doing the same thing. John J. Audubon said he was told that Indians sometimes emasculated a young bull so that the animal would grow "large and fatter."[1] But the likelihood of truth in these tales seems to be slim. First of all, those engaged in such acts would have had to wait three or four years to see the results, and it is doubtful that most men on the plains, white or Indian, would or could have waited around just to see what would happen. Stockmen today know that when beef cattle are castrated at a very young age they rarely grow larger than they would under normal circumstances. It is doubtful that buffalo are any different.

Yet one contemporary account takes some of the legend out of the 19th century buffalo-ox stories. E. Franklin Phillips, director of the Stamford Museum and Nature Center in Connecticut, told me in March 1968 about a buffalo he had purchased in 1962 from a Massachusetts animal dealer. The buffalo bull had been castrated. Six years after buying the buffalo, Phillips sold the animal back to the dealer. By then the buffalo was eight years old and a huge animal. "It took a couple of weeks to make a crate big enough to ship him," said Phillips, "and after he got to the dealer's place it took three days to get him out of the crate and into a barn. The next day the bull walked through the wall of the barn. The dealer shot the animal for fear he would turn over a school bus, of which several were passing the farm." Phillips said the castrated bull weighed two tons when killed.[2]

The plains buffalo cow is not as impressive a sight as the bull. With a less shaggy head, she does not have the noble appearance of the bull nor does she grow to be as large as the male. Then, too, her hump is not as large as the bull's. An average cow might weigh 700 pounds, although there are accounts of some weighing 1,300 to 1,500 pounds. But like the 3,000 pound bulls, such large cows were and still are difficult to find.

A wild plains buffalo of a hundred or more years ago was heaviest, actually fat, in the fall of the year. The animal had grazed all summer long on the rich grasses, but by midwinter, when grass became scarce and sometimes hard to find, the animal would rely upon the stored-up fat and what little grass could be found. By late winter a buffalo would tend to be lean, but with the spring grasses the animal would again increase in weight. In spite of such seasonal changes the weight of a buffalo never varied more than about 100 or 200 pounds.

In many ways the wild plains buffalo were like domestic cattle of today. They were gregarious, living together in loose herds, and like cattle they would travel together to water. After a period of grazing, the buffalo would slowly travel single file, following a trail to a nearby river or creek. On October 1, 1877 Colonel Nelson A. Miles' forces were facing the Nez Perce at Snake Creek—waiting, stalemated, each group hoping for arrival of allies. Major Henry Remsen Tilton wrote of that day:

> A cheerless morning, with clouds and winds and mist, succeeded by rain and, finally, snow. Early in the day we discovered in our rear two long lines of cavalry marching toward us on either flank. Were these Gen. Sturgis' troops, or the warriors of Sitting Bull? Many anxious moments were spent before we determined that they were buffalo marching in single file, with all the regularity and precision of soldiers.[3]

Though buffalo usually went to water once, sometimes twice, a day, if water was scarce they could and often did go without watering for several days, much longer than domestic cattle. How long they went without water depended upon their physical condition, their age, and the weather. Water was often hard to find on the plains and prairies. During very dry years, the small creeks would often dry up in the hot summers. Then the buffalo depended upon the rivers, whose water usually kept flowing. Then they might travel en masse fifty or even a hundred miles to find water. In such cases it was not uncommon for the animals to gorge themselves.

In nearly all the stories told about thirsty buffalo, the animals became drowsy after drinking large quantities of water. George Ruxton related one such story. During the 1840s, Ruxton was about to make camp one night on the plains when three buffalo bulls came up from a nearby river where they had watered. They leisurely walked in front of Ruxton on the trail. The animals moved slowly and stopped frequently. One of the bulls kept lying down on the ground whenever the two other bulls stopped. At this Ruxton became curious. Since this performance was most unnatural for buffalo, he followed them on his mule. When two of the buffalo stopped again and looked back at Ruxton, the third animal lay down again. He lay on the grass until

Ruxton rode up to within a few feet. Then the animal got up and followed slowly after the others. Ruxton decided to see just how close he could get to the drowsy buffalo. Taking his rifle, he dismounted and slowly moved toward the bull, who by then had stopped again. The bull never looked around as Ruxton walked up and placed his hand on the animal's rump. The bull paid no attention to Ruxton and again lay down on the grass. At this point Ruxton killed the buffalo. When he butchered the carcass later that evening, he was surprised to find the animal's stomach completely full of water. "Another pint would burst it," said Ruxton. Otherwise the bull was in perfect health and good condition.[4]

Another observation of this phenomenon was made in recent times by Edwin Drummond in southwest Oklahoma. When buffalo "get full of water and start up a creek bank where there is no mud, they stumble over a rock or trip in a ditch and fall down; with their stomachs full of water they can not get up and they die there," Drummond said.[5]

Summers in Kansas can get very hot and the summer of 1868 was no exception. Hot temperatures and dry conditions prevailed throughout Kansas and most of the other plains states from Texas northward to the Dakotas. Vegetation was sparse and water was scarce. Most of the small streams dried up for lack of rainfall. Only the largest rivers continued to flow. One day that summer a large herd of buffalo was seen moving across the northern edge of McPherson County in central Kansas. Settlers who saw the animals said the herd stretched for thirty miles. The buffalo apparently had been grazing to the east, but lack of water had driven them back westward toward the Smoky Hill River. When the first animals reached the river, they plunged in and filled themselves as fast as they could. Other buffalo crowded in and pushed the earlier arrivals out of the river onto the opposite bank. According to one man who saw the watering, "Hundreds, thousands, and even hundreds of thousands of buffalo watered at the river that hot day, and the buffalo drank the river dry."[6]

A similar story was told by Jeff Durfey, an old buffalo hunter, who in 1911 recalled his buffalo days as he sat on the front porch of his Osborne County, Kansas farm home. Durfey remembered the time he was camped on the banks of Beaver Creek in western Kansas. He said, "The stream was six feet wide and about six inches deep, with swiftly running water. A buffalo herd came to the creek above our camp and drank it dry. For hours the creek bed was dry until the great herd passed on." When one of his listeners asked Durfey whether that was the only time he had seen buffalo drink a stream dry, he replied, "No," and went on to relate how in 1872 he saw another herd of buffalo drink the Solomon River dry and the river was twenty-five feet wide and a foot deep before the buffalo came.[7]

Buffalo often could smell water miles away. Just how far would, of course, depend upon the wind. Charles Goodnight, the father of the Texas Panhandle, thought a buffalo could smell water eight or ten miles away if the wind were right.[8] Another early plainsman who observed both wild and captive buffalo, C. J. "Buffalo" Jones, thought the wild buffalo's sense of smell was "so keen" that the animal could tell "where a rich bunch of grass is, though buried a foot deep under the snow."[9]

In the wild state, buffalo relied more on their nose than on their eyes for warning. In 1800 Alexander Henry, an early trader in the Red River area of the Dakotas, saw a large herd of buffalo cows moving at full speed toward the south. When the animals came to the trail on which Henry's party had just passed, the buffalo stopped. Then suddenly "as if they had been fired at," wrote Henry, they turned and hurried off toward some mountains. The animals had undoubtedly picked up the trader's scent. Henry observed that if buffalo smell the track "of even a single person in the grass," they will run off in the opposite direction. "I have seen large herds, walking very slowly to pasture, and feeding as they went, come to a place where some persons had passed on foot." The buffalo, said Henry, would stop, "smell the ground, draw back a few paces, bellow, and tear up the earth with their horns."[10]

The buffalo's acute sense of smell was also mentioned by Joel Palmer, another early traveler on the plains. In June 1845 Palmer and his party were camped on the Platte River in what is today Nebraska. "At daylight a herd of buffalo approached near the camp; they were crossing the river, but as soon as they caught the scent, they retreated to the other side," wrote Palmer.[11]

Another early plains traveler, Edwin James, secretary and historian to Major Stephen H. Long's plains expedition of 1819-1820, saw a large herd of buffalo near the Big Bend of the Arkansas River in August 1820. When the wind carried the scent of the white men

Sniffing the wind, common practice especially during rutting season. *Courtesy Oklahoma Historical Society.*

"Buffalo Hunt, Approaching in a Ravine" (detail), George Catlin lithograph, from his *North American Indian Portfolio.*

"Head of American Buffalo,
to show the divergent Hair about the Eyes,"
Ernest Thompson Seton drawing, from his
Studies in the Art Anatomy of Animals.

John Palliser commented: "During my stay at Fort Union [winter 1847-48], I was frequently surprised at the friendly relations between our domestic cattle and the buffaloes, among
whom they mingled without the slightest hesitation. . . . I was still more astonished . . . to
see our little calves apparently preferring the companionship of the bison, particularly that
of the most colossal bulls." Palliser watched the calves closely and discovered the reason
for their preference. "The bison has the power of removing the snow with his admirably
shaped shovel-nose, so as to obtain the grass underneath it. His little companions, unable
to remove the frozen obstacle for themselves, were thankfully and fearlessly feeding in his
wake; the little heads of two of them visible every now and then, contesting an exposed
morsel under his very beard." *(Solitary Rambles)*

to the buffalo, they went "into a full bounding run."[12] There are many similar accounts of the keenness of nose of the buffalo in any season of the year. When newcomers to the plains saw buffalo two or three miles away take flight for no apparent reason, they concluded the animal had excellent eyesight, but the "oldtimers" knew differently. It was the wind blowing across the open country that usually betrayed man's presence to the buffalo.

Because the wild plains buffalo used his nose and not his eyes, some early white men on the plains thought the animal was almost blind. George Catlin was one who attributed the poor eyesight to the "profuse locks of hair that hang over their eyes."[13] Another, "Buffalo" Jones, thought it was due to the position of their eyes. From where their eyes are situated in their head, said Jones, "they cannot see directly in front; neither can they look backward on account of their immense shaggy shoulder."[14] But in truth neither Catlin nor Jones was correct. The hair had little effect on the animal's eyesight, and they were not blind. Their eyesight was as good as that of most domestic cattle today. Hoofed animals, as a rule, have poor eyesight. With the lateral set of the eyes there is no focusing power of binocular nature. Grazing in the close confines of large herds, the wild buffalo had no need to evolve sharp vision. They simply relied more on their sense of smell than on their eyesight.

Interestingly, several modern day buffalo men have observed that captive buffalo, those kept in close quarters such as in zoos, rely more on their eyes and less on their sense of smell, except when it comes to eating. There are many different smells found in zoos. Thus the animal has found his eyes more reliable than his nose. Captivity seems to have a similar effect on other large wild animals kept in zoos.

The wild buffalo's sense of hearing seems to have been as good as his sense of smell. In August 1810 Alexander Henry observed twelve buffalo cows crossing the North Saskatchewan near an Indian camp. The noise of dogs and Indian children turned the buffalo, reported Henry.[15] And, according to George Bird Grinnell, the Cheyenne Indians could frighten buffalo away from their camps by beating loudly on drums. "The Cheyennes believe that buffalo are afraid of a drum, but say that they do not mind singing," wrote Grinnell.[16]

Like domestic cattle, buffalo are cud chewers. They have a compound stomach and temporarily store partly chewed food in the first compartment, or paunch. The food is regurgitated in small pellets to the mouth to be thoroughly milled, and is passed on to the second compartment to start digestion. Cud chewing animals, in the class of *Ruminantia,* have no teeth in the front of the upper jaw, but they do have broad-crowned sharp-edged molar teeth to grind and mill food. If you watch a buffalo eating you will see him swallow and after a pause bring up a new pellet to chew. This is held in the cheek, causing a slight swelling. When the buffalo chews, it does so with a rotary motion, first on one side of the mouth, then on the opposite side where the upper and lower teeth do the job of milling. The buffalo chews only on one side of the mouth at a time.

Unlike domestic cattle today, the wild buffalo on the plains were very hardy animals. They lived and thrived when other animals, especially cattle, might have died. When winter blizzards hit the plains and prairies, the buffalo did not drift with the storm like cattle. Instead, they faced into the storm, either standing still waiting for the storm to pass or slowly heading into it. In this way the storm passed faster for the buffalo than it did for cattle, who would drift with the storm and frequently die from the elements.

Robert M. Wright thought buffalo would stand still or move against a storm because they were "much more thinly clad behind than in front." The animal would "naturally" face the storm, said Wright.[17] Others, for the lack of anything better to say, simply credited the buffalo with "better sense" than cattle. Yet, there is still some question whether buffalo did, in fact, "always" face into winter storms. I have never had the opportunity to observe a large herd of buffalo during a winter storm, but naturalist Tom McHugh says that all the present-day buffalo he has observed faced in random directions during storms.[18] This would seem to cast some doubt on the accuracy of stories by oldtimers. Yet, it is difficult to discount many such accounts. They are numerous. Have perhaps the present-day buffalo changed or have the winters of recent decades not been as severe as those of the late 1800s when most oldtimers made their observations? James Hertel and Bruce Poel own a small herd of buffalo at Fremont, Michigan. In 1973 Hertel told me that he had watched the animals during three snowy winters and they seemed invariably to face into a blizzard or a cold wind.

A wild plains buffalo did carry a lot of natural protection on his back. Besides his heavy robe, the massive growth of hair on his head, shoulders, and hump offered good protection from winter storms. But buffalo often died in blizzards. In January 1872 a very bad blizzard swept across Nebraska. The temperatures were below zero and icy north winds formed snow drifts several feet deep. Visibility was cut to zero. In the western part of the state a Union Pacific train was trapped between two huge snow drifts. The telegraph wires were tapped and calls for help sent out. As the persons aboard the train waited for aid, they huddled around the potbelly stoves in the train cars. Outside, someone noticed a herd of buffalo pushing their way up toward the train on the lee side. The animals were seeking shelter.

When some of the passengers became fearful that the buffalo might push the train cars over, the engineer blew the whistle several times, but the buffalo only pushed in closer to the train and the protection it offered from the blizzard. When revolver shots and yelling did not move the animals, everyone just sat back, waited, and watched. By the time the relief train arrived the storm had passed on. Many of the buffalo moved back onto the snow-covered prairie, but on the lee side of the train, where the buffalo had huddled, from one end of the train to the other there were huge, shaggy frozen carcasses of dead buffalo. Some were in standing position, others lying on the ground. They had frozen to death.

The Beatrice, Nebraska *Express,* which reported the incident a few days later, observed, "We think a buffalo robe a luxury in winter, but imagine the severity of a winter when the animal who furnishes the robe freezes to death under its natural protection."[19]

After such a blizzard ends, the plains and prairies often look as if a huge white bedsheet had been spread across the land. Then, if the clouds vanish as they usually do, and the sun comes out, the blanket of white will glisten like a million diamonds. It is then that the buffalo begin to search for food, and it was in times like this, a century or more ago, that buffalo were compared to hogs. Unlike horses who paw through snow to get at grass, buffalo root like a bunch of hogs, pushing their noses down through the snow until they find grass. Unless the snow is several feet deep, the distance to grass is of little concern to the animals if their noses pick up its scent. One early observer, Dr. James Hector, saw a herd of wild buffalo root through a twelve-inch snowfall to get to grass.[20] That was in 1858 on the northern plains, and there are numerous other accounts of buffalo doing the same thing.

History records at least one instance in which the animal's ability to root through snow was of no use. It was during the winter of 1844-45 on the Laramie Plains in what is now southeastern Wyoming. That winter, buffalo gathered on those plains as they had for years. Surrounded by mountains, the Laramie Plains had for centuries offered buffalo and other wild game shelter during the cold and windy winter months. But that winter a windless snowstorm covered the surface of the plains with nearly four feet of snow. After the snow stopped and the clouds disappeared, a bright sun warmed the top of the white surface. A few hours later at dusk the temperatures dropped and the melted snow on top froze solid. There was such a heavy crust of ice on the top of the snow that it was weeks before any of the buffalo, even the larger and more powerful animals, could make any headway. Few were able to root through the ice to get at the grass below. Thousands died of starvation. And as late as 1868 buffalo skulls and bones could still be seen dotting the Laramie Plains, grim reminders of the great snowstorm and the tragedy that followed.[21]

Winter storms were not the only natural disasters that plagued the buffalo. As winter snows melted, the rivers and streams would rapidly fill and often flood. It was not uncommon for buffalo attempting to swim a river to be pulled under by the fast current. Others sometimes found themselves mired or bogged in the muddy banks they had to cross to reach the water. The more they tried to escape, the deeper they sank.

Perhaps one of the most unbelievable stories about buffalo trapped by nature was related by John McDonnell, a partner of the Northwest Company. In 1795 a large herd of buffalo stampeded across the ice-covered Qu'Appelle River which flows eastward across the plains of southern Saskatchewan. Many of the animals broke through. When McDonnell reached the place in May of that year, he saw large numbers of dead buffalo in the river and along the banks. From his boat he counted the animals. By nightfall he had totaled 7,360 drowned and mired buffalo, and there were still more to be seen. "It is true," he wrote in his journal, "in one or two places I went on shore and walked from one carcass to the other, where they lay from three to five files deep."[22]

A few years later, in 1801, Alexander Henry saw a similar sight on the Red River of the north. In his journal he noted, "Drowned buffalo continue to drift by in whole herds throughout the month, and toward the end for two days and nights their dead bodies formed one continuous line in the current." Henry added that thousands of the dead animals were grounded along the river bank. The stench was so strong, he noted, that sometimes he could not eat.[23]

Still another natural enemy of the wild plains buffalo was the prairie fire which might sweep across the open grasslands for hundreds of miles before stopping at a river or other natural barrier. It was not uncommon for such fires to trap buffalo and other wild animals. The flames moved at a high rate of speed when fanned by the wind. There was nothing to stop the fire and it had no mercy for man or beast.

In the late 1850s, Henry Hind, after exploring the South Saskatchewan River in Canada, reported:

> Blind buffalo are frequently found accompanying herds, and sometimes they are met with alone. Their eyes have been destroyed by prairie fires, but their quickened sense of hearing and smell, and their increased alertness, enable them to guard against danger and make it more difficult to approach them in quiet weather than those possessing sight.[24]

Sometimes prairie fires were started by lightning which also killed buffalo. Indians sometimes set prairie fires, as was the case in November 1863, when a band of Cheyenne set fire to the prairie on the south side of the Solomon River not too far from where Minneapolis, Kansas stands today. The Cheyenne were trying to burn out a group of Delaware Indians trapping beaver and otter along the Solomon. The Delaware managed to escape, but the strong northwest winds swept the flames through the very dry prairie grass, suffocating buffalo. The following day nearly twenty dead buffalo, mostly cows, were found in the area swept by the fire. The buffalo were lying headed in the southeasterly direction, just as they had been running before the wind when overtaken by the rush of fire and smoke.[25]

An even more striking story of a prairie fire and buffalo was recorded by Alexander Henry in his *Journal* as he traveled across southern Canada on November 25, 1804. Wrote Henry:

> Plains burned in every direction and blind Buffalo seen every moment wandering about. The poor beasts have all the hair singed off, even the skin in many places is shrivelled up and terribly burned, and their eyes are swollen and closed fast. It was really pitiful to see them staggering about, sometimes running afoul of a large stone, and other times tumbling down hill and falling into creeks, not yet dead. The fire having passed only yesterday these animals were still good and fresh, and many of them exceedingly fat. Our road was the summit of the Hair Hills [Pembina Mountain] where the open ground is uneven and intercepted by many small creeks running eastward. The country is stony and barren. At sunset we arrived at the Indian camp, having made an extraordinary day's ride, and seen an incredible number of dead Buffalo. The fire raged all night toward the S.W.[26]

If there was anything a buffalo bull liked to do besides eat or drink, it was to roll on the ground or rub up against a tree, large boulder, or other hard object big enough to accommodate him. Rolling on the ground or "wallowing" was primarily enjoyed by bulls, but cows often joined in. With the coming of summer the buffalo shed their coats in great broad flakes or wads of mothy-looking felt, till the hinder portion of their bodies was positively bare. Then as the hot summer sun beat down on their nearly naked bodies and the insects moved in, the buffalo would search out a damp or muddy area in which to wallow. If they could not find a mud hole they would simply pick a spot on the prairie and make a wallow. George Catlin eloquently described how the buffalo did it:

> Finding in the low parts of the prairies a little stagnant water amongst the grass, and the ground underneath soft and saturated with moisture, an old bull lowers himself upon one knee, plunges his horns into the ground, throwing up the earth and soon making an excavation into which water trickles. It forms for him in a short time a cool and comfortable bath, in which he wallows like a hog in the mire. Then he throws himself flat upon his side,

and then, forcing himself violently around with his horns, his feet and his huge hump, he ploughs up the ground still more, thus enlarging his pool till he at length becomes nearly immersed. Besmeared with a coating of the pasty mixture, he at length rises, changed into "a monster of mud and ugliness," with the black mud dripping from his shaggy mane and thick woolly coat. The mud, soon drying upon his body, forms a covering that insures him immunity for hours from the attacks of insects.[27]

Many old narratives, particularly the reminiscences of oldtimers on the plains, imply that wild buffalo, especially the bulls, rolled completely over while wallowing. This is doubtful. In my conversations with buffalo men from Texas to Canada, I found only one instance in which a man had seen a buffalo roll completely over while wallowing. In that case the animal, a bull, was wallowing on a slope and to his own surprise gravity pulled him on over. He jumped up, amazed at what he thought he had done, looked around, and hurried off. I have watched many buffalo wallow on the prairie in private and government herds and I have never seen an adult buffalo roll completely over. After rolling on their side, they sometimes will push their feet into the air, but each time their hump stops them from rolling completely over.

Buffalo wallows made by the wild plains buffalo can still be seen today in many areas of the Dakotas, Wyoming, Montana, Kansas, Nebraska, Oklahoma, and the Texas Panhandle on grazing land that has yet to be touched by the plow. In some of these areas, rubbing stones are also visible. These are large boulders standing alone on the prairie against which the buffalo would rub themselves. Around some are depressions, several feet deep, made by the buffalo. The animals literally "walked the ground away" in a circle around the boulders as they rubbed.

Where trees were available on the plains, buffalo would rub against the bark, sometimes causing the sides of trees to be smooth. It was not uncommon for the animals, particularly the large bulls, to rub so hard that they would knock younger trees to the ground. And when civilization moved onto the plains, man gave the buffalo still other things to rub against. The corner of a sod house was perfect for rubbing, as was the telegraph pole.

As telegraph poles were set across the plains, buffalo found them to be a new source of rubbing delight. The buffalo daily shook down miles of wire scratching themselves on the poles. Then someone had the bright idea to stop the rubbing with bradawls, a small boring tool. He sent to St. Louis and Chicago for all the bradawls he could buy. They were driven into poles with the idea that they would wound buffalo as they rubbed against the poles. Thousands and thousands of bradawls were installed on poles in buffalo country, but the hoped-for results never materialized. The genius who thought up the idea did not know buffalo, nor how tough the animal's hide was, nor what really would happen with the bradawls. In March 1869 the Leavenworth, Kansas *Daily Commercial* reported there never had been a "greater mistake" made than with the bradawls. The buffalo loved them.

> For the first time they came to scratch sure of a sensation in their thick hides that thrilled them from horn to tail. They would go fifteen miles to find a bradawl. They fought huge battles around the poles containing them, and the victor would proudly climb the mountainous heap of rump and hump of the fallen, and scratch himself into bliss until the bradawl broke, or the pole came down. There has been no demand for bradawls from the Kansas region since the first invoice.[28]

Most men who have had anything to do with buffalo agree that the animals live longer than most domestic cattle. From the meager accounts this writer has found, it seems likely that the wild plains buffalo perhaps set longevity records unmatched by buffalo in captivity. On the plains and prairies the wild buffalo found conditions suited for a long life, provided natural disasters or man did not end their lives early. Bob Wright once noted, "I have heard it on the best authority that some of them live to be seventy-five or eighty years old, and it is quite common for them to live thirty or forty years. In fact, I think I have seen many a bull's head that I thought to be over thirty years old."[29]

"Buffalo" Jones once observed that he had seen buffalo "so old their horns had decayed and dropped off, which indicated that they live to a patriarchal age."[30] Still another, Charles Goodnight, who was not known to always agree with Jones, did agree on the question of the buffalo's longevity. Goodnight flatly asserted that some buffalo lived to be forty or fifty years old.[31]

Old Stub Horns (left), buffalo bull about 30 years old when this photo was taken. *Courtesy Division of Manuscripts, University of Oklahoma Library.*

IV

Those of the Mountains and Woods

Old hunters of the west recognized a difference between the buffalo that lived in the mountains and those down on the plains.

George Bird Grinnell 1904

Colonel Richard I. Dodge liked to tell the story about a sportsman friend who in the late 1800s decided to go into the Rockies in search of the so-called "mountain buffalo." As Dodge told the story, a few days after his friend took a small hunting party into the mountains, they came upon fresh buffalo tracks. Because the terrain was so rocky, they had to dismount and follow the tracks on foot. Not knowing how long they would be gone, they left a man to watch the horses.

When night came they still had not seen any buffalo. The party made camp, enjoyed a good meal, and had a good night's sleep. Early the next morning they started again after the buffalo. About 11 o'clock in the morning the party stopped to rest on a high ledge overlooking a deep mountain valley and spotted a herd of about twenty buffalo lying peacefully on a rocky slope several hundred feet below. Moving close to the edge, Dodge's friend picked out a full grown bull, aimed his rifle, and fired. The shot echoed through the valley like a cannon blast. The hunter was sure his bullet had hit its mark, but a moment later he began to think otherwise. In an instant the buffalo were on their feet and running. A few seconds later they vanished over what appeared from above to be a precipice. At first the hunter and those in his party thought the frightened animals had, in sudden panic, jumped to their death. But when the men made their way down to the spot, they found a steep incline down which the buffalo had slid in their escape. One of the animals, however, had gone no farther than the bottom. Lying dead on the rocky bottom was the large bull the hunter had shot. The bullet had finally taken effect.[1]

The journals of early explorers, trappers, and travelers tell us something of buffalo in the mountains, but there are many more accounts of buffalo in the plains. There are two reasons for this. First, there were fewer buffalo in the mountains than on the plains; the mountainous terrain simply was not conductive for the development of large herds such as on the open plains. Second, the white man had little contact with the mountain buffalo. During the winter months, trappers and travelers usually moved along the rivers and streams while the mountain buffalo wintered in well-protected valleys. During the summer months, as the buffalo grazed high in the mountains, the trappers were at the rendezvous or preparing for another season of trapping. Travelers going through the Rockies usually stayed close to the rivers and easy passes near firewood and meadows for grazing. Thus they saw very few buffalo in the mountains.

What may be the earliest account of mountain buffalo was made by explorer Zebulon Pike. In January 1807 he found buffalo not far from what became known as Pike's Peak. The animals were among the snow, but Pike did not refer to them as mountain buffalo, nor did anyone else until around 1850. Several writers before then did, however, note the difference between buffalo in the mountains and those on the plains. One of these was Washington Irving. In his book, *Adventures of Captain Bonneville,* first published in 1837, Irving wrote, "The buffalo on the Pacific side of the Rocky Mountains are fleeter and more active than those on the Atlantic."[2]

Another writer, Col. John C. Frémont, who traveled through the Rockies during the early 1840s, also saw buffalo in the mountains. On June 17, 1844, Frémont reported:

> We continued our way among the waters of the Park over the foot-hills of the bordering mountains, where we found good pasturage, and surprised and killed some buffalo. We fell into a broad and excellent trail, made by buffalo, where a wagon would pass with ease; and in the course of the morning we crossed the

summits of the Rocky Mountains, through a pass which was one of the most beautiful we had ever seen.

A few days later, about June 21, 1844, Frémont and his men saw buffalo as they neared Hoosier Pass, 11,541 feet. Wrote Frémont, "We surprised a herd of buffalo, enjoying the shade at a small lake among the pines, and they made the dry branches crack, as they broke through the woods."[3]

George Frederick Ruxton, who also traveled through the Rockies in the 1840s, reported the existence of buffalo in the mountains. Ruxton said they were harder to find than those on the plains. They roam in small bands "and require no little trouble and expertness to find and kill." Ruxton noted that in the mountains "one may hunt for days without discovering more than one band of half a dozen."[4]

None of these men, however, referred to the buffalo as "mountain" buffalo. The name appears to have been given by hunters around 1850 and to have been first used in print in 1871 when W. H. Brewer published *Animal Life in the Rocky Mountains of Colorado*. Wrote Brewer, "They were frequently described to me as a marked variety known to the hunters as *Mountain* buffalo, and quite unlike the buffalo of the plains, smaller in size, the hair longer, more shaggy, and blacker, with other well marked differences."[5]

Although Colonel Richard I. Dodge never did see a live mountain buffalo, he heard many stories about the animal and in his *The Plains of the Great West* devoted several pages to this non-plains creature. He concluded:

> [The mountain buffalo] bears about the same relation to the plains buffalo as a sturdy mountain pony does to a well-built American horse; his body is lighter, whilst his legs are shorter, but much thicker and stronger, than the plains animal, thus enabling him to perform feats of climbing and tumbling almost incredible in such a huge and apparently unwieldy beast.[6]

Colonel Dodge probably gathered more information about mountain buffalo than any other writer of the 19th century, including William T. Hornaday of the Smithsonian. Hornaday relied heavily on Dodge's material in discussing mountain buffalo in his *Report of the National Museum* in 1887 dealing with the extermination of the buffalo. At some point before 1877, Dodge hunted for mountain buffalo in the Rockies but with no success. Later he wrote, "I have wasted much time and a great deal of wind in vain endeavours to add one of these animals to my bag. My figure is no longer adapted to mountain climbing and the possession of a bison's head of my own killing is one of my blighted hopes."[7]

The lack of knowledge concerning mountain buffalo is probably one reason why today there is still an air of mystery, a certain strange fascination, about the animal. Most authorities believe they lived in small groups numbering between five and thirty, rarely more. And it is generally believed that mountain buffalo annually moved about altitudinally. Also, several observers including Dodge have written that the mountain buffalo were wilder than their plains cousins, but there is little evidence to back up these conjectures. There are only stories and tales.

One of the best literary examples of the apparent wildness of mountain buffalo is a story told by Dr. R. W. Shufeldt in June 1888. Shufeldt came upon eight mountain buffalo on the northern slope of the Big Horn Mountains in the autumn of 1877. These mountains extend from northern Wyoming into southern Montana. Shufeldt wrote:

> We came upon them during a fearful blizzard of heavy hail, during which our animals could scarcely retain their feet. In fact, the packer's mule absolutely lay down on the ground rather than risk being blown down the mountain side, and my own horse, totally unable to face such a violent blow and the pelting hail (the stones being as large as big marbles), positively stood stock-still, facing an old buffalo bull that was not more than 25 feet in front of me. . . .
>
> Strange to say, this fearful gust did not last more than ten minutes, when it stopped as suddenly as it had commenced, and I deliberately killed my old buffalo at one shot, just where he stood, and, separating two other bulls from the rest, charged them down a rugged ravine. They passed over this and into another one, but with less precipitous sides and no trees in the way, and when I was on top of the intervening ridge I noticed that the largest bull had halted in the bottom. Checking my horse, an excellent buffalo hunter, I fired down at him without dis-

mounting. The ball merely barked his shoulder, and to my infinite surprise he turned and charged me up the hill. Stepping to one side of my horse, with the charging and infuriated bull not 10 feet to my front, I fired upon him, and the heavy ball took him square in the chest, bringing him to his knees, with a gush of scarlet blood from his mouth and nostrils.

Upon examining the specimen, I found it to be an old bull, apparently smaller and very much blacker than the ones I had seen killed on the plains only a day or so before. Then I examined the first one I had shot, as well as others which were killed by the packer from the same bunch, and I came to the conclusion that they were typical representatives of the variety known as the "mountain buffalo," a form much more active in movement, of slighter limbs, blacker, and far more dangerous to attack.[8]

In September 1882 G. O. "Coquina" Shields, a well-known sportsman and writer, L. A. Huffman, the Miles City, Montana, photographer, and Jack Conley, a packer, had an experience similar to Shufeldt's, also in the Big Horn Mountains. After spending one night camped in Wyoming, the three men set out the next morning along the eastern side of the mountains into Montana. From time to time they came to small parks or meadows covered with rich grass and surrounded by thick timber. Near one they found fresh tracks of a buffalo. Shields and his companions dismounted, picketed their horses, and followed the buffalo tracks on foot. Soon they could tell that the animal was feeding as it moved along. At that point the hunters split up. Conley went to the left of the trail, Shields to the right, and Huffman stayed with the trail. About a quarter mile ahead of him, Shields saw a steep rugged rock. No sooner had he reached its top than he saw the mountain buffalo "quietly grazing and browsing on some weeds among the rocks." The animal was about 150 yards away. Shields dropped to one knee, drew a bead, and fired. "I distinctly heard the dull 'whack' of the ball as it struck him," wrote Shields. The animal's tail twitched quickly over his back as if he were shooing a fly away. Then he turned and plunged madly over the rocks in the opposite direction. Meanwhile, Conley and Huffman spotted the animal and both fired. The buffalo became furious and turned and lunged toward Shields, who was closest, but as the buffalo neared the hunter he staggered and suddenly fell to the ground. The bullets had taken effect.

Recalling his adventure, Shields described the animal as a "mountain buffalo, a large, finely-formed, noble-looking animal; his fur is finer, darker and curls more than that of the plains buffalo."[9]

To men like Shields the mountain buffalo was considered a real prize. Killing one was not easy and hunting them in the mountains was considered "real sport." During the 1870s and the early 1880s many hunters went after mountain buffalo. Just how many animals were killed is unknown, but by the middle 1880s they were becoming scarce through the Rockies, from Colorado northward into Montana. By 1893 only five small groups of mountain buffalo were reported in North America. Four were in Colorado, the other in Yellowstone National Park. Those in Colorado comprised a small number in Middle Park, another group of perhaps twenty in the Kenosha Range, another ten or fifteen at Hahn's Park in Routt County, and a handful around Dolores.[10] But by 1900 these animals were gone. Man had completed the job in Colorado of removing the mountain buffalo, and he was working toward that end at Yellowstone to the north.

The buffalo found in Yellowstone by John Colter in the early 1800s were mountain buffalo. By the time that rugged area of northwestern Wyoming was made our first National Park in 1872, several hundred mountain buffalo could be found within its boundaries. But by 1900, the number had been reduced considerably by poachers. In 1902 a group of plains buffalo were brought to the park to restore the area's buffalo population. Unfortunately, the cross-breeding that resulted apparently destroyed the true mountain buffalo. The new arrivals, however, did not live as they had on the plains. Instead, they adopted the habits of Yellowstone's mountain buffalo and spent the winters in the valleys and the summers high in the mountains.[11]

Today the only living descendants of the mountain buffalo can be seen in Yellowstone National Park. They act much like the mountain variety of seventy years ago. Rarely are they spotted by tourists during the summer months, but when winter comes they are frequent visitors to the meadows. In 1973 there were approximately 800 buffalo in Yellowstone National Park.[12]

As mentioned earlier, the mountain buffalo is classified by zoologists as *Bison athabascae.* The wood buffalo of the Canadian far north also fall under this classification, and they look and act much like the mountain buffalo of a hundred years ago in spite of their different homes. Samuel Hearne was probably the first white man to see a wood buffalo. He saw the animal in the early 1770s and described one bull which was killed as "so heavy" that six to eight Indians skinning the animal did not even attempt to turn him over. Instead, they first skinned one side and cut it up to lessen the weight before turning the carcass over.[13]

Between the late 1700s and about 1890, all accounts of wood buffalo indicate "an uneven distribution and a general scarcity" of the animal, writes W. A. Fuller, Canadian zoologist. Fuller adds, "During that time almost the only Europeans in the country in addition to the explorers were a few fur traders, so the scarcity cannot be ascribed to the influence of the white hunters as it was in Southern Canada and the United States."[14]

By 1893 fewer than 300 wood buffalo could be located in Canada. They were found in the lower Peace River and Slave River areas of Canada's Northwest Territories, only about 400 miles south of the Arctic Circle. These areas were very difficult for man to reach during the summer months and almost impossible to traverse during the winter.

That year, 1893, the Royal Canadian Mounted Police got the job of protecting the wood buffalo, and under their watchful eyes the animals prospered and multiplied. Then, in 1911, six "buffalo rangers" were appointed to patrol the buffalo range and to protect the animals from poachers. By 1922, when the wood buffalo's range became Wood Buffalo National Park, there were about 1,500 wood buffalo in the area. But about then officials of the Canadian government made the same mistake officials at Yellowstone National Park had made about twenty years earlier. To increase the buffalo population in the newly established Canadian park, they decided to transport some plains buffalo from government herds in Alberta to Wood Buffalo National Park. Between 1925 and 1928, almost 7,000 plains buffalo were shipped by train and by barge to the far north and released among the wood buffalo. Scientists today regret that transplant of buffalo because nearly all the buffalo born since then in the park have been hybrids of the two original species.

Wood buffalo family, mounted at National Museum of Canada, Ottawa. *Courtesy Museum.*

This illustration showing Indians hunting mountain buffalo in the winter is by S. Eastman and appears in H. R. Schoolcraft's *Indian Tribes of the United States, 1860.*

"Buffalo Hunt, on Snow Shoes," George Catlin lithograph, from his *North American Indian Portfolio*.

Mountain buffalo, from H. R. Schoolcraft's *Indian Tribes of the United States*, 1860.

But not all of the wood buffalo had contact with the plains buffalo. W. A. Fuller suspected that some pure wood buffalo stock might have remained untouched by the plains variety in an inaccessible area of the park's northwest corner. In 1959, N. S. Novakowski explored that region and found about 200 buffalo. Five were killed and taken to Ottawa, where they were identified as true wood buffalo, not hybrids. When Canadian authorities learned that the true wood buffalo still existed, the Canadian Wildlife Service undertook a program to move some of the animals into an area where there would be absolutely no chance for contact with the hybrids. During the early and middle 1960s more than a hundred wood buffalo were captured in two separate winter roundups and placed in isolation. More buffalo might have been taken, but the roundups were held in February, perhaps the coldest month in the far north. Only then was the area where the animals lived accessible to ground vehicles. Temperatures ranged down to near 40° below zero and heavy snow made the job all but impossible, but the task was completed. Today the Canadian Wildlife Service closely watches over a small number of pure wood buffalo, the last of a kind.[15]

V

For The Indian

I go to kill the buffalo.
The Great Spirit sent the buffalo.
On hills, in plains and woods.
So give me my bow; give me my bow;
I go to kill the buffalo.

Sioux song

The lush grass, blanketing the sloping plains for as far as the eye could see, rippled in the gentle wind. In the distance, grazing peacefully in the warm sun, were perhaps 200 buffalo. To the north, downwind at some distance, a few figures moved slowly through the tall grass. Indians. As they moved closer to the buffalo, they dropped to their hands and knees. Slowly they crawled toward the unsuspecting animals.

Soon, on a prearranged signal, they jumped to their feet. Yelling, the Indians ran toward the buffalo. Surprised and alarmed, the buffalo started to run south, away from the Indians. Over a small rise the buffalo went, their hooves sounding like thunder as they trampled the earth. Suddenly the lead animals came to an arroyo hidden by the tall grass. Quickly they tried to change course, but it was too late. Other buffalo struck their leaders from behind and those in front went crashing into the arroyo. The others behind followed. Buffalo upon buffalo filled the arroyo. In a matter of a minute or so nearly all the herd had tumbled over. Some were stunned by the fall. Others struggled unsuccessfully in the dust and blood to regain their footing. They could not escape. Only a handful of buffalo at the back of the running mass were able to miss the jump, and they ran off on a course parallel to the arroyo.

As the first buffalo tumbled into the arroyo, Indians, hiding below, jumped to their feet and began throwing lances from point-blank

52

range into the moaning and helpless creatures on the arroyo floor. In a short time the slaughter was over. Then the butchering began. The Indians must have laid in a good food supply that spring day.

This hunt occurred more than 8,000 years ago on the rolling plains, not far southeast of where the town of Kit Carson, Colorado stands today. The facts, as they are known, have been pieced together by archaeologists who in 1957 unearthed the remains of 193 buffalo and forty-seven Indian artifacts, including twenty-seven flint projectile points.[1]

At least 10,000 years ago the Indians who then lived on the plains hunted buffalo. The early type of projectile points found in and near the fossil remains of extinct buffalo buried on the plains leaves little doubt of this.[2] But just when the first Indian hunted buffalo on the plains or how many Indians first resided on that vast expanse of open country is unknown. Estimates of the Indian population of North America at the time of Columbus' landfall here have ranged from 850,000 to 9,800,000.[3] Whatever the actual figure, probably about one-sixth of them lived on the millions of acres of plains and prairies stretching from the Missouri River to the Rockies and from the Rio Grande north to the Canadian plains.

To those Indians living on the plains and prairies, the buffalo was an important commodity. The animal provided not only food but also shelter, clothing, and other necessities. When Alvar Muñez Cabeza de Vaca crossed Texas in the early 1500s, he became the first white man to see the extent to which the Indians depended upon the buffalo. At a spot not far south of modern Corpus Christi, Texas, Cabeza de Vaca noted:

> The cows [buffalo] reach here, and I have seen them three times and eaten of their meat. . . . They come from toward the north forward through the country . . . and spread themselves all over the land more than four hundred leagues; and on all this road along the valleys through which they come, the people dwelling along there descend and live upon them; and they take inland many skins.[4]

And a decade later, when Francisco Vasquez de Coronado went in search of the golden city of Quivera, he too saw with his own eyes how the Indians lived off the buffalo.

But such dependence was not the case for Indians living east of the Mississippi. Though they killed buffalo, they never placed such dependence upon the animal. Their land was far more productive than the plains and prairies; there was much wild game, and more tribes fished, trapped, and raised crops in the woodland environments east of the Mississippi.

The buffalo dominated all phases of the Plains Indians' life: their thinking, their philosophy, their religion. The Indians followed the buffalo within their tribe's generally accepted territorial limits, for during the late summer and fall the animal transformed some tribes into nomads as they hunted the buffalo.[5]

At about the time the white man first began to penetrate the plains and prairies, there were about two dozen Plains Indian tribes dependent, some more than others, upon the buffalo.[6] Many of these tribes originally came from a woodland environment, but only a few maintained their old arts of fishing, trapping, and planting crops. The Mandans, Gros Ventres, Pawnee, and Osages, for example, raised some crops and at times were not as dependent upon the buffalo as were the other tribes, but when crops failed they were forced to rely almost entirely upon the buffalo for survival.

Some Plains Indian tribes believed that buffalo came from underground. Tribes on the southern plains thought there was a huge cave somewhere on the Staked Plains of northwest Texas which was the buffalo's home. It was from there, they believed, that each spring some benevolent spirit sent huge numbers of buffalo out onto the plains for the Indians. Colonel Richard I. Dodge, writing in 1883, reported:

> One Indian has gravely and solemnly assured me that he has been at these caverns, and with his own eyes saw the buffalo coming out in countless throngs. Others have told me that their fathers or uncles, or some other of the old men have been there. In 1879 Stone Calf assured me that he knew exactly where these caves were, though he had never seen them, that the Good God had provided this means for the constant supply of food for the Indian, and that however recklessly the white men might slaughter, they never could exterminate them. When

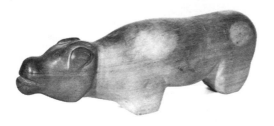

a

b

c

Above and on the following page are pictured a few ways in which the American Indian has employed a buffalo motif in various ritual, play, or utilitarian items.
a. Fox Indian, 1875-1900. Carved wood (17¾" long). *Courtesy Museum of the American Indian, Heye Foundation.*
b. Cheyenne Sun Dance buffalo, 18th century. Carved wood (8¾" long) with fiber-tuft tail and inset carved wooden horns. *Courtesy Philbrook Art Center, Tulsa.*
c. Sioux toy buffalo (14" long). *Courtesy Museum of the American Indian, Heye Foundation.*
d. Sioux painted hide shield (20½" dia.). *Courtesy Museum of the American Indian, Heye Foundation.*
e. Pueblo (Zuni) buffalo head (10¾" high), from altar of Buffalo Clan; used during Shalako Dance. Horsehair and carved/painted cottonwood. *Courtesy Taylor Museum, Colorado Springs Fine Arts Center.*
f. Cherokee buffalo mask, used in Booger Dance. Carved wood (11¼" high). *Courtesy Museum of the American Indian, Heye Foundation.*
g. Arapaho painted buffalo skull, used in Sun Dance. *Courtesy Museum of the American Indian, Heye Foundation.*
h. Sioux club. Head (7¼") is buffalo head carved from stone; handle (19½") is wooden, bead-wrapped. *Courtesy Denver Art Museum.*
i. Sioux pipe, c. 1870. Bowl (5") is buffalo carved from catlinite; stem (17½") is carved wood. *Courtesy Ralph T. Coe, William Rockhill Nelson Gallery of Art, Kansas City, Missouri.*

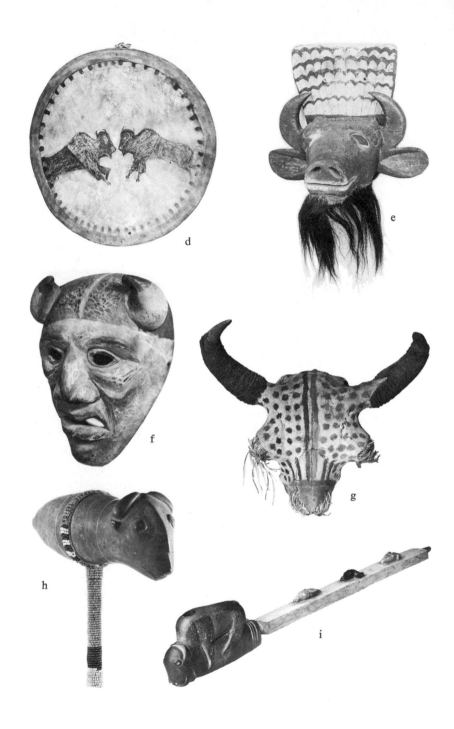

last I saw him, the old man was beginning to waver in this belief, and feared that the Bad God had shut up the openings and that his people must starve.[7]

For the tribes in the north there was a similar explanation. Indians in southern Canada thought their buffalo came from under a lake not far south of what is now Battleford, Saskatchewan.[8] Several hundred miles away the Sioux believed their buffalo came from a large cave. The Sioux legend recalled that in "olden times" an Indian traveling over the land came upon a hill with many caves in it. He explored them, and in one he found buffalo tracks on the floor and buffalo hair on the sides where the animals had rubbed against the walls. The Indian believed that cave was the home of buffalo that lived underground. "These buffalo had many earth lodges, and there they raised their children," he said.[9]

It is very doubtful that any other animal in the world has ever matched the buffalo in providing so many commodities of prime importance to any one people. The number of non-food uses of the buffalo has been placed as high as eighty-seven.[10] However, Red Cloud, a former Sioux chief, once told a white man that he knew of only twenty-two uses of the animal by his tribe:

> His meat sustained life; it was cut in strips and dried, it was chopped up and packed in skins, its tallow and grease were preserved—all for winter use; its bones afforded material for implements and weapons; its skull was preserved as great medicine; its hide furnished blankets, garments, boats, ropes, and a warm and portable house; its hoofs produced glue, its sinews were used for bowstrings and a most excellent substitute for twine.[11]

On the woodless plains even buffalo droppings were valuable to the Indians, and later to the settlers. Buffalo chips often were the only fuel available. One early plainsman said buffalo chips burned "like peat, producing no unpleasant effects."[12] Another reported, "When dry, it catches easily, burns readily, and makes a hot fire with but very little smoke, although it is rapidly consumed."[13] Still another pioneer noted with a straight face that buffalo chips even had a cooking advantage. "A steak cooked over them requires no pepper," he said.[14] Buffalo chips were used for more than fuel. When stones or rocks could not be found on the plains, buffalo chips were some-times stacked in piles as markers by Indians and white men alike.[15]

On occasion, Indians on the plains would use the hide of a newly killed buffalo to cook the animal's own meat. Old Indians have told how a hunting or war party on the move would, after killing a buffalo, turn the carcass on its back and butcher the animal. Then, digging a hole in the ground, they would push the fresh hide into the depres-sion. After stones were heated over a nearby fire, they would be tossed into the hide. Water would then be poured over them. The cut-up portions of buffalo meat would then be dropped into the water, and the meat would slowly boil. Sometimes the same feat would be performed above ground. The buffalo's paunch would be hung from wood stakes above a fire. The results were the same, a well-cooked piece of buffalo meat to ease the pains of a long day on the trail.[16]

In the Mandan tribe it was the woman's job to cure and preserve buffalo meat. It was not considered a job for men. George Catlin wrote that the meat "is all cured and dried in the sun without the aid of smoke or salt." The women first would cut the choicest parts from the animal. These pieces were then cut across the grain "so as to secure alternate layers of lean and fat in strips of about one inch thick." The women then hung the meat over poles "resting on crutches, out of the way of dogs and wolves; there it is exposed to the sun for several days." Catlin observed, "It is by that time so effectually dried that it can be carried everywhere, even during the hottest months of the year . . . without damage." Catlin felt that the "purity of the air" on the plains might have had something to do with preserving the meat. He noted that there were many salt springs in Mandan country but none of the Indians used salt "in any manner." Catlin also noticed that Indian women cooked their meat longer than white men did. "I have found that meat thus cooked can be eaten and relished without salt or any other condiment," he said.[17]

Long before Indians on the plains began using horses, they were able hunters of buffalo. Probably the first method they used to kill buffalo was the "surround." They would form a cordon around a herd, circling on foot only as many buffalo as they felt they could handle. Then running in circles around the terrified animals, yelling loudly, the Indians would slowly close the circle, making it smaller and smaller. Then at an appropriate time they would let go with their lances and arrows. As the first buffalo would fall near the outside of the circle, it would act as a roadblock for the animals still in the ring. Soon many such roadblocks would trap the remaining live buffalo. Unwilling to pass, the animals inside would eventually meet their fate.[18]

One Indian surround was witnessed in the early 1800s by John Dunbar, a Presbyterian missionary. He went to live with the Pawnees in 1835. His first November with them, he joined the Indians on a buffalo hunt along what is today the Platte River in Nebraska. Dunbar later reported that the Indians killed about 300 buffalo in the surround.[19] Colonel Dodge reported that a white hunter friend saw some Indians kill the same number of buffalo in another surround. "The whole affair occupied less than ten minutes after the signal was given, and . . . not a single buffalo escaped," he said.[20] In 1872 Charles J. "Buffalo" Jones claimed he saw one of the last surrounds conducted by the Pawnee Indians. He reported that out of a herd of about 2,000, "only forty-one" buffalo were killed.[21]

Another buffalo hunting method used by Indians was called "impounding." It was more difficult than the "surround," and most authorities seem to think that for that reason it probably developed after the "surround." Impounding was used more widely in partly wooded, rolling, and rocky country than on the flat plains. Where wood or rock was available, Indians would build a strong corral, drive buffalo inside, and kill them.[22] It was not uncommon for them to use the same enclosure over and over provided they could continue to find buffalo in the same area. One such pound is about ten miles above the junction of the Elbow and the Bow Rivers near Calgary, Alberta. It was discovered in 1875 and contained a large deposit of buffalo bones buried under several feet of dirt.[23] There is another pound about ten miles from Fort Union in eastern Montana. Thousands of buffalo bones have been found there.[24]

Still another method used by Indians to kill buffalo was the buffalo jump or *piskun,* as it was often called. Indians would find a cliff, then locate a herd of buffalo and stampede the animals over the precipice. Indians waiting below would finish off any buffalo that might survive the fall.

The Crow Indians may have been the first tribe on the northern plains to use buffalo jumps. They were one of the first tribes to migrate from an eastern forest area out onto the plains of southern Montana and northern Wyoming. The Crow had previously used the "jump" technique in the east to kill buffalo and other animals. They also used other techniques such as crowding animals into snow banks and surrounds, and they also slaughtered buffalo trapped on ice in the winter. When the Crow Indians came west they found terrain ideally suited for jumps and soon began using the technique more and more to kill buffalo.

Joseph Medicine Crow, a Montana Crow Indian, said his people credited Old Man Coyote, who they claimed was the creator of the world and the teacher of the people, with giving them the idea of crowding buffalo over cliffs. Joseph Medicine Crow tells the legend:

> Old Man Coyote was hungry one day and so were the rest of the people, and he decided to go look for some meat. Soon he found a herd of buffalo upon a bench. He decided to trick them over a nearby cliff hidden by a thick cloud or fog. In his usual bragging style, he challenged the head buffalo to a race. Well, as the story goes, immediately his challenge was accepted and the race was on. Naturally Old Man Coyote selected the course, was ahead, and when they approached the cliff, he disappeared quickly. The buffalo herd went over the cliff. Old Man Coyote jumped around to the lower end and returned with his clothing all messed up, if he was wearing clothing; maybe he changed himself into a buffalo. Anyway, he made his nose bleed and pretended like he was hurt. The inference of this little legend was that buffalo were driven over cliffs a long, long time ago by the Crow Indians.[25]

History records that other Indian tribes on the northern plains and in the foothills of the northern Rockies also used buffalo jumps. Today, from northern Wyoming northward through Montana into Alberta and Saskatchewan, at least a hundred buffalo jumps can still be seen. One is called Old Women's Buffalo Jump. It is near the town of Cayley, Alberta, about fifty miles south of Calgary. Legend says that one day Indian men journeyed to a village close to the jump to take women of the village as their wives. All of the Indian men except

DRIVING BUFFALO OVER A BLUFF.

Theodore R. Davis drawing,
from *Harper's,* January 1869.

"Buffalo Hunt, Surround," George Catlin lithograph, from his *North American Indian Portfolio.*

"Buffalo Hunt by Indians,"
C. F. Wimar painting, 1861.
*Courtesy Kansas State
Historical Society.*

"Indian Hunting Buffalo," Felix O. C. Darley drawing, 1844. *Courtesy Kansas
State Historical Society.*

one, "The Oldman," found a wife. He became so enraged by the turn of events that he changed himself into a pine tree. As recently as sixty or seventy years ago there was a lone pine tree standing at the site of Old Women's Buffalo Jump. It is not there today.[26]

Another buffalo jump is in Big Horn County, Montana, at the junction of Hoodoo Creek and Dry Head Creek. It was used by the Crow Indians and is still called by them the "place of many buffalo dried heads." The area around it is called "Dry Head Country." The story or legend of how "Dry Head Country" got its name began with an Indian chief called Tip of the Fur. He and part of his tribe were camped where Beauvais Creek and the Big Horn River meet, across from the present-day town of St. Xavier, Montana. Tip of the Fur and another chief, Red Bear, were rivals. Both were medicine men or sorcerers who claimed they had black magic powers. One day both men became involved in a misunderstanding of some sort. Tip of the Fur told Red Bear, "Now I'm going to leave you behind. As soon as I leave you're going to get hungry, and you'll be hungry for days. In the meantime, I'll have lots of buffalo with me."

Tip of the Fur and his party then moved out of camp, leaving Red Bear and his followers alone. In a matter of hours a thick fog settled over the area. Light drizzle began falling. It was bad for Red Bear. The fog remained for days and days. Red Bear's hunters could not even see their horses. They could see only a few feet in front of them. And they could not find any game. They were hungry for a week, maybe two. In the meantime, Tip of the Fur went into the Beauvais country, the Dry Head Basin area, with all of the buffalo in front of him. They seemed to be in a mass movement down into Basin country. Since so many buffalo were nearby, Tip of the Fur thought it wise to get his fall supply of meat. He and his hunters found a cliff at the junction of the two streams. As some of the buffalo moved close to the cliff, the Indians stampeded the animals over. When the kill ended, and as the hunters prepared to butcher the buffalo, Tip of the Fur ordered his braves to "cut off the heads of the buffalo and pile them in one spot." This was done.

Afterwards, whenever Indians made a buffalo kill at that spot, the heads of all dead buffalo were always cut off and piled in one large heap. In time there was a massive pile of dried buffalo skulls. After the wild buffalo on the northern plains were no more, Indians would still restack the skulls each year. The last time this was done was in the early 1900s. Since then souvenir hunters have carried away the skulls at the spot which is still called by many persons the "place of many dried heads."[27]

Another technique used in killing buffalo was to encircle the animals with fire. Indians, particularly the Miami and Santee and a few other tribes of the upper Mississippi country, would set fires on all sides but one of a buffalo herd. At the open end the hunters would wait in ambush. The buffalo would try to escape through the opening, only to be killed.

In June 1971 a team of scientists headed by Waldo Wedel, senior archeologist at the Smithsonian Institution in Washington, found the tooth of a horse while digging at the site of an old Indian village near Lyons, Kansas. That tooth, estimated by Wedel to be from around the middle 1600s, about 100 years after Coronado, is a clue that horses were used by Indians of the early plains culture. Until the Kansas discovery, many authorities thought that Indians on the southern plains did not obtain the horse until the 18th century.

"The tooth we recovered is not proof, in my judgment," said Dr. Wedel, "that the Indians who lived on the site actually used the horse. The piece in question bore several cuts and may have served as an ornament or talisman of some sort, quite possibly received in trade or otherwise from contemporary people who had horses."[28]

Just how the Indians came to have horses is still a matter of conjecture and controversy. Hypotheses have ranged from the Indians' having the horse before the Spaniards arrived to their first obtaining the animals from wild herds formed from horses lost or abandoned by early Spanish explorers. Many authorities today believe that the Indians of the plains first obtained horses in trade at Spanish settlements in New Mexico and Texas; in time, horses changed hands between tribes, and eventually Indians from Texas northward into southern Canada had them.

With the horse, the Plains Indians had mobility. No longer was a whole village needed to hunt buffalo. A few skillful Indians on horseback could kill many animals and provide more than enough meat for a village. But when no buffalo could be found near camp, the Indians usually turned to "medicine." Often the rites were held even when buffalo were plentiful. Every plains tribe had its own magical rite which they performed for some benevolent spirit who they thought would produce more buffalo. Such a ceremony was usually directed by a medicine man, and it usually included mimicry of a desired event. Often the medicine man made it appear that he had indeed caused the buffalo to return, but what appeared to be magic was in truth usually a natural occurrence. The medicine man was a student of nature. He knew that the same result would normally follow a specific occurrence. He knew a rain storm would always end. Spring would always come. Winter would always bring colder weather. Heavy objects would always sink in water, and so forth. And he knew that if he waited long enough the buffalo would return. Experience had taught him that.

During the 1830s George Catlin was living with the Mandan Indians on the Upper Missouri. When the buffalo wandered away from the village, he would watch the chiefs and medicine men sit in solemn council to decide on what course of action to follow. Nearly always, wrote Catlin, they decided upon the "old and only expedient which 'never had failed'," the buffalo dance.

> The chief issues his order to his runners or criers, who proclaim it through the village—and in a few minutes the dance begins. The place where this strange operation is carried on is in the public area in the centre of the village, and in front of the great medicine or mystery lodge. About ten or fifteen Mandans at a time join in the dance, each one with the skin of the buffalo's head (or mask) with the horns on, placed over his head, and in his hand his favourite bow or lance, with which he is used to slay the buffalo. . . .
>
> It never fails, nor can it, for it cannot be stopped (but is going incessantly day and night) until "buffalo come." Drums are beating and rattles are shaken, and songs and yells incessantly are shouted, and lookers-on stand ready with masks on their heads, and weapons in hand, to take the place of each one as he becomes fatigued, and jumps out of the ring.

While all of this was going on, Indians were posted on nearby hills to watch for buffalo. When they discover the animals, wrote Catlin, the signal is given—the "throwing their robes"—and word is passed to the village. There everyone thanks the Great Spirit, the medicine men, and the dancers for what they have done. Catlin said that buffalo dances sometimes lasted "two or three weeks without stopping an instant, until the joyful moment when buffaloes made their appearance."[29] So the dance "always" worked.

The Blackfeet Indians, before a buffalo hunt, would worship the *Iniskim* or buffalo stone. It was a fossil type of marine shell which in a crude sort of way resembled a buffalo. Probably because of its appearance the Blackfeet thought the stone had some magical powers that lured the buffalo to their hunting grounds.

There are several versions of the legend of how the Blackfeet first began using the buffalo stone. Each describes how the stone revealed itself to an Indian woman one day. The stone also gave the woman a number of songs to cast spells upon the buffalo. In the Blackfeet ritual, a man and a woman, usually husband and wife, would join together to carry out the ceremony. Sometimes they were assisted by friends. There would be a series of songs and sometimes imitative dances. The songs usually took up most of the night and early morning before the village started out on a buffalo hunt.[30]

A strange tale about a buffalo head was told by an adopted Blackfeet, Chief Buffalo Child Long Lance, who is said to have been a North Carolina mixture of Negro and Indian. One night a band of Crow Indians crept into the Blackfeet camp on the northern plains and stole all the horses. The next morning the Blackfeet decided to chase the Crows, recapture the stolen ponies, and maybe add some Crow ponies to their remuda. As was the custom, before starting out, the Blackfeet chief consulted the medicine man. After much discussion the medicine man gave his approval for the chase. For three days the Blackfeet pursued the Crow. The fourth night the medicine man announced that he wanted to see whether the spirits still favored the chase; he said to find a buffalo and to bring back its head. Some men left. When they returned, they reported they could not find a buffalo. The medicine man, White Dog by name, told them to wait. He retired to his teepee. A little while later, White Dog came outside and described a dream, a vision, he had just had. He ordered the men to follow a nearby river for a distance of six buffalo arrows, one buffalo arrow being about 350 yards or the longest distance at which an arrow could kill a buffalo. "Then," said White Dog, "cross the river and climb the butte on the other side. In the distance you will see a lone buffalo."

Again the men left camp. It was after dark when they returned, and swinging from a pole was the head of a buffalo. White Dog examined the shaggy head and announced it was perfect for what he needed. He then directed that the head be placed just outside the camp and a big fire built. All the men formed a semicircle around the roaring blaze and faced the direction in which they had been chasing the Crow Indians. The medicine man moved in and took up a position in the center of the other Indians and began singing. In a few moments some coyotes outside the camp joined in with their mournful howls. After White Dog sang one song, he told everyone to watch the buffalo head for signs that would mean they should continue the chase. As all turned their heads toward the buffalo's shaggy head, White Dog began a second song. When he reached a part that mentioned the buffalo's head, sparks began to shoot out of the left eye of the head. In a moment the head began to sway from side to side and smoke rolled out of its nostrils. It even panted. "The spirits favor the Blackfeet," said White Dog. Within a few days the Blackfeet recaptured many of their ponies and killed some of the enemy.[31]

The northern Cheyenne Indians felt there was great medicine in a "buffalo cap" or sacred hat. The *is si wun,* as they called it, was perhaps their most sacred object. It exemplified the buffalo as subsistence. The hat was a head covering made from the hide of a buffalo cow's head. Two handpainted buffalo horns were attached. It was kept by the medicine man or chief priest along with the medicine arrows. Like the hat, the arrows were considered sacred objects. There were four arrows in all, each stone-headed and of very fine workmanship. Two of the arrows were for war, the other two for hunting. The arrows and the medicine hat were prayed to, sacrificed to, and sworn by. And when the Cheyennes went into battle their buffalo hat and arrows went along.[32]

Some Plains Indians were not above using their most precious commodity, the buffalo, both as a decoy and as a shield in battle. In one tale told by an old cavalry trooper, seventy-six recruits under the command of Lieutenant William B. Royall set out from Fort Leavenworth in the late spring of 1848 to reinforce the Santa Fe battalion. On their journey west they were called upon to perform numerous functions. One was to escort a paymaster to Fort Mann, a rest stop and repair station midway between Fort Leavenworth and Santa Fe, about four miles west of present-day Dodge City, Kansas. When the soldiers reached Coon Creek, about forty miles east of Fort Mann, near where Kinsley, Kansas, is today, they set up camp near the creek's edge. Tandy Giddings, an old plainsman who was traveling with the soldiers, walked up to Lt. Royall and said, "You'd better double the guards tonight."

"Why?" asked the lieutenant.

"Well sir, we haven't seen a buffalo for two days, and that's a sign there are Indians around."

Lt. Royall took Giddings' advice and posted an extra guard that night, but nothing happened. However, at daybreak, just as everyone was about to have breakfast, the soldiers heard what sounded like wolves. There were howls from the far side of Coon Creek, then from down the creek, and finally from the uplands on the side where the soldiers were camped.

"Look out, boys, I've heard them wolves many a time," shouted Giddings. "Them's Indians howling."

Moments later, an immense herd of buffalo appeared along the creek stampeding straight for the soldiers' camp. For a moment, the soldiers forgot Giddings' warning. Some grabbed their rifles and prepared to "get some fresh meat for breakfast." But Giddings yelled, "Hold on, boys, the Indians are behind them buffalo." The soldiers held their fire. The buffalo kept coming, heading straight for the camp. As they neared, for some unknown reason they veered to one side, passing the camp at a safe distance. Behind them, just as Giddings had predicted, came Indians mounted on horses. As they came into full view, others appeared from two other directions charging toward the soldiers. It was a war party of about 800 Comanches and Apaches that converged upon the seventy-six troopers, a paymaster, and one old plainsman. Holding shields made from the necks of bull buffalo, the Indians merged into one charging force. The soldiers opened fire.

These Indians had fought the U.S. Army before. They were used to drawing the soldiers' fire while at a pretty safe distance. Then, before the soldiers would have time to reload their muzzleloaders, the Indians would swoop down and lance the troopers to death. But this time things were different. New breech-loading carbines had been issued the recruits at Fort Leavenworth. These rifles could be loaded and fired five times a minute. As the Indians charged in for what they thought would be the kill, the soldiers opened fire again. Many Indians fell to the ground. Others, surprised, withdrew. But again the Indians attacked, and again the soldiers fired. Not understanding how the soldiers could load and fire their rifles so fast, the Indians withdrew about a mile from the camp. The soldiers could hear them shouting and yelling at each other. Fifteen minutes passed. The soldiers waited. Then they could see that the Indians were preparing to charge again. When the charge came, the Indians were spread out in a line about a hundred yards wide and eight-ten yards deep. In front of their line was a woman, an Indian Joan of Arc, who urged them on. She was dressed in red.

This time the Indians apparently thought they could drive the soldiers back to the creek bed, where other Indians were waiting. The ground between the soldiers and the Indians was clear. The Indians charged in their new formation, shouting with unearthly yells, their lances held high in the morning sun. At 400 yards the soldiers fired, then again at 300 yards, at 200 yards, and at 100 yards. But Indians still kept coming. Then someone yelled, "Shoot their horses." Everyone dropped his carbine to the level of the horses and fired. In an instant many Indian ponies fell to the ground. They were either dead or wounded. Those Indians behind the front ranks stopped, looked at the fallen horses and their horseless comrades scrambling to their feet. Suddenly, almost in union, they turned and fled. Some Indians were seen riding double, having picked up a horseless brother. Those who were not wounded ran away. And the woman in red vanished among the fleeing Indians. The soldiers were ordered to mount their horses and follow. They did, and they killed more Indians before returning to camp. Thus was the battle of Coon Creek, which began with a stampeding herd of buffalo.[33]

When the white man introduced guns on the plains and prairies, he did not readily displace the bows and arrows and spears used for centuries by the Plains Indian. The white man's muskets were difficult to load on horseback. They could not be fired as fast as arrows, and at close range a bow and an arrow in the hands of a skillful Indian could be just as accurate as a musket. Not until the repeating rifle was introduced in the 1870s did the Indians begin to discard their bows and arrows.

The bows used by Plains Indians were of two general types: the sinew-backed or compound bow and the plain wooden bow. They were made from wood or horn and were shorter after the Indians began to use the horse. George Catlin, in his *North American Indians,* said the standard length of bows among the Crow and Blackfeet was three feet. The arrows were proportionate in length to the bows. Indians made them by passing a piece of wood through a hole drilled in a piece of horn and rubbing the wood between two grooved stones. This straightened and rounded the wood. To increase the arrow's accuracy, three feathers were generally attached to the butt end of the shaft, which was usually notched. For shooting small game and birds, most Indians on the plains did not attach heads to the arrows. But for buffalo and other large game an arrowhead of bone, horn, or stone was usually fastened to the end. Later metal points were used.

Did Indians herd buffalo much like cattlemen herd domestic cattle? There is evidence that at least one tribe did. Father A. McG. Beede observed:

> They say that from the year 1864 onward till after the Sitting Bull-Custer battle they—the Western Sioux Indians—kept the Buffalo herded back into Montana or the extreme western part of the present North Dakota. This statement, strange as it may seem, is confirmed by numerous white frontiersmen. . . . Old Indians say, and other evidence shows, that these Sioux Indians extended their herd line as far north only as the Kill-deer Mountains—never called Killdeer Mountains by the Sioux—and to the point on the Missouri River just north of these mountains. Meanwhile, the Buffalo, now quite safe from ruthless depredation by white men, increased greatly in numbers; and in the summer of 1875, many of these Buffalo crossed the Missouri from the south side to the north side—from the present McKenzie County

to the present Williams County and Mountrail County—and that escaping the Indian herd line, they roamed up and down the Missouri often swimming the river forth and back, and going as far down the river as Painted Woods, sixty miles up river from Bismarck, N.D.

All white frontiersmen know well that from 1864 onward till after the Sitting Bull-Custer battle, no Buffalo were to be found on the vast plains west of the Missouri River and east of this Indian herd line, except a very few that occasionally slipped through the herd line and were not followed by Indians and killed for food. But from 1875 on, large numbers of Buffalo were sometimes found in the north country which was beyond this Indian herd line, and some of them wandered as far east as the Turtle Mountains country.

The Buffalo herds that crossed the Missouri in the late summer of 1879 or 1880, and those that drifted east from Montana after the Custer fight, mingled together in the territory west of the Missouri River, and for three years after 1879 there were so many Buffalo in that country that only a few thoughtful men supposed that the time should ever come when Buffalo hides could not be had in plenty.

After the Sitting Bull-Custer battle, which was the end of this careful attention given to the Buffalo by the Western Sioux Indians, the Buffalo that had thus increased in numbers came freely eastward, and seemed, at times, almost to fill the plains country west of the Missouri River in North Dakota. In old times, Indians never ruthlessly slaughtered Buffalo, or other animals. They used what was necessary and preserved the rest. The Buffalo were the Indians' Cattle.

The Western Sioux Indians regarded Buffalo stealing about as a Dakota cattle-rancher regards Cattle stealing—as a crime deserving death. This Buffalo stealing by white men in the Dakotas was many times more the cause of trouble with the Sioux than all the gold in the Black Hills. I know this from hearing old Indians freely talk among themselves. That the trouble was chiefly caused by the Black Hills gold is an erroneous notion of the matter originated by white men. This notion has become fixed in written history, but it is incorrect.[34]

If America had never been discovered by white men and if the Indians of the plains had had no horses, untold numbers of buffalo might still be roaming the plains and prairies of North America. Before the white men brought horses, which gave the Indians mobility, the Plains Indians were not excessively wasteful with the buffalo. Sometimes they did kill more than they needed, but at times Indians also starved for the lack of buffalo. The number killed by Indians before the white men came was only a very small percentage of the total number of buffalo on the plains and prairies. Frank Gilbert Roe, after many years of study on this matter, concluded, "There is not a fragment of evidence that I have been able to discover to suggest that the Indians would ever have exterminated the buffalo herds."[35]

Under the influence of white men and horses, however, the Plains Indians did contribute to the buffalo slaughter. With the horse the Indians prospered for a time. They were able to kill more buffalo and raise their standard of living. Indian populations increased. But more Indians required more buffalo meat. And as contact with the white man grew, the Indian began to kill more buffalo to obtain robes to trade for the white man's goods.

John D. Hunter, a white man who spent his childhood and young manhood with the Indians, noted in 1832, "I have never known a solitary instance of their wantonly destroying any of these animals, *except* on the hunting grounds of their enemies, or encouraged to it by the prospect of bartering their skins with the traders."[36]

Thus the white man was responsible, directly and indirectly, for the slaughter of the buffalo. As Frank Gilbert Roe once told me, "The true slayers en masse were the white 'buffalo butchers' who swarmed on the Buffalo Plains."[37]

Perhaps the words of Red Cloud, the Sioux chief, best exemplify the situation. In the 1880s he told a white friend that the Sioux never killed buffalo wantonly—"Where the Indian killed one buffalo, the hide and tongue hunters killed fifty."[38]

Indians preparing for a buffalo hunt on the 101 Ranch near Ponca City, Oklahoma, c. 1909. *Courtesy Division of Manuscripts, University of Oklahoma Library.*

Three American Indian Movement members at Wounded Knee, South Dakota, March 1973. *Courtesy Wide World Photos.*

Title page of W. E. Webb's 1872 book, *Buffalo Land,* drawn by Kansas artist
Henry Worrall.

VI

When the Killing Began

*No intelligent excuse can be offered for the senseless slaughter
and awful waste of a valuable and harmless animal purely for
personal gain or to satisfy a blood-lust to kill.*

Martin S. Garretson 1938

The near extermination of the American buffalo did not happen overnight, nor was one generation of human beings fully responsible for clearing the plains and prairies of this most noble animal. The process was slow. It started with the Indians. Then came the white man, and as he developed the fur trade west of the Mississippi, the momentum of the killing increased.

Before the 19th century began, Americans were not active in the fur trade of the north. The French and French Canadians were first. They entered the trade in the 1600s. Then came the English who formed the Hudson's Bay Company around 1670. It was more than 138 years later that Americans actively entered the fur trade. That was after the United States purchased the Louisiana Territory from the French in 1803 and after Lewis and Clark explored the northern part of the new territory between 1804 and 1806. America's active involvement in the fur trade began in 1808 when the St. Louis Missouri Fur Company was formed by Manuel Lisa, Pierre Chouteau, William Clark, and others. It was followed a year later by the American Fur Company, which was to make its owner, John Jacob Astor, a fortune. These companies hired men to live in the wilderness and to trade with the Indians.

The War of 1812 disrupted further American expansion of the fur trade into the northwest. By 1820, however, the prospects for increased trade appeared good. The price of furs was holding steady in the East in spite of a depression. There was better profit in fur trading than had existed for years. At St. Louis new trading companies were formed and by the early 1820s, even though Indians had killed several traders in the north, expeditions succeeded in enlarging the fur trade along the Upper Missouri. From the beginning, the fur traders of the northwest did not engage in trapping animals themselves. They built trading posts, called them forts, and obtained furs trapped by Indians. Most of the Indian tribes of the north, with the exception of the Blackfeet, traded the pelts of beaver and other small game and buffalo robes for beads, tobacco, hand mirrors, axes, blankets, calicoes, powder, lead, and whiskey. Beaver pelts were the most prized during the early 1800s. They were what the traders wanted, but other pelts and buffalo robes were accepted. It was good business, even though buffalo robes traded cheaply.

To the south the American involvement in the fur trade of the southern Rockies and the southern plains did not get fully under way until 1821, after Mexican Independence. It was then that the Santa Fe Trail was opened between Westport (now Kansas City) and Santa Fe. Soon Taos and Santa Fe became bases for trapping and trading with the Indians. In 1822, a year after the Santa Fe trail opened, William Henry Ashley joined forces with Andrew Henry in the fur trade at St. Louis. Instead of depending upon the Indians to supply furs, Ashley decided to have white men catch the beaver. In the *Missouri Gazette,* a newspaper published at St. Louis, he advertised for "enterprizing young men." Those he hired were paid according to the number of skins they obtained. These were the men who became known as "mountain men." Most were self-sufficient, confident individualists who more often than not lived off the land like the Indians. There were Kit Carson, Osborne Russell, John H. Weber, Hugh Glass, Thomas Fitzpatrick, Philip Thompson, Jedediah Smith, Thomas "Pegleg" Smith, to name only a few.

And there was Jim Bridger. Bridger was born in 1804 and he joined Ashley's fur company at the age of eighteen. Two years later he crossed the Rocky Mountains and viewed the Great Salt Lake, on whose shores buffalo could then be found. Bridger, somewhat of a legend in his own day, had the reputation of being quite a storyteller. He is credited with a yarn about a great snow storm that supposedly killed vast numbers of buffalo in the Salt Lake Valley. As the story goes, Bridger pickled the buffalo's meat in the Great Salt Lake, thereby preserving it. For many years, as Bridger told the story, he and his Ute Indian friends lived on the plentiful supply of pickled buffalo meat.[1]

For about fifteen years these mountain men reigned supreme in the Rockies. Concentrating mostly on the central Rockies, they procured vast numbers of beaver pelts. In 1825 William Ashley introduced the first of sixteen "rendezvous" where caravans from the East would annually meet the mountain men during the summer months in or near the Rockies and exchange supplies and Indian trading items for pelts taken by the mountain men during the previous season.

The first rendezvous was held on Henry's Fork, along the present-day boundary between Wyoming and Utah. Ashley's diary relates that 120 men were at that first meeting, which saw much drinking, gambling, singing, and other fun and games. When Ashley returned east in October 1825, he carried with him nearly 9,000 pounds of beaver worth about $50,000. Few, if any, buffalo robes went east with Ashley. In what was then the United States the trade in buffalo robes was not yet established, but to the north across the international boundary in Canada, buffalo robes were already items of commerce along with pemmican. Pemmican was dried buffalo meat pounded fine. Sometimes berries were mixed with it. One recipe for pemmican reads:

> Cut the buffalo's flesh into large lumps, then into thin slices. Hang up in the sun or over a fire and dry. When it is thoroughly dry, place upon rawhides spread out upon the ground. Pound until the meat is reduced to a pulp. Then place pulp into a strong container. Pour boiled tallow, boiled separately, over pulp. Stir together and thoroughly mix until dry pulp is soldered down into a hard, solid mass by the melted fat poured slowly over it. Then, after mixture cools, place in airtight container.

This recipe was obtained by G. O. Shields during the late 1800s. Just where he obtained it is unknown, but it is similar to an old Indian recipe that I obtained from a Crow Indian friend in Montana. The only difference is that the Crow, and probably other tribes as well, used bags about the size of flour sacks, made of buffalo hide with the hair on the outside, as containers in which the meat and tallow were mixed. Once the mixing was finished, the bags were sewed up as tightly as possible. Even in buffalo hide the pemmican did not spoil. There was no limit to how long it would keep. Really, it kept itself, requiring no spices or seasoning or anything else for its preservation.

In Canada, Alexander Henry was, in part, responsible for the early trade in pemmican and buffalo robes. He had led a band of voyageurs and hunters for the North West Company to the Red River area of southern Canada in 1800. He found countless buffalo on the surrounding plains. By the following year Henry had sent 5,000 pounds of pemmican and thirty-one buffalo robes east. He also had built a permanent trading post on the Pembina River, a tributary of the Red.

During the next eight years the North West Company shipped buffalo robes and pemmican to the east. In 1812 competition developed. Thomas Douglas, the fifth Earl of Selkirk, purchased land from the Hudson's Bay Company and settled a colony of Scottish Highlanders on the Red River a few miles below the mouth of the Pembina. Until they could grow crops, the settlers hunted buffalo. This alarmed the North West Company, which felt it was entitled to the hunting rights in the area, and soon the bloody Pemmican War began. It lasted nearly nine years until the North West Company was merged with the Hudson's Bay Company in 1821. The Hudson's Bay Company dominated the trade in what is today Canada for the remainder of the century.

The settlers along the Red River were able to return to buffalo hunting by 1820. In 1821 they formed a joint-stock company called the Buffalo Wool Company. It was to provide "a substitute for wool, which substitute was to be the wool of the wild buffalo, which was to be collected in the Plains, and manufactured both for the use of the colonists and for export, and to establish a tannery for manufacturing the buffalo-hides for domestic purposes." A capital of £2,000 sterling was raised, and orders were sent to England for machinery, implements, dyes, and skilled workmen. Two groups of workers arrived, including "curriers, skinners, sorters, wool dressers, teasers, and bark manufacturers, of all grades, ages and sexes." But soon the workers became disillusioned. "The wool and the hides were not to be got, as stated, for the picking up; the hides soon costing the company 6s each, and the wool 1s 6d. per pound." The workers and even the officials of the Buffalo Wool Company turned to "the bottle and the glass."

They have been described as "wallowing in temperance." The hides were allowed to rot and the wool to spoil, and the tannery proved a complete failure. Besides using up their capital, the company found itself in debt. Bankruptcy followed.

Some samples of buffalo cloth had been made and sent to England but sold for a loss. It cost two pounds ten shillings per yard to make along the Red River. In England it could be sold for only four shillings and sixpence, less than 10 per cent of the cost.[2]

Because of the extraordinary demand for buffalo robes when the company began, the settlers had organized buffalo hunting on a scale about as large as anything ever seen on the plains. Although the company failed in 1822, the hunts lasted for two more decades.

During the 1820s, the mountain men had little contact with buffalo in the Rocky Mountains. Although mountain buffalo existed there, encounters were rare. Only when the mountain men ventured out onto the plains did they find many buffalo. On the high plains east of the Rockies, the trade in buffalo robes and buffalo tongues began to increase, and during the early 1820s about 200,000 buffalo were killed annually on the plains. But only about 5,000 of these were killed by white men.[3]

The trade in buffalo robes and tongues was particularly heavy along the Upper Missouri. There Indians were encouraged to kill the animals for robes and tongues and to trade these items at the growing numbers of trading posts. Buffalo were often killed for their tongues alone. This was particularly true during the summer months, when the buffalo's robe was poor. In 1804 Charles Mackenzie, after visiting some Mandan villages, reported, "Large parties who went daily in pursuit of the buffalo often killed whole herds, but returned only with the tongues."[4] And George Catlin made a similar observation in the early 1830s. After he had arrived at Fort Pierre on the Missouri, he learned that only a few days earlier 500 or 600 Sioux had gone buffalo hunting. When the Indians returned, they had 1,400 fresh buffalo tongues, for which they were given a few gallons of whiskey.[5]

Long before the white man came to the plains and prairies, Indians had considered buffalo tongue a great delicacy. Tongues had even played a role in some Plains Indian rituals connected with the buffalo. When the white man discovered tongues, they soon became popular and the market developed. By the early 1830s, large numbers of buffalo tongues, packed in salt, were shipped down the Missouri River. In the summer of 1831, a newly built steamboat named *The Yellowstone,* built at Louisville by Kenneth McKenzie, then the leading trader on the Upper Missouri, delivered to St. Louis a full cargo "of buffalo robes, furs and peltries besides ten thousand pounds of buffalo tongues."[6]

Though buffalo tongues could be obtained at any time of the year, good buffalo robes could not. The buffalo carried a good robe only between November and March, the months when nature gives the animal a thick and full robe to protect it from the snow and bitter cold. And then only buffalo cows provided really good robes. Only young bulls, less than three years old, had full robes. The hair on the hind quarters of older bulls was only as long as that of a horse. The cow's robe, however, had hair that was uniformly the same thickness.

The Indians, prime suppliers of robes, needed much time and labor to properly prepare and dress each robe. Most Indian families could produce only a limited number of robes each year. John B. Sanford, an official of the American Fur Company, observed, "It is seldom that a lodge trades more than twenty skins a year."[7] For a buffalo robe, the skin first had to be fastened by its edges with rawhide thongs to a frame, much like an old-fashioned quilting-frame, made of four stout poles tied together at right angles. The only tools used were a hoe-like device and plenty of muscle. The underside was scraped, a few inches at a time, until the back was soft and pliable. This often took several days. The final quality of any robe was determined by how good the fur was to begin with and how much effort went into working the robe.[8]

As the 1830s began, the business of fur trading was beginning to undergo change. Heavy, unrestricted trapping began to deplete the supply of beaver in the Rockies. By 1835, for example, only about 25,000 beaver pelts were shipped down the Missouri River from Fort Union, compared to nearly 50,000 buffalo robes. Meantime, in the East, hat manufacturers were already beginning to substitute seal and South American nutria for beaver. Then something else happened. The United States began to trade with China. New England ships brought cargoes of silk, among other things, from the Far East. As attention in the United States suddenly focused upon China, the hat makers turned to silk for hats, joining other industries capitalizing on products from the Far East. Beaver, which had sold in the eastern United States during the early 1830s for between $4 and $6 a pound, dropped in price to between $1 and $2 a pound by 1840. That was the last year the rendezvous system of disposing of pelts in the Rockies was used.

It was also the last year for what had become known as the Red River hunts of southern Canada. Jean Baptiste commanded the last hunt in June 1840. For twenty years the hunters had held an annual buffalo hunt using Red River carts patterned after the *charette* or country cart of France. Each cart had two 5-foot-high wheels. They were pulled or pushed by men and women. In that last Red River hunt in 1840, more than 1,200 carts were used by the 1,630 persons who took part. Nearly 3,500 buffalo were killed and skinned, and when the hunters started for market in August, their carts carried not only the 3,500 hides, but also 240 bales of dried buffalo meat and 375 bags of pemmican. The hunts ceased to exist because buffalo were becoming hard to find in the large number necessary for the Red River colonists to make a profit[9]

To the south of Canada, along the Upper Missouri, American Indian traders frequently hunted buffalo for meat but never on the scale of the Red River hunts. John J. Audubon witnessed such a hunt in July 1843, while he was staying at Fort Union, about seven miles above the mouth of the Yellowstone River in what is today eastern Montana. Audubon saw traders kill three buffalo near Fox River. He noted the traders' uncivilized habits when he reported that Alexander Culbertson, a young Englishman in charge of Fort Union, broke open one buffalo's head and ate part of the animal's brains raw. Other traders did the same. "The very sight of this turned my stomach," wrote Audubon.[10]

George Ruxton had a similar experience during his trip through the Rockies in the early 1840s. "The powers of the Canadian voyageurs and hunters in the consumption of meat strike the greenhorn with wonder and astonishment, and are only equaled by the gastronomical capabilities exhibited by the Indian dogs, both following the same plan in their epicurean gorgings," said Ruxton. In slaughtering a fat buffalo cow, he said:

The hunter carefully lays by, as a titbit for himself, the "boudins" and medullary intestine, which are prepared by being inverted and partially cleaned (this, however, is not thought indispensable).

The depouille or fleece, the short and delicious hump-rib and "tender-loin," are then carefully stowed away, and with these the rough edge of the appetite is removed. But *the* course, is, *par excellence,* the sundry yards of "boudin," which lightly browned over the embers of the fire, slide down the well-lubricated throat of the hungry mountaineer, yard after yard disappearing in quick succession.

Ruxton told of the time he saw two French Canadian trappers eating the intestines of a buffalo in what might be called a contest. The two men were seated on the ground. Between them, on a rather dirty saddlecloth of buffalo skin, was a huge coil of buffalo innards looking something like a giant snake. Each man began eating one end of the coil.

As yard after yard slid glibly down their throats, and the serpent on the saddlecloth was dwindling from an anaconda to a moderate-sized rattlesnake, it became a great point with each of the feasters to hurry his operation, so as to gain a march upon his neighbor, and improve the opportunity by swallowing more than his just portion; each, at the same time, exhorting the other, whatever he did, to feed fair, and every now and then, overcome by the unblushing attempts of his partner to bolt a vigorous mouthful, would suddenly jerk back his head, drawing out the same moment, by the retreating motion, several yards of boudin from his neighbor's mouth and stomach (for the greasy viand required no mastication, and was bolted whole), and snapping up himself the ravished portions, greedily swallowed them, to be in turn again withdrawn and subjected to a similar process by the other.[11]

The 1840s saw a continued increase in the trade of buffalo robes along the Upper Missouri River. There trading posts were used to capture the trade. To the south, however, where trading posts had not been necessary because mountain men had obtained the furs and disposed of them through the rendezvous system, new trading posts were built. And traders began to obtain buffalo robes and other items from the Indians on the plains east of the Rockies.

In 1840 traders at Forts St. Vrain and Bent in present-day Colorado shipped 15,000 buffalo robes east to St. Louis over the Santa Fe Trail. A year later, they sent 1,100 large bales of robes weighing nearly fifty tons.[12] But the robe trade over the Sante Fe Trail and on the southern plains never did surpass the robe trade along the Upper Missouri and Yellowstone Rivers. During the first half of the 19th century, the vast majority of buffalo robes sold in the East were obtained from the northern plains and from the fur companies' trading posts.

When Captain P. St. George Cooke visited Fort Laramie in 1845, he reported, "The trade at this post is principally for buffalo robes; nine thousand were lately sent off by the American Fur Company and how many by the other company I do not know."[13] The following year a St. Louis newspaper, *Missouri Republican,* reported that eight Mackinaw boats had arrived from Fort Laramie with "1100 packs of buffalo robes, 10 packs of beaver, and 3 packs of bear and wolf skins." In 1850 about 100,000 buffalo robes reached the St. Louis market alone, more than any previous annual total.

Other than fur company records, newspapers are the best source for information on the buffalo robe trade, particularly those newspapers in St. Louis, which for many years was the fur trading capital. As an example, the *Missouri Republican* gave the following account of steamboat arrivals and cargo for June 1854 at St. Louis, which will give the reader some idea as to the trade in buffalo robes:

> The El Paso brought down Harvey, Primeau & Co., upper Missouri fur returns (the cargoes of three Mackinaw boats); and aboard the Highland Mary, reaching St. Louis June 14 were the contents from three more of this company's boats (786 bales of, and 440 loose buffalo robes, 99 packages of furs and skins, also a lot of buffalo tongues). Upper Missouri fur returns (1,024 buffalo robe bales, and 44 deer skin packs) of Pierre Chouteau, Jr. & Co., were brought down from Council Bluffs by the Honduras, which left there June 20 and reached St. Louis on the 25th.

Three years later, in 1857, the trading posts along the Upper Missouri were by themselves shipping 75,000 buffalo robes annually to eastern markets such as St. Louis.[14]

In his history of the American buffalo, J. A. Allen provides good information regarding the buffalo robe trade in 1857. He recalls meeting F. F. Gerard, a Cree interpreter, who spent twenty-five years as an agent for the American Fur Company in the Upper Missouri country. From Gerard, Allen learned that the trade in buffalo robes at the principal posts on the Upper Missouri in 1857 was as follows:

> At Fort Benton, the number received amounted to 3,600 bales, or 36,000 robes; at Fort Union, 2,700 to 3,000 bales, or about 30,000 robes. At Forts Clarke and Berthoud, 500 bales at each post, or about 10,000 robes; at Fort Pierre, 1,900 bales, or 19,000 robes; giving a total for one year of about 75,000 robes, which he informed me was about the annual average at that period. Allowing that the Indians retained only as many more for their own use, and estimating . . . that one robe represents the destruction of three buffaloes, gives four hundred and fifty thousand as the number killed by a portion only of the Upper Missouri Indians in one-third of a year, or over a million and a third annually.[15]

During the late 1840s and early '50s, many emigrants crossed what is today Kansas and Nebraska, the area the Santa Fe and Oregon Trails traversed. The emigrants saw many buffalo and frequently killed the animals for meat. By 1855 civilization was forcing the buffalo farther west in Kansas, almost 150 miles west of the Missouri and the north-south boundaries of eastern Kansas and Nebraska.

One emigrant, Peter Bryant, who came from Illinois to settle in Kansas near Holton, about forty miles west of the Missouri River, wrote his brother in August 1859, "Buffaloes can be found 100 miles west of here on the Republican Fork." Bryant added that he had talked with a Pawnee Indian who had watched one party of white hunters kill 300 buffalo, but had seen no Pawnees killing the animals. Bryant said the Indian was most concerned about the white men killing "his" buffalo.[16]

Many sportsmen traveled to Kansas during the 1850s to shoot buffalo. Of all who came west during that decade, perhaps the most famous was Sir St. George Gore of Ireland. Gore outfitted his hunting expedition in St. Louis in the spring of 1854, then headed west. By the time his party returned to St. Louis nearly three years later, Gore had spent more than $500,000 and traveled at least 6,000 miles with forty servants, about fifty hunting dogs, numerous wagons and horses, and a number of companions, including scientists and mountain men.

Accounts of Gore's expedition tell of their killing 105 bears, 1,600 elk and deer, and more than 2,000 buffalo. Gore hired the old mountain man Jim Bridger to guide the expedition up the North Platte River and eventually to the mouth of the Tongue River in eastern Montana. There Gore and his party camped the winter of 1855-56 and killed so many buffalo that the Indians complained to their agent. The federal government ordered Gore to stop hunting buffalo in that area. This was the first and only time that the U.S. government moved to stop the wholesale slaughter of the buffalo while large numbers of the animals still roamed the plains and prairies. But it appears the government was more concerned about Gore's safety among the Indians than it was about the killing of so many buffalo. Whether Gore himself actually shot many buffalo is doubtful. Accounts of his almost unbelievable slaughtering expedition indicate he was a bad shot and had to have help to bring down even one shaggy.[17]

More typical of the buffalo hunts of the late 1850s and early 1860s is one in which five men from Wyandotte, Kansas Territory—now Kansas City, Kansas—participated. The five—John P. Alden, T. J. Darling, I. D. Heath, John Blachly, and Alanson Reeve—set out from Wyandotte in late September 1860, taking with them two ox teams, each with two yoke of oxen, a pair of ponies with a light wagon, a small rat terrier, a big dog,

> a tent, three Sharp's and one muzzle-loading rifle, two shot guns for small game, three Colt's revolvers, Navy size, plus five lbs. powder, ten lbs. shot, twenty-five lbs. lead, five sacks flour, two sides bacon, one bushel onions, one bushel potatoes, seven bushels corn meal, four cwt. salt, fifty lbs. sugar, ten lbs. coffee five gallons sorghum, one gallon common molasses, table salt, pepper, ginger, pipes and tobacco for three smokers, &c., &c., &c. . . .

The men traveled west to the bottom lands of the Solomon, Saline, and Smoky Hill Rivers. From the start they kept a log of their trip. The first week in buffalo country was logged as follows:

> *Friday, 12th,* drove on up Solomon, and camped on Hard-crossing creek (fitly named). There the buffalo carcasses were very numerous, showing that there had been great slaughter among them a few weeks previous. A good many Irish and German settlers in this neighborhood. Shot at more buffalo.
>
> *Saturday, 13th,* drove on across Sand creek, a beautiful soft water stream, and camped on Solomon. Shot at more buffalo, but brought none down yet.
>
> *Sunday, 14th,* forded Solomon and camped on Salt creek. Saw a few wild turkeys. The timber consists of cottonwood, burr oak, white oak, black walnut and elm.
>
> *Monday, 15th,* killed three buffalo, and brought two, nicely dressed into camp before sundown. Lost the other from not being able to dress it soon enough. Buffalo must be dressed immediately after being killed, or the meat will spoil. Two of these were killed each with a single ball, while the third was so tenacious of life that he refused to give up till he had nearly a pound of lead under his skin.
>
> *Tuesday, 16th,* spent the day cutting up beef, and commencing the process of "jerking." Cut up the hind quarters in thin slices, across the grain, which are then dipped in hot brine, or allowed to lie in cold brine all night, and afterwards the meat is spread upon small strips of wood or upon wire, and dried by smoke, sun and wind.
>
> *Wednesday, Oct. 17th,* Darling, Blachly and Reeve went out and brought in one buffalo, nicely dressed. Some one remained in camp all the time.
>
> *Thursday, 18th,* killed a fat young buffalo, and "jerked" the whole.
>
> *Friday, 19th,* killed one buffalo.

When the five buffalo hunters returned to Wyandotte on November 22, 1860, they brought 5,000 pounds "of as fine buffalo meat as ever tickled the palate of a hungry man." The hunt was considered a success.[18]

Inexperienced buffalo hunters would often take out after buffalo. Some had success, if they were lucky, while others did not. One such unsuccessful hunting expedition consisted of twelve men and boys from Alma, Kansas Territory, in 1860. They headed west from Alma toward buffalo country, about sixty miles farther west. With five ox-drawn wagons the group traveled twenty miles a day until they found buffalo near present-day McPherson, Kansas. There were many buffalo in the vicinity, but not being experienced at hunting the animal, the group were able to kill only one old bull whose meat was too tough to eat. They finally settled on buying buffalo meat at 50 cents a carcass from some professional hunters they met.[19]

Occasionally, buffalo would wander east of their normal range. If their wanderings led them to settlements in eastern Kansas, they were almost always killed on sight. Edward Secrest remembered such an incident. He settled in the Fancy Creek area not far north of present-day Manhattan, Kansas, in the late 1850s. One warm June day in 1857 he was working in the fields not far from his cabin when a neighbor brought the news: buffalo had been seen moving east off the prairie and down into the valley where Secrest and other settlers lived. Secrest quit work and headed for home to get his rifle. But as he approached his small cabin, there stood a huge buffalo bull rubbing his shaggy shoulder against a corner of the cabin. The bull was standing there "calmly and unconcerned as if he owned the whole shebang in fee simple," recalled Secrest, who could do nothing but wait and watch. After a while, the buffalo, satisfying his itches, took off along Fancy Creek going farther south into the valley. Secrest ran into his cabin for his rifle, but by the time he got back outside the buffalo was out of range. Later he learned that a neighbor had killed the animal. Though Secrest saw no other buffalo that day, he learned that other settlers in the area had killed several. The animals were skinned and butchered and the meat divided up. Secrest got a share. Those wandering buffalo were welcomed with open arms by Secrest and his neighbors. As he recalled, the people in the Fancy Creek area

lived in clover the summer, fall and winter following. Their larders were replenished as never before. We jerked Indian fashion, and hung it up well cured, in long rows and festoons on the joists in the back part of the cabin. When our Mill Creek Hoosier boys would come up occasionally in their lumber wagons on a visit during the winter days, as the Old Boreas howled and drove the snow like sifted flour under the clapboarded roof, we sat around a roaring fire. As we tired of telling Hoosier yarns, one after another would get up and cut down a chunk of buffalo bull beef, return to the fire and cut slice after slice of the juicy steak and talk of the halcyon days a coming when Kansas would be a Free State with the wild territorial sunflower as its glorious emblem.[20]

Not too much is recorded about the hunting of buffalo on the plains and prairies during the Civil War. Buffalo were killed by both Indians and white men, and the trade in buffalo robes and meat did continue because the buffalo range was far to the west of where the battles were being fought. During much of the 1860s, wagon load after wagon load of buffalo robes, obtained from Indians, was shipped east by freight wagons across the southern plains. One Kansas freighter, Theodore Weichselbaum, a close friend of my grandfather, Archie Long, hauled freight back and forth across Kansas from 1861 until 1869, when the railroads forced him out of business. Weichselbaum, a big man who nearly always wore a bandana tied around his neck and a large hat, was typical of the men who freighted buffalo robes to market during this period. Some freighters only carried the robes to market for a fee. Weichselbaum, however, would buy them from traders in western Kansas who had traded them from Indians. As he recalled, "I hauled thousands of buffalo robes to Leavenworth with my teams. I sold them mostly to W. C. Lobenstein, for from five to six dollars apiece, cash."[21]

During the winter of 1861-62, Nebraska trader John Nelson also relied upon the Indians to do the killing:

> I took a couple of wagons and two teamsters for the Gilmans, to trade with the Indians on Big Turkey Creek, about one hundred and fifty miles away. There I found the Oglalas, Brules, and a few Cheyennes, and amongst them my old friends, Spotted Tail and Two Buck Elk. Spotted Tail had been released from prison

and returned to his tribe. Both were glad to see me again, and I passed a very pleasant three months with them, returning to the ranche with the whole of my stock sold out and laden with a good supply of furs. I believe I had four hundred and seventy-five of the best buffalo robes alone, besides other skins, raw hide lariats etc. The robes cost about one and a quarter dollar each, and sold at five dollars; the lariats cost twenty-five cents, and sold at four dollars. In addition, I had two thousand dried buffalo tongues, which I purchased for thirty-six cents a dozen and resold to the emigrants at twenty-four dollars a dozen.[22]

For the Indians on the plains, the 1860s were a decade of change: the white man's encroachment upon Indian lands, the white hunter's killing of the Indian's buffalo for sport and meat, the failure or inability of Washington to enforce the 1861 treaty of Fort Wise, giving the Arapahoe and Cheyenne Indians of the Upper Arkansas a new reservation. The Indians became restless, resentful of the whites, and generally unhappy.

In the spring of 1863, S. G. Colley, the Indian agent for the Upper Arkansas, reported that some young Kiowas had demanded or forcibly taken goods from small wagon trains in the area. Colley reported, however, that starvation caused most of the depredations committed by the Indians. He noted that there was not a buffalo within 200 miles, adding, "Thousands and thousands of buffalo are killed by hunters during the summer and fall merely for their hides and tallow, to the displeasure and injury of the Indians."[23]

In late 1863, H. T. Ketcham, special Indian agent, was sent west to vaccinate the Indians against smallpox, a disease brought to the plains years before by the white man. Ketcham reported that the Indians were poor, sick, and starving on the Arkansas River, Pawnee Fork, and Walnut Creek in Kansas Territory. Ketcham said many Indians were living on cattle that had belonged to emigrants who had either died or turned back, leaving their animals to roam the land. He also noticed that buffalo were very scarce along the Arkansas and that the Indians were bitter against the white hunters and traders who had traded them whiskey for what few buffalo robes they had.

By May 1864, the Cheyennes on the South Platte were becoming hostile toward whites. By early June of that year, Indian raids were becoming frequent occurrences. On one summer day—June 29, 1864 —a train of thirty wagons was attacked on the road along Lodgepole Creek between Julesburg and Fort Laramie. That same day, many miles to the south, a coach was attacked on the Arkansas River between Fort Larned, Kansas and Fort Lyon, Colorado Territory. In the months that followed, the Sioux and the North Arapahoes joined the Cheyennes, who had invited them, in an all-out war against the whites.

As the Civil War continued in the East, Washington had its hands full. Few troops could be spared for the Indian Wars in the West. But when the Civil War ended in April 1865, more troops were sent west. New Army posts were established to secure the plains from the Indians. To feed the soldiers, the Army made contracts with local men to hunt buffalo and to supply buffalo meat to the military posts. The railroads, which began pushing westward after the war, also soon became part of the buffalo's story.

When the Union Pacific began crossing Nebraska in the late 1860s, and the Eastern Division of the Union Pacific Railway (later the Kansas Pacific) started across Kansas, these railroads, like the Army, contracted with hunters to supply buffalo meat for the railroad construction crews. Later the Atchison, Topeka & Santa Fe did the same thing.

For a time, killing buffalo for their meat became a profitable business for such men as William F. Cody. He entered into contract with the Goddard Brothers, who had in turn contracted to feed the Eastern Division of the U.P. (Kansas Pacific) construction crews. According to one source, Cody, who became known as Buffalo Bill, killed 4,280 buffalo in eighteen months. But in May 1868 his job ended. The railroad construction stopped at Sheridan, Kansas, and Cody's services were no longer needed.[24]

Today, Cody is probably the most famous "killer" of the buffalo in history, even though many other men killed far more buffalo than Cody. Cody gained the publicity and fame.

As the railroads reached buffalo country, the iron horse and the buffalo met. For the animal, the trains brought trouble. A Kansas newspaper, the *Junction City Union,* reported on November 2, 1867:

We learn that about ten days ago, while the passenger train was

"The Robe Press," Theodore R. Davis drawing, from *Harper's,* January 1869.

"Curing Hides and Bones," Paul Frenzeny and Jules Tavernier drawing, from *Harper's Weekly,* April 4, 1874. *Courtesy Kansas State Historical Society.*

Drawing by Theodore R. Davis, who wrote: "It would seem to be hardly possible to imagine a more novel sight than a small band of buffalo loping along within a few hundred feet of a railroad train in rapid motion, while the passengers are engaged in shooting, from every available window, with rifles, carbines, and revolvers. An American scene, certainly." (*Harper's,* January 1869) *Courtesy Richard Fitch, Santa Fe.*

coursing the plains, on the track of the Union Pacific railway west of Ellsworth, a buffalo was discovered ahead, and failing to get beyond rifle range was handsomely shot by a marksman on the locomotive. A few days afterwards, in about the same locality, the train was intercepted by a whole herd of buffaloes and compelled to halt until they had crossed the track! This is the statement as made to us. The Kansas City Journal of Commerce says that for three miles the buffaloes pushed along parallel with the train, heedless of many shots fired among them, and finally swept across the track, ahead of the locomotive, fairly worsing the iron horse by bringing him to a halt.

Though the buffalo did cause some delays on the railroads in buffalo country, it was exciting to ride a train through or near a buffalo herd, and soon some of the railroads began advertising sightseeing excursions and buffalo hunts as added enticements to get people to ride the roads. The advertising worked, and hunting buffalo by rail became a favorite pastime for many.

One of the best accounts of such an excursion was given by E. N. Andrews. He was one of about 300 persons who left Lawrence, Kansas in October 1868 on the Kansas Pacific Railway in an excursion party organized for the benefit of a church. Their train consisted of five passenger coaches, one smoking car, one baggage car, and one freight car. The party had about thirty women and 250 men with seventy-five or eighty guns. They spent the first night sleeping in their seats at Ellsworth, Kansas. At dawn the train headed west again and by evening the party reached the 325 milepost west of Kansas City.

In the distance, as if upon a gentle slope, we beheld at least a thousand buffalo feeding, though this was not a circumstance to that which followed. For ten miles vast numbers continued in sight; and not only for this short (?) distance, but for forty miles, buffalo were scattered along the horizon, some nearer and others more remote. In estimating the number, the only fitting word was "innumerable;" one hundred thousand was too small a number, a million would be more correct. Besides, who could tell how many miles those herds, or *the* herd, extended beyond the visible horizon? It were vain to imagine!

By the time the train reached the 365 milepost the party saw buffalo near the train.

> Three bulls were on the left of the track, though nearly all that we had seen were on the right, or north of that barrier, while now on their southward course, feeding in their slow advance. They kept pace with the train for at least a quarter of a mile, while the boys blazed away at them without effect. It was their design to get ahead of the train and cross over to the main body of their fellows; and they finally accomplished their object. The cowcatcher, however, became almost a bull-catcher, for it seemed to graze one as he passed on the jump. As soon as the three were well over upon the right, they turned backward, at a small angle away from the train, and then it was that powder and ball were brought into requisition! Shots enough were fired to rout a regiment of men. Ah! see that bull in advance there; he has stopped a second; he turns a kind of reproachful look toward the train; he starts again on the lope a step or two; he hesitates; poises on the right legs; a pail-full of blood gushes warm from his nostrils; he falls flat upon the right side, dead. . . . The engineer was kind enough to shut off the steam; the train stopped, and such a scrambling and screeching was never before heard on the Plains, except among the red men, as we rushed forth to see our first game lying in his gore. The writer had the pleasure of first putting hands on the dark locks of the noble monster who had fallen so bravely. Another distinguished himself by mounting the fallen brave. Then came the ladies; a ring was formed; the cornet band gathered around, and, as if to tantalize the spirits of all departed buffalo, as well as Indians, played Yankee Doodle. I thought that "Hail to the Chief," would have done more honor to the departed.
>
> And now butcher-knives and butchers were in requisition. "Let us eviscerate and carry home this our first captive without further mutilation, that we may give our friends the pleasure of seeing the dimensions of the animal." This seemed a good plan, and we proceeded to carry it out. After the butchers had done their work, a rope was attached to the horns, and the animal, weighing about fifteen hundred pounds, was dragged to the cars

and thence lifted on board the freight-car, a few of our party climbing upon the top of the car, the better to pull on the rope.[25]

Civilization had arrived on the Kansas prairie.

As the railroads pushed westward across buffalo country, from the south came the Texas cattlemen driving their cattle to the railroads in Kansas. Like everyone else, the trail drivers saw buffalo. B. A. Barroum of Del Rio, Texas, an oldtime trail driver, wrote in 1920 for J. Marvin Hunter's *The Trail Drivers of Texas* that he saw many buffalo in 1871 while driving a herd of longhorns north to Kansas:

> On a plain about half way between the Red Fork and the Salt Fork we had to stop our herds until the buffalo passed. Buffalo, horses, elk, deer, antelope, wolves, and some cattle were all mixed together and it took several hours for them to pass, with our assistance, so that we could proceed on our journey. I think there were more buffalo in that herd than I ever saw of any living thing, unless it was an army of grass-hoppers in Kansas in July, 1874.[26]

Another cattleman, George W. Saunders, the organizer of the Old Time Trail Drivers Association, recalled for Hunter how in 1870 at the age of seventeen he first went up the trail to Kansas with a herd of longhorns:

> At Pond Creek [Indian Territory] we encountered our first buffalo. The plains were literally covered with these animals, and when we came in sight of them all of the boys quit the herd and gave chase. It was a wonderful sight to see these cowboys dashing after those big husky monsters, shooting at them from all angles. We soon learned that it did no good to shoot them in the forehead, as we were accustomed to shooting beeves with our pistols, for the bullets would not penetrate their skull. We would dash by them and shoot them between the eyes without apparent effect, so we began shooting them behind the shoulder and that brought them down. I killed two or three of the grown buffaloes and roped a yearling which I was glad to turn loose and let him get away with a good rope. I soon became satisfied with the excitement incident to killing buffalo, swimming streams, being in

stampedes, and passing through thunder storms, but I still longed to be mixed up in an Indian fight, for I had not yet had that sort of experience.[27]

The coming of the railroads to buffalo country saw not only the development of eastern markets for Texas cattle but also an increased demand for buffalo robes. The publicity given railroad excursions into the land of the buffalo made more people in the East want to own a buffalo robe. A robe was not only a symbol of the excitement on the frontier but also a practical thing to have. Buffalo robes were used as covers by drivers and passengers on sleighs and buggies and other forms of open transportation during the winter months. Some people had coats made from buffalo robes.

The largest supply of buffalo robes in the 1860s continued to be from the northern plains, from traders along the Upper Missouri and Yellowstone. In the south, Indian troubles had reduced the number of robes that traders could obtain. During the '60s, Leavenworth, Kansas became a trading center for robes. On the west side of the Missouri River, Leavenworth was ideally situated for river travel from the north and overland travel from the west.

The man responsible for Leavenworth's status in the robe trade was E. H. Durfee, an Indian trader. Durfee came to Leavenworth in 1861 from Marion, Wayne County, New York. Within a few years he built an immense and lucrative trade with the Indians, both on the southern plains and along the Upper Missouri. In 1868 Durfee was interviewed by a newspaperman with the *Leavenworth Daily Conservative*. The story that follows was published in that newspaper on May 8, 1868, and is an excellent description of Indian trading in the West during the 1860s:

Probably there is no business carried on in this country of which so little is known by the public generally as Indian trading. We yesterday had a very interesting chat with Mr. E. H. Durfee, one of the oldest and most widely known Indian traders who have ever been in the West. We are indebted to him for a great many interesting items about the business, which we have decided to lay before our readers.

The Southern Trade

He is the sole proprietor of the establishment here, which is the headquarters for the traffic with the Southern Indians. The posts on the upper

Missouri are owned by Durfee & Peck. The Southern Indians, or those south of the Arkansas, supplied, are the following, with their estimated numbers: Comanches, 23,600; Apaches and Cheyennes, 3,500; Osages and Kaws, 4,000. The larger tribes, as nearly everybody knows, are divided into bands, under various names, which we will not give here.

The Northern Trade

The Indians of the North with which they trade are all Sioux, numbering, it is estimated, upwards of 70,000. They are located in Dacotah and Montana. The Sioux are divided into twelve or fifteen bands. Some of their trade comes from the British Possessions, and the whole extent of it is from there to Texas. The only rival of Durfee & Peck is the Northwestern Fur Company. The competition is sharp, and is carried on with all the energy which characterizes the Yankee everywhere, whether in Wall street or in a log cabin a thousand miles from civilization.

The Posts, and Men Employed

Durfee & Peck have employed at their posts, in all, about one hundred men. A large number of these are fitted out every season by them with arms and traps with which they get their furs and turn them over to their employers, receiving therefor goods, which they in turn sell to the Indians.

They have on the upper Missouri seven posts, at which are stored and kept for sale all kinds of goods which the Indians want to buy, and where they come in with their skins. The houses used are all built of logs, with mud roofs, saw mills being scarce up that way.

The Hunting and Trapping Season

The season in which furs and peltries are secured by the hunters and trappers is from October to February. After that time the shedding of the coat commences and the hair fades and becomes worthless. The animals most sought for and which produce the most desirable skins are the following, placed in the order of value: Otter, beaver, buffalo, wolf, elk, bear, fox, deer, and coon. Mink is considered too small game, among the Indian trappers in particular.

How They Are Killed

The buffalo are killed mostly by arrows, as they are not only less expensive, but can be withdrawn and used again. These animals are generally hunted in the following manner: A large herd is surrounded and gradually driven in together. And here is exhibited a piece of strategy thoroughly Indian. The stragglers on the outside of the main herd are shot in the liver and will bleed to death internally in going four or five miles. The hunters will keep on driving them, and the carcasses at the close of the chase are not scattered over so large an extent of ground as they would be if the stragglers were shot dead. When the circle is well closed in, the hunters begin to shoot at the heart. Their ponies are all trained and will not enter the herd, but keep always around the outside, though the rider does not draw a rein on them after the main herd is reached.

The wolves are all poisoned in the following manner: A quarter of buffalo is either taken in a wagon or dragged over the prairie; at the distance of about 40 rods apart, numerous stakes are stuck in the ground, on the top of which is impaled a small piece of the meat, which has been poisoned with strychnine. The wolves strike the trail and follow it up, taking the pieces as they go. Next morning the hunters go along the line and skin the dead animals.

Dressing and Tanning

The Indians use the brains of the animal to tan it with. They first stretch the skin over a frame. They then rub on the brains, mixed with juices obtained from certain roots and plants. They are then scraped with various implements, hoes being used. They say the brains draw out the grease. After they are dry, they are painted and ornamented. The paint used is of the very finest qualities of Chinese vermillion and chrome yellow and green. These are imported by Durfee & Peck.

Bringing in the Skins

As soon as the season is over the Indians put the hides and furs on poles, which are dragged by ponies, sometimes a distance of 300 miles, to the nearest trading post. The whole band generally comes in with them. At the posts are opposition runners, in the employ of the Northwestern Company and Durfee & Peck. They keep on the watch, and as soon as a band comes in sight they mount their ponies and start off to secure the customers. Those with whom they decide to trade are compelled by custom to give the band a great feast, which lasts one day. Then business commences.

What the Indians Buy

The articles most in demand by the red men are coffee and sugar, of which they are very fond. In dry goods they want blankets, cloth, prints; a few of them buy saddles and bridles. An ornament called an Iroquis shell, which is picked up on the seashore somewhere in Europe, is in great demand. Mr. Durfee says he has seen an Indian sell fifteen out of twenty buffalo robes for these shells.

Big Canoes

The Indians know the boats which are loaded with goods for them by the tops of the smokestacks being painted red. They call them "big canoes," and as soon as they get into Indian country the news is carried ahead by runners, and they all know when the boat will arrive. They never molest them, and Durfee & Peck have never met with any loss at their hands.

Miscellaneous

Mr. Durfee has sent off one boat load of goods this season, per steamer *Benton,* which will be back in June, loaded with furs, robes and peltries. She took up 250 tons. The *Big Horn,* which has gone up with Government freight, will also bring down a cargo. The *Benton* will also make another trip this season. The farthest that boats go up from Leavenworth is 2,700 miles, by the river. The proceeds of the stock to be brought down by the *Benton* this year will be about $150,000. They have sutler's stores at Forts Sully, Rice and Stevenson, which are entirely separate from the Indian business.

Durfee & Peck handle yearly from 25,000 to 30,000 buffalo robes, which average about $8.00 apiece. The furs are, of course, much higher, and the whole business comprises an enormous trade.

There is a popular idea that some of the buffalo robes which we find in market are tanned by white men. This is not so. The Indians do it all. White men have tried it, but failed.

Mr. Durfee has, during his various trips to the mountains secured a large number of pets; among them he has kept the following animals, which are at his New York residence: one bear, one antelope, one deer, one badger, a red fox and two American eagles. He had two buffalo but they died.

Not quite two years after this article, Theodore R. Davis, a writer for *Harper's Monthly,* reported that Durfee and Charles Bates of St. Louis were the "two great gatherers or collectors of buffalo-robes" and that their combined collections during a single year amounted to more than 200,000 robes. "During good years," wrote Davis, "nearly a quarter of a million of skins" found their way to the New York market. There they were classed as first, second, third, and calf on the wholesale market. In 1869 the prices paid by New York wholesale dealers, according to Davis, were something like "$16.50, $12.50 and $8.50, this being the prices for first, second and third rate skins; calf-skins bring from $3.50 to $4.00, and are not much dealt in." Davis noted that Bates and Durfee sometimes held their robes for market as long as "three or four years, this being done when the market does not range to suit them, though one would think that controlling the trade as they do they might dictate the prices of the robe." Davis contended that a Comanche robe was

> perhaps the best in its dressing, but the fur is not likely to be so good as that of the Sioux dressed robe. The only way of accounting for this is the fact of climate, the Comanche being a southern Indian, and the Sioux ranging far to the north. The Sioux robe is not, however, so well dressed as either the Comanche or Iowa robes. What is known as the split robe—that is,

a robe which has been divided into two parts and is sewn together after it has been dressed—is uncommon among the southern Indians, but frequently met with in trading with the Sioux.[28]

Through 1871, buffalo robes were the major commercial product in dollar value obtained from the buffalo by the white man. But that was about to change.

"Scalped by the white man" was the title given this buffalo carcass picture by the photographer who took it on the plains of western Kansas during the early 1870s. *Courtesy Kansas State Historical Society.*

VII

The Slaughter

A cold wind blew across the prairie when the last buffalo fell . . . a death-wind for my people.

Sitting Bull 19th century

A few centuries ago, the Argentine pampas was stocked with Spanish cattle. In time the animals multiplied until vast numbers roamed the open expanse of sometimes level, sometimes rolling grassland. Cattle became so cheap that a gaucho would think nothing of killing a bull in the evening just for the animal's horns, which would be used as a picket pin for his horse.

By the late 1850s, there were so many cattle on the pampas that thousands had to be killed just to make room for future increases. The land could support only so many animals. The meat of those cattle that were slaughtered was of little value to anyone except the gauchos. There was no modern process of refrigeration. Except during the winter months, the meat was a total loss. But the animal's hide did not spoil. No matter when it was taken—summer or winter— it could be used, if properly cared for. And when so many cattle were killed, the bones and hides were the only things that were of value and could be sold. In time the cattle of the pampas became a source of hides as were cattle in California during the 1830s and later—that California story is vividly told by Richard Henry Dana in *Two Years Before the Mast*. By the late 1860s the cattle of the pampas were one of the world's major sources of leather. Boat loads of hides were regularly shipped to Europe and America, but sometime around 1870 the supply began to dwindle. Too many cattle had been killed.

Aware of this, tanners on both sides of the Atlantic began eyeing other possible sources of leather. They were already using hides from cattle in the United States, but the supply was not sufficient. Several

93

tanners began experimenting with the hides of different animals to see which would produce the best leather.[1]

In the winter of 1870, an English tannery made a contract with William C. Lobenstein, a dealer in pelts, hides, and leather at Leavenworth, Kansas, to purchase 500 buffalo hides. The tanners wanted to experiment with them. Lobenstein did not have the hides in his warehouse, but having dealt in buffalo robes for many years, he knew where to get them. He got in touch with Charles Rath and A. C. Myers at Fort Dodge, Kansas, which was then about the center of the buffalo range. Both men had been prime suppliers of buffalo robes to Lobenstein. What hides Rath and Myers did not have in their warehouse, they bought from other hunters, including J. Wright Mooar. Then they shipped 500 hides to Lobenstein, who in turn shipped them to England.[2]

At Fort Dodge, J. Wright Mooar had fifty-seven hides left over which Rath and Myers did not buy. Suspicious of what was in the wind, Mooar decided to ship his extra hides to his brother who lived and worked in New York City. There, Mooar's brother, John Combs Mooar, took the hides to Bates Hide House on Broadway. He had no trouble selling them to some tanners from Philadelphia who were anxious to experiment on buffalo hides.[3]

At about the same time, now early in 1871, J. N. DuBois, a fur and hide dealer at Kansas City, filled an order and shipped several bales of buffalo hides to tanners in his native Germany.[4] Like the American and English tanners, the Germans wanted to see what kind of leather they could make from buffalo hides.

Whether the Americans, the English, or the Germans were first with their findings is unknown. But soon DuBois, Lobenstein, and the Mooars learned the news. Buffalo hides could be used to make fine leather. The English even went so far as to say that the hide of the buffalo was much more pliant and elastic than cowhide. The English tanners immediately ordered more buffalo hides to make leather for army accoutrements.[5]

Orders to buy buffalo hides were sent to Lobenstein and DuBois, who in turn spread the word across the plains and prairies that they would buy the hides of buffalo killed at any time of the year, with or without hair on them. Lobenstein advertised for hides in the growing number of frontier newspapers, while DuBois, in addition to some newspaper advertising, flooded the small towns in buffalo country with printed circulars stating that he would buy buffalo hides. DuBois offered to pay $2.25 for cow hides and $3.25 for the larger buffalo bull hides *if* his directions were followed. Those directions were simple. A buffalo hunter, after killing and skinning a buffalo, was to take the hide and stretch and peg it down on the ground with the skin-side up. This kept the hides straight for handling, hauling, and shipping. DuBois also introduced the hunters to a hide bug poison that had been used on cattle hides in Argentina. When sprinkled on dry hides it kept moths and bugs away.[6]

The news that tanners had found a way to make good leather from buffalo hides spread like a prairie fire across buffalo country. It meant not only a new market for the buffalo runners but also year-around work for those wanting it. The hunters were jubilant. By the summer of 1872, the plains of eastern Colorado, southwestern Nebraska, and western Kansas were alive with buffalo hunters. One Wichita, Kansas newspaper reported:

> Thousands upon thousands of hides are being brought in here by hunters. In places whole acres of ground are covered with these hides spread out to dry. It is estimated that there is, south of the Arkansas and west of Wichita, from one to two thousand men shooting buffalo for their hides alone.[7]

About 150 miles west of Wichita was the new town of Dodge City. Buffalo hunting was the town's chief industry. As a matter of fact, the community was originally called Buffalo City. However, in applying to the government for a post office, it was discovered that there was already a Buffalo Station out on the Union Pacific line and a little town of Buffalo in Wilson County. So, not wanting too many Buffalos in Kansas, the postmaster-general suggested the name Dodge City, in honor of nearby Fort Dodge. Dodge City it was, and almost overnight this tiny village became the hide capital when the Atchison, Topeka & Santa Fe Railroad reached there in early September 1872. During the three months that followed, the railroad shipped 43,029 hides and 1,436,290 pounds of buffalo meat east. One observer concluded that these figures represented about 50,000 dead buffalo and did not include those animals killed in "wanton cruelty, miscalled sport, and for food for the frontier residents."[8]

On August 29, 1872, just before the Atchison, Topeka & Santa Fe reached Dodge City, the Newton *Kansan* had reported:

> G. M. Richards, the "buffalo excursionist," has made arrangements with the A.T.&S.F. . . . for buffalo excursions to the Arkansas valley; the excursions to run every two weeks until the weather becomes too cold, commencing October 1st, and to run from Atchison to Fort Dodge and return via Wichita, thus giving the excursionists a ride of 800 miles and a view of the most beautiful valley east of the Rocky Mountains, for the small sum of $10.

From Dodge City and other railroad towns in Kansas and Nebraska, in or near buffalo country, hides, robes, and meat were shipped east. Though statistics on the total volume of the railroad's buffalo trade are incomplete, it is known that the Santa Fe in Kansas transported about 19,000 Indian-tanned robes annually during the early 1870s. As for hides, meat, and bones shipped east from Kansas, there are conflicting reports. In his account of his days as a buffalo runner, Frank Mayer says that in Dodge City and surrounding country a total of 3,158,730 buffalo were killed during 1872-74. This figure, said Mayer, did not include about 405,000 buffalo killed each of those three years by Indians.[9]

But Colonel Richard I. Dodge provides figures obtained, in part, from the railroads:[10]

Railroad Buffalo Shipments

Product	Year	Atchison, Topeka & Santa Fe	Union Pacific, Kansas Pacific, and all other RRs	Totals
Hides,	1872	165,721	331,442	497,163
number	1873	251,443	502,886	754,329
of	1874	42,289	84,578	126,867
		459,453	918,906	1,378,359
Meat,	1872	none	none	none
lbs.	1873	1,617,600	3,235,200	4,852,800
of	1874	632,800	1,265,600	1,898,400
		2,250,400	4,500,800	6,751,200

As can be seen, Dodge's figures for the total number of hides shipped east during the three-year period are low compared to Frank Mayer's figures for the total number of buffalo killed around Dodge City during the same period.

The Atchison, Topeka & Santa Fe and the Kansas Pacific—both running through Kansas—and the Union Pacific—running through Nebraska—carried the bulk of the buffalo trade during the 1870s. Though these railroads undoubtedly kept records, their representatives have reported and still report that they are unable to find such records, if they even exist. I, and many others, have attempted diligently but unsuccessfully to find these records. It is doubtful whether they ever will be found.

A reporter for a Topeka newspaper, A. P. Baldwin, reported from Dodge City in late November 1872:

> Every ravine is full of hunters, and camp fires can be seen for miles in every direction. The hides and saddles of fourteen hundred buffalo were brought into town today. Twenty miles west of Dodge an immense herd of creatures, covering an extent of country two miles in width and ten miles in length, was passed by the construction train. Fourteen were run over and killed by the engine. Two hours were consumed by the construction train in endeavoring to get through the herd. Several calves were run over and injured, and the construction men, while in the act of capturing some of them, were charged upon by several hundred buffalo and barely escaped with their lives.[11]

About a month later, on December 26, 1872, the Newton *Kansan* reported:

> It is estimated that there are about two thousand buffalo hunters now pursuing game in Western Kansas, and that they average bringing down about fifteen buffalo daily. One man near Dodge City killed 100 in a day. The hides and meat bringing him a handsome sum of $300. At Dodge City the hams are worth 1½ to 2 cents a pound and the hides from $1.50 to $2.50 a piece.

Until about 1873, fresh buffalo meat could be shipped east only during the cold winter months, when it would remain frozen. After 1872 when refrigeration cars came into use on the railroads—they were crude by today's standards—the railroads began shipping some buffalo meat east during the warmer months of the year. Still there was great spoilage, and about the only thing that could safely be shipped year-around was buffalo tongues, salted and shipped in barrels, and buffalo hams that had been cured.

The Newton *Kansan* reported on January 9, 1873 that the preserving and shipping of hams at Dodge City

> is now nearly as extensive a business as that of caring for the hides, and that the town of Dodge City, containing some five hundred inhabitants, is supported almost entirely by this trade.
>
> For the purpose of curing buffalo hams and preparing them for shipment east, several houses have been erected and are now in successful running operation in that place. The tongues are also treated in proper manner for the trade, and one of the recent Acceptance Party to the State line [Colorado line] informed us that he saw at one station two thousand of them in one pile awaiting shipment to Dodge City. . . . As to the number of buffalo still at large, they are undoubtedly becoming more scarce and the supply must soon have an end, and we doubt not every butcher shop in Kansas will welcome the day of its cessation, as many people are now relying altogether upon the buffalo for their meat.

Though many newspaper accounts of buffalo hunting implied that it was a fun-thrilled "get rich quick" way of life, such was not the case. There was money to be made in hunting buffalo for their hides and hams, but the work was dirty and the life was rough and sometimes dangerous, particularly for those hunters going after buffalo in Indian country or during the cold months of winter.

The vivid account of one hunter's life, from his own diary, is perhaps the best example of this. Henry Raymond, a native of Illinois, came to Kansas in 1872 to join his brother Theodore, and for more than a year they hunted buffalo with Bat and Ed Masterson in western Kansas. One week of Henry Raymond's diary reads as follows:

Dec.   Sunday 8.   1872
Shaved with pocket knife. Ed Jim and me went to Indian village. Saw the squaws tanning robes. Arapaho tribe.

Dec. Monday 9. 1872
Skinned 30 buffalo. Very cold day. Cloudy and windy at night. I killed
3 of buffalo.

Dec. Tuesday 10. 1872
first snow, last night. warm today. wounded buffalo this morning. Big
John here. Jim and I went aft turkys. Abe came at night brough
[word illegible] letter from Liza Lane.

Dec. Wednesday 11. 1872
Bat, Abe, Ed, Jim Rigny and me went to Indian camp to trade. The
[his brother Theodore] went to Dodge with Nixen. Killed my first
grouse today. Saw Indians eating lice.

Dec. Thursday 12. 1872
Sat up nearly all night last night. So cold did not work. hauled load
wood, went to Yahoos camp at eve.

Dec. Friday 13. 1872
butchered 4 buffaloes. Snow on ground. quail hunting. Abe tried to
kill some. I shot a kyote on run, grazed his back.

Dec. Saturday 14. 1872
terrible windy and cold. did not work, only pegged a few hides.
The [Theodore] came in at night.[12]

During the 1860s, men like William Cody, those who were the first
professional buffalo hunters on the plains, used a variety of rifles to
kill buffalo. Henrys, Ballards, Spencers, and other rifles were used,
but when a hunter had a choice, he usually chose the .50-70 Springfield Army Musket, sometimes called the "needle gun." These .50
caliber Army rifles were Civil War muskets converted to breech load-
ers by the Allin system and relined to .50 caliber for centerfire cartridges. The cartridge used 70 grains of powder and a conical 450
grain bullet. They were accurate and had much "knock-down" power.
William Cody used such a converted Army rifle which he named
"Lucretia Borgia." It could throw a ball double the weight of the ordinary carbine and could kill a buffalo at 600 yards provided it hit
the animal in a vital spot.

After buffalo hides became valuable for leather in 1872, the growing number of hide hunters on the plains began to demand a better rifle. Two eastern companies, first Remington, then Sharps, answered the call.

Remington produced a "rolling-block" single-shot rifle using the breech mechanism developed by Joseph Rider during the Civil War. Remington cartridges were the .44-77 and .44-90 bottlenecked calibers. The .44-90 was advertised in the '70s as Remington's .44 "Special" cartridge, but others soon became available, including the .40-70, .45-70, .50-70, and the .58 Govt.

Sharps began experimenting, after being convinced by some buffalo hunters, including J. Wright Mooar, that a larger rifle was needed. They sent samples to the buffalo range for hunters to try. W. Skelton Glenn, a Texas buffalo hunter, told about the efforts of Sharps to produce a buffalo gun:

> Sharps manufactured an octagon shape barrel of various lengths from 26 to 30 inches, long, 50 calibre, using 380 grains of lead and burning 120 grains of powder, with a reloading outfit including a bullet mold to make bullets on the range. After experimenting sometime, it did not give the satisfaction desired in Nebraska and Kansas, because the wind was much stronger than it was in Bridgeport, Conn. [home of Sharps Rifle Company from 1876 to 1881]. The merchants wrote that the bullet would catch too much wind drift, so Sharps decided to make another calibre of smaller size; so made a 44, burning the same amount of powder and lead, but using a bottle-neck shell. After experimenting a long time this one did not give satisfaction, as it leaded too badly, and there was no certainty to the marksmanship. He next went to making a 40 with less powder and lead. This proved like the others to be a failure, as it would not even knock the buffalo down, not having enough lead. He then began to make a 45, using 380 grains of lead and 120 grains of powder. This took a straight shell [case], and he thought it would overcome the leading, giving the weight and being a smaller calibre than the 50 and would not wind drift. [It worked.][13]

The Sharps was a side-hammer, single-shot rifle adapted from the the Civil War and pre-war percussion Sharps. The .50 caliber model, mentioned by Glenn, appeared about 1866, but a variety of loads and calibers did not develop until later. Probably the two most popular Sharps cartridges were the .40-90-420 (.40 caliber bullet weighing 420 grains propelled by 90 grains of black powder) and the .45-120-550. These cartridges had nearly 2,400 foot-pounds of muzzle-energy, and a buffalo hit in a vital spot by one of these bullets would almost always drop in its tracks.

One Kansas buffalo hunter who used a Sharps, George W. Brown, recalled:

> It shot a hundred and twenty grains of powder and the bullets were an inch and a quarter long. When one of these big leads would hit a buffalo, whether it hit the right place or not, it would make him sick. It wouldn't be long till I put another into him. If that wasn't enough I'd put still another into him. I have often shot a buffalo ten or fifteen times before I got him down.[14]

By modern standards the 19th century buffalo cartridges were slow, being in the 1,400 to 1,500 feet-per-second class, compared to the present-day 30-06 with a 2,700 to 3,370 feet-per-second velocity. But the soft bullets of the 19th century were heavy. When they struck, they hit with much force. The low velocities naturally gave high trajectory, but the hunters, or "buffalo runners," as they were sometimes called, quickly learned to estimate distances accurately.

A good buffalo hunting rifle cost money. Frank H. Mayer recalled that one rifle would cost $100 to $150, "not including the necessary telescope sight." Many hunters used such telescopic sights of 10, 20, or sometimes 30 power. When they did, they usually used a support for their rifle, sometimes formed by two hardwood or iron strips bolted in the middle to form an "X" when unfolded. The bottom points of the "X" were pushed into the ground, and the rifle rested in the crotch or upper angle of the cross sticks. Sometimes, hunters might use a "Y" shaped branch from a tree and rest the rifle in the crotch.

Frank Mayer purchased his first Sharps second-hand from Colonel Richard Irving Dodge. It was a practically new .40-90-420:

> It was a beautiful piece, with its imported walnut stock and forearm, its shiny blue 32-inch barrel. At $125 I considered it a bargain. This Sharps weighed 12 pounds. On the barrel I mounted a full-length one-inch tube telescope, made by A. Voll-

mer of Jena, Germany. Originally the scope, a 20-power, came
with plain crosshairs. These I supplemented with upper and
lower stadia hairs, set so they would cover a vertical space of
thirty inches at 200 yards.

I was proud of that first Sharps of mine. . . . At first it used a
320-grain bullet, but I experimented with one a hundred grains
heavier, and thereafter used the 420-grain projectile. It killed
quicker. In making this change I didn't sacrifice anything in
velocity, because by then I had begun to use the English powder . . . and it added 10 to 30 percent efficiency to my shooting.
After a year or two, having plenty of buffalo dollars in my jeans,
I talked myself into believing that I needed an extra rifle in reserve—so I bought two. One was a .40-70-320—a light little
gun for deer and antelope but too impotent for buff. The other
was another .40-90-420. Both used bottle-necked cartridges;
don't ask me how I fell for that sort of thing after vowing I was
off bottle-necks for life. I paid $100 for the .40-70, $115 for the
.40-90—current prices then. Prices on Sharps declined rapidly
after the buffalo years, and I saw plenty of Sharps in John
Lower's gun shop in Denver at $35, $40 and $50—identical
with the guns I paid $100 to $125 for a couple of years earlier.

It cost Frank H. Mayer about $2,000 to become a buffalo runner
(he never called himself a buffalo hunter—that was the mark of a
tenderfoot). Mayer was the last surviving buffalo runner when he
died in 1954 at Fairplay, Colorado, at the age of 104. He said, "A
buffalo outfit was simple, and I could have made mine simpler but I
wanted to do things up brown. All you needed was horses or mules,
wagons, camp equipment, and firearms, and you were in business."
In Mayer's reminiscences, a classic in the buffalo slaughter story, he
described his "running" outfit:

> I bought two wagons in St. Joseph, Missouri. The big one, drawn
> by twelve mules, we used in hauling hides; the small one, drawn
> by six mules, was our camp wagon. Both were equipped with
> nine-inch tread flat iron wheels and steel boxes or bed of ⅛-inch
> steel. I remember what I paid for the big wagon—$650. I re-
> member what I paid for the small one—$400. I already had a
> couple of good saddle horses, which I went to much pains to

train. I taught them to lie flat while I was shooting at game, so as to avoid detection by roving Indians. We always used American horses because they were bigger, stronger, more dependable than the mustangs or Indian ponies which the Indians used because they didn't have anything better. A good buffalo horse, though, was worth real money on the range, anywhere from $250 to $500.

For over half a decade Mayer was a buffalo runner. To him it was a business as it was to most of the runners.

When I went into the business, I sat down and figured that I was indeed one of fortune's children. Just think! There were 20,000,000 buffalo, each worth at least $3—$60,000,000. At the very outside cartridges cost 25 cents each, so every time I fired one I got my investment back twelve times over. I could kill a hundred a day, $300 gross, or counting everything, $200 net profit a day. And $200 times thirty, would be, let me see, $200 times thirty—that would be $6,000 a month—or three times what was paid, it seems to me, the President of the United States, and a hundred times what a man with a good job in the '70s could be expected to earn. Was I not lucky that I discovered this quick and easy way to fortune? I thought I was. . . .

We never killed all the buff we could, but only as many as our skinners could handle. Every outfit had its quota, which was determined by the ambition and the number of skinners. My regular quota was twenty-five a day, but on days when my crew weren't tired, I sometimes would run this up to 50 or even 60. But there I stopped, no matter how plentiful the buff were. Killing more than we could use would waste buff, which wasn't important; it also would waste ammunition, which was.[15]

During the early 1870s, Fort Wallace, along the Smoky Hill River in far western Kansas, joined Dodge City as a supply center for buffalo hunters, especially those hunting along the Kansas-Colorado border. Although it never developed into a large trading center like Dodge City, Fort Wallace did fulfill the needs of many hunters, as did several other small frontier towns and military posts on the southern plains. Nearly one thousand hides were shipped from Wallace in the winter of 1872-73, besides shipments of meat.

In December 1872 Capt. Louis T. Morris, stationed at Fort Wallace, accompanied a party of eastern friends on a buffalo hunt, and in 1873 Capt. John A. Irwin, Company D, Sixth Cavalry, with a detachment of the Third Infantry, conducted a hunt for some visiting British army officers. In both of these hunts, Homer W. Wheeler, post sutler at Fort Wallace, acted as guide; he later authored a book called *Buffalo Days*.

One old hunter who worked out of Fort Wallace was known only as "Kentuck." He had a camp on Punished Woman's fork, forty miles to the south. He made about $10,000 from hides and meat before 1875, killing about 3,700 buffalo. Kentuck and other hunters did much business with Peter Robidoux, a French Canadian, who had set up a trading post at Fort Wallace during the late 1860s.

"First a little stock of drugs, then general merchandise, where I kept for sale *everything,*" recalled Robidoux years later. "Everything, that was my sign I had painted over the door. I sold everything from a postage stamp to the real old stuff; from a jew's harp to the big Sharps rifle which was used to kill the buffalo and a real menace to the Indians." Robidoux traded merchandise to the hunters, scouts, and Indians for buffalo hides:

> These fresh hides were piled in big piles in a stockade I had built back of my store. When I had accumulated a car or more of hides I would ship them to Leavenworth, where I shipped hundreds of carloads of buffalo hides. Sometimes a sneaking, thirsty Indian would scale the walls of the stockade from the rear, steal a hide, bring it in the front door and trade it for bottled merchandise. Times were good. The soldiers at the Fort spent their money freely while it lasted.[16]

In January 1872 soldiers from Fort Wallace participated in a royal buffalo hunt along the Kansas-Colorado border. Two months earlier, Alexander II, then emperor of Russia, had sent his third son, the Grand Duke Alexis, as a special ambassador to President Grant. While in Washington, Alexis met General Philip Sheridan, who invited him to go buffalo hunting in the West. Alexis accepted eagerly. Aboard a luxurious private train Alexis and General Sheridan traveled west in early January, 1872. At Omaha, Lt. Colonel George A. Custer joined the party, and they headed for North Platte. There, "Buffalo Bill"

Cody joined the party. After a few days of hunting, Alexis traveled to Cheyenne and south to Denver. Then he decided he wanted to hunt buffalo again.

On January 20, his private train stopped at Kit Carson, Colorado, 140 miles southeast of Denver. Using Army horses brought in from Fort Wallace, the hunters spent several hours chasing a large herd of buffalo on the prairie a few miles southeast of Kit Carson. About fifty buffalo were killed. Alexis downed five. The animals were skinned and the meat, hides, and trophies loaded aboard the eastbound train. When the train reached Topeka not quite two days later, the Kansas legislature recessed to join in the celebration. There was a big parade and a banquet at the Fifth Avenue Hotel to honor the Grand Duke, who had killed less than a dozen buffalo on his royal hunt.

The year 1872 was the best year for hide hunters in Kansas. It is remembered as the year of the "big hunt." One summer day in '72, George W. Brown, a hunter, returned to Dodge City with about 600 hides which he sold immediately to a Kansas City buyer, Colonel E. W. Eskew. Eskew was then a partner with J. N. DuBois in the hide business. Brown recalled:

> He paid three dollars for the bulls and two dollars for the cows. While I was tending saloon, and before I got my hides delivered, a hunter drove up in front of my door and says, "Brown, are you buying hides?" I says, "Yes." He says, "I've got two hundred bull hides on my wagon." I asked him what he wanted for them. He says, "Two dollars." I says, "You go down and bring them up, I'll take them." After he was gone, I went on and saw old man Eskew; I told him I had two hundred more bull hides to add to our deal already made. He says, "That's all right; the more you've got the better." The young man drove up with the hides and old man Eskew counted them. I received the hides from the young man at two dollars apiece and old man Eskew received them from me at three dollars apiece and it was all done at one count.

Brown told how Eskew was very anxious to have him buy hides after that incident. Eskew pulled out his check book and began to discuss having Brown buy hides for him. But Brown declined the offer. "I told the old man I'd rather hunt," said Brown.[17]

Jim Lamb, another buffalo runner, is wrongly credited by a Kansas newspaper as having discovered the hunting method which became known as a "stand." The method was already well known among many buffalo hunters and especially the Indians when in the fall of 1872 Jim Lamb left Wallace, Kansas to hunt for shaggies. The newspaper account read:

> He had a fine outfit, two fine teams, wagons and complete hunting equipment. He had with him three men as skinners, campmen, etc. After getting out about 12 miles they came to a branch of Brush Creek having some timber in sight, such as a few cottonwoods and willows. Knowing there was water there, Lamb sent his men ahead to make camp. He said he would go to a bunch of buffaloes they had seen laying down at a short distance and see if he could get one for a supply of fresh meat. [Lamb crawled up on the buffalo and] picked out a fine plump-looking young cow and made a good lung-shot, hitting her about midway in the body and just back of the shoulder. Buffalo shot this way die a frantic death, jumping and plunging and finally coughing and discharging blood from their nose. Lamb, not wanting any more buffalo lay perfectly still watching this performance when, to his great astonishment, on looking up he saw the herd had stopped fleeing and were slowly coming back. They began milling around the fallen buffalo and seemed stupid. Lamb began shooting and his men in camp, hearing so many shots, brought him a couple of fresh guns and a lot of ammunition. When it was all over he had 56 dead buffaloes and numerous cripples wandering around. This was the biggest killing up to then from one position and to this very day [1925] the spot is known as "Jim Lamb's Slaughter Pen" by the people in and around Wallace, Kansas.[18]

What Jim Lamb had discovered for himself in 1872 had been practiced earlier by numerous other buffalo runners and even earlier by the Indians. The Indians, however, called their method "still hunting." George Douglas Brewerton, who crossed overland during the summer of 1848, wrote that in still hunting "the hunter must take advantage of every favorable peculiarity of the ground as he crawls

Buffalo hide hunters' camp on Evans Creek near Buffalo Gap in the Texas Panhandle, 1874. George Robertson photo. *Courtesy Division of Manuscripts, University of Oklahoma Library.*

"A Still Hunt," James Henry Moser painting, at National Collection of Fine Arts. *Courtesy Smithsonian Institution.*

Buffalo hides staked out at Dry Ridge, a watering station on the Santa Fe line, near the Kansas-Colorado border, early 1870s. *Courtesy Kansas State Historical Society.*

Charles Rath's buffalo hide yard in Dodge City, mid-1870s. Rath is sitting on the pile of hides. The man in the white shirt is D. W. Auchutz, who ran the hide bailer for Rath. During the peak of the buffalo slaughter in western Kansas, this yard handled as many as 80,000 hides at one time. *Courtesy Kansas State Historical Society.*

cautiously upon his prey; for should the buffalo 'wind' him, even though he may have been as yet unseen, the alarmed animal will carry his hump steaks far beyond the reach of even a Jake Hawkins [Hawken] rifle in double-quick time."[19]

What has often been reported to be the world's record for buffalo killing on the southern plains was set by Thomas C. Nixon during the early 1870s. One September day in Dodge City, Nixon invited his friends to see him set what he boasted would be "a record for buffalo killing which would last for all time." Whether he had had a few too many drinks or was just feeling good, history does not record, but as the story is told, Nixon's friends agreed to judge the event. Nixon got his rifle and plenty of ammunition and headed south of Dodge with his "judges." After crossing the headwaters of Bluff Creek in Meade County, they saw a buffalo herd which Nixon said was large enough for his record. He left the judges to watch from a nearby hill, took his rifle and ammunition, and crawled up as close as he could to the herd. Nixon signaled his friends on the hill to keep track of the time. Then he began to kill. As buffalo dropped to the ground, those still standing began bellowing and pawing the earth. Some moved from one downed animal to another, sometimes butting the dead with their big shaggy heads. Others would poke a fallen animal with their horns. Soon Nixon stopped firing and stood up. The few buffalo still alive apparently saw or smelled him as he started walking toward them. They took off running in the opposite direction. Shortly Nixon's judges arrived and counted the dead buffalo. In forty minutes Nixon had killed 120 buffalo.[20]

As the "big hunt" was going on in Kansas in 1872, the first permanent buffalo camp in Texas was being established at Big Lake, near what is today the boundary line between Knox and Haskell Counties. Before 1872 ended, more than 20,000 buffalo were killed on the Staked Plains of West Texas. Thousands of pounds of meat were cured by the hunters and the following spring shipped along with the hides by ox team to Denison, Texas. From there they were shipped north by rail.[21]

By 1873, as fewer and fewer buffalo were found north of the Arkansas in western Kansas, some buffalo hunters headed south to Texas. Most, however, did not go directly south from Dodge City; there were unfriendly Indians in western Indian Territory and in what is today the panhandles of Oklahoma and Texas. The Medicine Lodge Treaties of 1867 had provided that white hunters would not hunt south of the Arkansas. It was Indian land. The buffalo and other wild game were for the Comanches, Cheyennes, Arapahoes, and Kiowas. Most hunters, more out of respect for the Indians than for the treaty, did not travel due south. Instead they went east from Dodge City to Wichita, then south through less hostile Indian Territory to Denison, Texas, from where they pushed westward to the southern buffalo range.

But in July 1873 John Webb and J. Wright Mooar, both buffalo hunters, decided to scout south from Dodge City into Indian country. "Buffalo, a solid herd as far as we could see, all day they opened up before us and came together again behind us"—that is the way Mooar later described what he and Webb saw. Both men soon returned to Dodge City, where they reported what they had seen. Nearly all the hunters wished they could head south across the Arkansas River and take the hides from those buffalo, but as Mooar recalled, "All contended it was not safe to cross the strip as it was Indian territory." According to Mooar, a young buffalo hunter by the name of Steel Frazier proposed that they go to Fort Dodge and ask the commanding officer, Colonel Richard I. Dodge, what to do. They rode out to the fort and asked Dodge what would be the penalty if they went south into Indian territory and killed some buffalo.

"Boys," Mooar quoted Dodge as saying, "if I were a buffalo hunter I would hunt buffalo where the buffalo are."[22]

J. Wright Mooar and the other buffalo hunters did just that.

In the months that followed, the buffalo hunters left Dodge City and headed south. Dodge City merchants, not wanting to lose the hunters' business, decided to head south with the hunters and establish a branch trading post in the new buffalo country. They chose Adobe Walls, about 160 miles south-southwest of Dodge City, as the site for their new trading center. Adobe Walls was about in the middle of the buffalo range in the south. Only ten years before, Colonel Kit Carson, commanding U.S. troops, had fought a band of Kiowa and Comanche Indians at the site. Much earlier, before 1840, William Bent had built an adobe trading post, called Adobe Fort, nearby, but it had been abandoned sometime in the 1840s.[23]

Large freight wagons carried building materials and supplies south from Dodge City. At Adobe Walls two stores, a blacksmith shop, and a saloon were constructed, stocked, and made ready for business by May 20, 1874. But a stockade nearby was not completed. "As spring advanced and the weather became warm, the work lagged," noted Robert M. Wright.[24] Along with Charles Rath and other Dodge City merchants, Wright had been responsible for setting up the new trading center, if it could be called that. By early June, buffalo hunters were finding many buffalo and were making Adobe Walls their headquarters. Because Indians began causing some problems, many of the hunters frequently came to Adobe Walls to spend the night, thinking there was strength in numbers.

Early on the morning of June 27, 1874, more than 500 Indians attacked the twenty-eight men and one woman at Adobe Walls. The Indians—Cheyennes, Kiowas, and Arapahoes—killed three white men and numerous horses and oxen before they were driven off. At least thirteen Indians were killed and others wounded. But instead of accepting defeat the Indians separated into tribal bands and went on the rampage in parts of Texas, New Mexico, and Colorado and along the southern border of Kansas. The merchants and buffalo hunters at Adobe Walls abandoned the site and most returned to Dodge City.[25]

Some hunters then decided to try their luck to the north in Wyoming, Montana, and Dakota Territories. Others gave up the business to settle down in one of the many new frontier towns. A few hunters found enough buffalo roaming eastern Colorado and far western Kansas to keep busy that summer and fall, but still others decided to try the far southern plains again. Those who went south took the long way around—to miss the Indians—and came up on the West Texas buffalo range from the east. By late fall of 1874, the frontier outpost of Fort Griffin, Texas was rapidly becoming the new headquarters in West Texas for buffalo hunters. On a bluff overlooking the Clear Fork of the Brazos River, Fort Griffin had been founded in 1867. Now from this small but growing military outpost, buffalo hunters fanned out to find the shaggies. They found them in abundance.

Southward to the Concho Valley of Texas the killing increased. Joe S. McComb, one of the first white men to hunt buffalo on the Texas plains, followed the herds into Nolan County near Sweetwater in the fall of 1875. Nolan, Mitchell, and Howard Counties were then in the heart of the southern buffalo range. McComb continued to follow the buffalo with other hunters and in 1876 moved his headquarters to

Morgan Creek, west of where Colorado City, Texas stands today. The following year, McComb and other hunters camped at Mossy Rock Springs, near Signal Mountain, and at Mustang Pond, in what is today Midland County, Texas. Soon they had killed nearly all the buffalo in the region.[26]

In the '70s it was a rare thing to find a woman in any buffalo hunter's camp. Married hunters nearly always left their wives at home in one of the frontier towns. But Tom Bird's wife was different. In December 1876 she went with her husband to hunt buffalo in West Texas. Bird, after spending time in the Texas Rangers, had resigned to hunt the shaggies. With his wife, he spent a few days near Fort Belknap in Young County, Texas, preparing for the trip. There they gathered what they needed, including "a large buffalo gun, Sharps 45, weighed 16 pounds, used a cartridge about five inches long, a reloading outfit with plenty of ammunition, several hundred cartridge shells, primers and keg of powder, 50 pounds of led, two I. Wilson skinning knives, one buoy knife, and a large steel . . . and a good pair of rest sticks to hold up the heavy gun while in action."

With a friend, Alec Johnes, Tom Bird and his wife set out for the lower panhandle region of Texas. Near where Dickens and Cottle Counties come together, about seventy-five miles east-northeast of Lubbock, Texas, the trio set up their camp. The first order of business was to build winter quarters. Johnes built a rock house. "He threw up the walls, chinked with mud and covered with dirt all complete in three or four days time," wrote Mrs. Bird years later.

> Our house was different. It was on the same order that most of the hunters used, they were called teepees made of buffalo hides. That was before dugout day. They were made as follows: first, built a frame of small china poles split some for rafters, no nails were used, raw hide strings instead. We took dried buffalo hides tied the legs together put them around the wall, wool side out, then another tier of hides over these in the same manner to break the joints of those under neath. The roof was made of the same like order as the walls, tying down the legs all around the edges. The door was made of a frame of split poles with a buffalo hide stretched over this, legs tied inside, then the little rock chimney with fire place which was crude of course came

next. Beside the door the floor was then carpeted with buffalo hides squared up to fit the wool side up, all was complete, and a more clean and comfortable little home you could not find in any of the eastern cities.

For three years this was the home of Tom and Ella Bird. During that time they killed many buffalo.

> I would often go out with Mr. Bird in the wagon on his hunts. It was wonderful to me to see them kill the buffalo. The method they used in shooting them was queer. The country was not altogether level prairie. They could usually slip up in three or four hundred yards of them. When all was ready they would shoot the one that seemed to be the leader. They never shot them behind the shoulder, in the heart or they would pitch, buck around and break the stand. Always shoot far back in the body behind the ribs. This made them sick. They would hump up, walk around and lie down, then wait a moment till another led out. Shoot it then, another leader, on and on till you had shot several, then they would begin milling around and around. You had a stand on them, then you could kill all you wanted.[27]

W. S. Glenn, who hunted buffalo in Texas during 1876-77, described how a buffalo hunting outfit functioned:

> They would range from six to a dozen men, there being one hunter who killed the buffalo and took out the tongues, also the tallow. As the tallow was of an oily nature, it was equal to butter; it was used for lubricating our guns and we loaded our own shells, each shell had to be lubricated and it was used also for greasing wagons and also for lights in camp. Often chunks as large as an ear of corn were thrown on the fire to make heat. This [the removal of the tallow] had to be done while the meat was fresh, the hunter throwing it into a tree to wind dry; if the skinner forgot it, it would often stay there all winter and still be good to eat in the spring and better to eat after hanging there in the wind a few days.
>
> There were generally two men to the wagon, and their business was to follow up the hunter, if they were not in sight after the hunter had made a killing, he would proceed in their direction

until he had met them, and when they would see him, he would signal with his hat where the killing was. If they got to the buffalo when they were fresh, their duty was to take out all the humps, tongues and tallow from the best buffalo. The hunter would then hunt more if they did not have hides enough to make a load or finish their day's work.

A remarkable good hunter would kill seventy-five to a hundred in a day, an average hunter about fifty, and a common one twenty-five, some hardly enough to run a camp. It was just like in any other business. A good skinner would skin from sixty to seventy-five, an average man from thirty to forty, and a common one from fifteen to twenty-five. These skinners were also paid by the hide, about five cents less than the hunter was getting for killing, being furnished with a grind stone, knives and steel and a team and wagon. The men were furnished with some kind of a gun, not as valuable as Sharps' rifle, to kill cripples with, also kips and calves that were standing around. In several incidents it has been known to happen while the skinner was busy, they would slip up and knock him over. Toward the latter part of the hunt, when all the big ones were killed, I have seen as many as five hundred up to a thousand in a bunch, nothing but calves and have ridden up to them, if the wind was right.

There was always a boss or head man for each camp, and he was often the hunter. If he was a successful hunter, he would soon run a camp of his own, but if of the Pat Garrett style and gamble his money away would never run one successfully, as the business men soon learned he gambled and wouldn't credit him.

There also was a camp rustler to each camp, and his business was to watch the camp and look after everything about the camp, and prepare the meals, early and late. The boss and cook generally slept together in order to get him up in time to get breakfast, and whose-ever time it was to feed the stock, the other would peg the hides killed the day before.

As soon as he got his hides in suitable shape, he would examine them to see if any would do to flesh. If the wind and weather were right a week or ten days was all it required; all surplus flesh was removed by him, after they were dry a little, the flesh coming off easily. It usually took from eight to ten days

for an ordinary hide to dry so they could be ricked and stacked.
Watching these hides, he could tell by trampling over them by
the rattle when they were right, and would go out about sundown
and pull the pins out of the dry hides, reversing them top for
bottom.

Glenn described what he considered to be the best method of skinning a buffalo:

Have a stick about the size of a chair post with one end sharpened and the other end with a spike in it sharpened to a point.
If the buffalo was lying on his side and if he was a large one,
the skinner would have to get down on his knees under the fore-
shoulder and raise it up far enough until the backbone showed on
the opposite side. When he got him up at the proper height, he
would jab the stick in the ground, catch the spike behind the
fore shoulder just below the brisket. If the ground was not soft,
the buffalo's weight would not force him back more than three
or four inches. He would commence at the head on the under-
jaw, and would run a straight line from the brisket to the root
of the tail. Next the foot, running a straight line to the tail. After
ripping both, he would commence at the hoof and run over the
knee, coming out a little below the brisket. In ripping in this manner, the hide was uniform to stretch, although a great many
skinners would rip straight down the foreleg on the inside and
the same with the hind leg, but this way would leave a gap and
the first way did not do so.

He would then begin peeling the hide off, taking the jaws running to the back of the neck. When he had peeled the hide over
the backbone, he would go to the other side and skin the two
legs down, when about halfway down the ribs he would pull out
the prod, that side being in good skinning shape, and if all had
been done right, and he had shoved the hide properly underneath, it would all be on one side and he could pull it out and
throw it in the wagon.

These scrubhorn bull hides were, when green, slippery and
slick, weighing some eighty pounds. [They were] hard to throw
into the wagon, and if he were skinning by himself, would spread
it out on the ground, roll the legs under and when each side was

properly wrapped, it would be some three feet wide. He would then commence at the jaw and roll to the tail and taking hold of this tail would throw it in the wagon the same as a sack of flour.[28]

By late 1877, when Glenn and other buffalo hunters turned to other kinds of work, there were only scattered bands of buffalo on the plains of West Texas, eastern New Mexico, and what is now the Oklahoma Panhandle. What probably was the last large buffalo hunt on the southern plains for purely mercenary motives occurred in late 1877 about a hundred miles north of Tascosa, Texas. Two parties of buffalo hunters found about 200 buffalo and successfully killed fifty-two. Ten of the skins were preserved for mounting. The heads of the forty-two others were cut off and preserved for mounting and the skins were prepared as robes. The mountable skins were sold at these prices: young cows, $50 to $60; adult cows, $75 to $100; adult bulls, $150. The unmounted heads of young bulls sold at $25 to $30; adult bull heads, $50; young cows, $10 to $12; adult cows, $15 to $25. A few of the better buffalo robes went for $20 apiece; two of the robes were sold to the Hudson's Bay Company for $350.[29]

That year, 1877, what had been the four main buffalo hunters' headquarters in Texas—Buffalo Gap in Taylor County, Hide Town (now Snyder) in Scurry County, Rath City in Stonewall County, and Fort Griffin in Shackelford County—were nearly deserted. Most of the buffalo hunters had left.[30] And by 1880 the business of buffalo hunting on the southern plains had ended. The buffalo were gone.

Of all the tales told by buffalo hunters on the southern plains, there is one that seemingly has outlived nearly all others. It has been told hundreds of times and in several different ways. It is the story of a buffalo hunter, a buffalo hide, a cold "blue norther," and wolves. This particular version, the one I like best, was told by J. Wright Mooar in 1928, many, many years after he retired from hunting buffalo in West Texas.

It was February 1877. Jim Innis, a buffalo hunter, was camped on what is today called Sweetwater Creek in the southern part of Scurry County, Texas. Getting up one foggy morning, he set off on foot to hunt buffalo, leaving his horse and his buffalo skinners to enjoy themselves in camp. He found some buffalo in the fog and killed and skinned two of them. He wounded a third, a big old bull, and began to move up on him to finish him off. But the fog was bad, and Innis could not judge distance in fog. He got closer to the bull than he should, and suddenly the animal charged. To see the buffalo coming through the fog startled Jim. Instead of standing his ground and firing, he dropped his rifle and ran for a tree. Beating the bull to the tree, Innis jumped up and lifted himself to its heights. The huge buffalo, instead of stopping, ran straight into the tree with full force. The nearly dead tree snapped off and crashed. Innis landed right on top of the buffalo and then hit the ground. Hit by the tree, the buffalo took off like a scared jackrabbit. Innis watched the buffalo vanish into the fog. He shook himself. All his bones seemed to be in place, but his whole body hurt. He thought it best to find his gun just in case the buffalo should return. He found it and checked that it was still in good working order. But which way was camp? Innis did not know. He had lost all sense of direction.

He wandered for a while and found one of the carcasses he had skinned earlier in the day. Since it was getting dark, he decided to make camp for the night. He built a fire and cut off some of the buffalo meat from the carcass and pitched it on the coals. It sure smelled good roasting on the red coals, and it tasted just as good. After he had eaten all he wanted, he tossed the rest away and spread out the freshly skinned buffalo hide—hair side up—on the damp ground. Lying down on one edge, Innis rolled himself up in the hide and soon fell asleep. A little while later, a "blue norther" blew in and froze the hide, sealing Jim Innis inside. But Jim did not know this. He was fast asleep like a bear in hibernation. There were a great many lobo wolves in the country in those days. Where Jim had made camp the wolves were plentiful, and sure enough, attracted by the smell of cooked meat, a pack of lobos came into Jim's makeshift camp. They finished off the meat he had thrown away at supper time, and then the lobos began to tackle the hide in which Jim was wrapped. As they began pulling off the scrappy pieces of meat still clinging to the outside, Jim woke up and let out a yell that might have been heard all the way up to Kansas. But Jim's yelling did not bother the wolves. They just kept stripping the meat off the hide, and Jim just kept yelling. Some of the animals by then had found the carcass nearby, but a few of the critters stayed with the hide and Jim as if they belonged together. As they dug into the hide for tough pieces of meat, they tore Jim's clothes and bit into his side. At last, as the first light of morning appeared in the eastern horizon, the lobos sneaked off to wherever lobos go in the daylight. At about 10 o'clock the sun finally thawed out the hide so that Jim could unroll himself and squeeze out.

The clear blue sky was a welcome sight to Jim Innis. And the sun felt good. Better still, he knew which way camp was. When he got there, his own skinners hardly recognized him. He looked like an old man, his face in seams and his hair practically white. Jim Innis had gone buffalo hunting a blackhaired young man and had come back looking old and grayhaired.[31]

As the supply of buffalo on the southern plains dwindled, some hide hunters turned toward the northern plains, but many hesitated. They were fearful of the Indians, especially the Sioux. Also, getting the buffalo hides to market was more difficult on the northern plains. Transportation was still primitive. In the early 1870s in Dakota Territory the railroads were just beginning, and then only in the eastern portion. Overland freighting was next to impossible except during the summer months. The rivers were the best means of transporting hides to such eastern markets as Leavenworth, Kansas City, and St. Louis.

The Northern Pacific Railroad, although founded in 1864, did not begin to push westward until February 1870. By 1871 it had crossed only the Red River of the north at Fargo, Dakota Territory. Slowly, under the protection of soldiers, it pushed across Indian country reaching Jamestown, Dakota Territory in the summer of 1872, and Bismarck, Dakota Territory, on the Missouri River, in the summer of 1873.[32] But there it stopped. The business panic of 1873 delayed further construction.

The western end of the Northern Pacific was still at Bismarck the following year when gold was discovered in the Black Hills a few hundred miles to the south. Lt. Colonel George A. Custer, on orders of General Sheridan, had led an expedition into the Black Hills, land that the government had promised to the Sioux Indians. When geologists found gold, the government tried to buy the Black Hills from the Indians, but negotiations broke down. Meantime, gold seekers swarmed into the Black Hills, defying the Sioux, the government, and

Skinning a buffalo in northern Montana, 1879. L. A. Huffman photo. *Courtesy Montana Historical Society.*

"Taking the Robe," from *Drawings by Frederic Remington,* 1897.

"Hide Hunters at Work," M. S. Garretson drawing. *Courtesy Kansas State Historical Society.*

John Douglas Campbell, the Marquis of Lorne, was Governor General of Canada from 1878 to 1883. In 1881 Lord Lorne and entourage made a well publicized tour of western Canada in order to promote the Canadian Pacific Railway, immigration and settlement, and investment in the area by skeptical British bankers. During the first week of September, at Red Deer River, about fifty miles northeast of present-day Calgary, one of the last buffalo hunts on the Canadian prairies took place when several members of the Governor General's party chased and killed three bulls. It was the first herd seen in years. The above drawing (pencil and watercolor) was done by Sydney Prior Hall, artist-reporter for the London *Graphic* and a member of the Lorne party. The officer shooting the wounded buffalo is Capt. Percival, aide-de-camp; the mounted man on the right is Chief Poundmaker of the Crees, guide to the party. *Courtesy Public Archives of Canada, Ottawa.*

the Army. By early 1876, the gold rush was fully under way and the Indians on the northern plains were ready to explode. When they did, the Army, the buffalo hunters, the traders, and other white men on the northern plains were kept busy. There was little time for hunting buffalo except for food.

On June 25, 1876, General Custer and more than 200 soldiers were killed in the battle of the Little Bighorn, but within a year the great Sioux uprising was over. Sitting Bull and a few followers fled into Canada. Another Indian chief, Crazy Horse, surrendered; and for the white men on the northern plains, the land was now considered safe.

Other white hunters then began to converge onto the northern buffalo range, joining those who had made it safely through the Indian troubles. By then, however, the northern buffalo range had shrunk to eastern Montana, northern Wyoming, and western Dakota. Only in that area could large numbers of buffalo be found.

Just how many buffalo were in that area was reflected in an interview with U.S. Marshall X. Biedler of Montana in August, 1879:

"How far is Poplar Creek agency from Miles City?" asked the reporter.

"Nearly 175 miles by the way we came," answered Biedler.

"Were there any buffalo in sight anywhere on your trip?"

"Buffalo!" replied Biedler, "Why for four days we saw nothing else but buffalo. For 70 miles we were in the center, it seemed, of a herd numbering into the millions. It was the largest herd I ever saw."[33]

Such sights were no longer common to the south. In Wyoming and eastward through Dakota and Nebraska most of the buffalo had been killed by 1879, as was the case in southern Canada, where the animal was already considered scarce.[34] There, Indians, instead of the white men, hunted the few remaining buffalo. Few white settlers were on the Canadian plains during the 1870s because of the Hudson's Bay Company's "deliberate and persistent opposition to settlement in western Canada."[35] It should be noted here that the Canadian Pacific Railroad, unlike her sister lines south across the border, played no part in Canada's buffalo slaughter. The Canadian rail line was built after the buffalo were all but exterminated in southern Canada.[36]

But south across the international boundary in Dakota Territory, the

Northern Pacific Railroad, which had stopped its westward push in 1873, began moving again in 1879. Construction of the line west of Bismarck resumed. After the unscrambling of the line's financial situation, the road laid track to a point about fifty miles west of Mandan. In 1880 it reached the Dakota-Montana boundary near the center of the northern buffalo range. When the railroad arrived, a silver spike was driven into the roadbed to commemorate an event that would also help bring about the downfall of the buffalo on the northern plains. By then, 1880, the killing of the buffalo in the north was beginning in earnest.

Glendive, Montana Territory soon became the Dodge City of the north. It was the headquarters for hundreds of buffalo hunters who brought their hides to sell and to trade for supplies. Almost daily, railroad cars loaded with buffalo hides were shipped east on the Northern Pacific to fulfill the needs of eastern and foreign tanners. West of the railroad in 1880, steamboats continued to carry large numbers of buffalo hides down the Missouri and Yellowstone Rivers. Reporting on the season's activity of 1880, the Sioux City, Iowa *Journal* said:

> Most of our citizens saw the big load of buffalo hides that the C. K. Peck brought down last season, a load that hid everything about the boat below the roof of the hurricane desk. There were ten thousand hides in that load, and they were all brought out of the Yellowstone on one trip and transferred to the C. K. Peck. How such a load could have been piled on the little Terry not even the men on the boat appeared to know. It hid every part of the boat, barring only the pilot house and smokestacks. But such a load will not be again attempted. For such boats as ply the Yellowstone there are at least fifteen full loads of buffalo hides and other pelts. Reckoning one thousand hides to three carloads and adding to this fifty cars for the other pelts, it will take at least three hundred and fifty box-cars to carry this stupendous bulk of peltry East to market. These figures are not guesses, but estimates made by men whose business it is to know about the amount of hides and furs awaiting shipment.[37]

When the Northern Pacific railroad reached Miles City, Montana Territory in 1881, more hide hunters than ever assembled on the plains of eastern Montana. Miles City soon joined Glendive as a hunter's headquarters.

The winter of 1881-82 thousands upon thousands of buffalo were killed for their hides. One hunter, Vic Smith, is reported to have killed "one hundred and seven buffaloes in one 'stand' in about an hour's time, and without shifting his point of attack." Smith's "stand" supposedly took place about a hundred miles northeast of Miles City.[38]

By January 1882, prices for buffalo robes and "green" hides had reached an all-time high. Hide hunters were making a good profit for their efforts, more than the hunters on the southern plains had made. And unlike the southerners of a decade earlier, the northern hunters avoided waste and used care and skill in removing and preserving hides. Some hunters may have realized that the supply of buffalo was near an end. One source said that every 100 hides sold in the north represented only 110 dead buffalo.[39] Earlier in the south, considerably more buffalo than that were killed for fewer hides.

During 1882, the Northern Pacific Railway carried 200,000 hides out of Montana and the Dakotas. That was the biggest year for the hunters. The following year, 1883, only 40,000 hides were reportedly sent east.[40] Ironically, 1883 was the year that Dakota Territory, where nearly all buffalo had been killed, passed a law protecting the animal. This was typical also of other states and territories to the south. They passed laws to protect the buffalo after the animal was gone.

By the summer of 1883, many hunters in eastern Montana refused to believe the buffalo was nearly gone. Some buffalo hunters packed up and left, realizing the truth, but there were many who were sure more buffalo would return in August, when the animals usually ranged south out of Canada to find food. The hunters stayed, but they waited in vain. August and September passed without any sign of the great buffalo herds that had been seen during the previous years. There were no great herds. Nearly all had been slaughtered.[41]

The few buffalo that had survived the white hunters on the northern plains were scattered in small bands. They were hardly worth the effort of the hunters to track down. And by 1884, hide hunting as a business on the northern plains had all but ended. What is believed to have been the last carload of buffalo robes and hides from the northern plains was shipped from Dickinson in far western Dakota in 1884.[42]

William T. Hornaday, in his Smithsonian report of 1887, gave these annual totals for buffalo hides shipped east from the northern plains during the early 1880s:

Buffalo Shipments

Year	Number of Buffalo Hides Shipped East
1881	50,000
1882	200,000
1883	40,000
1884	300

Hornaday wrote that he found no records to indicate that any carloads of buffalo hides were shipped east from the northern plains in 1885.[43]

By then the American and foreign tanners had already begun to search for another animal to supply hides with which to make leather. The great slaughter of the American buffalo had ended, as had the slaughter of cattle on the pampas of Argentina not quite two decades before.

VIII

Early Attempts to Save the Buffalo

There is no calamity which a great nation can invite which equals that which follows a supine submission to wrong and injustice.

Grover Cleveland 1895

As early as 1776, travelers crossing what is today the southeastern United States became concerned about the killing of buffalo "for the sake of perhaps his tongue only."[1] During the next twenty-five years, others saw buffalo being killed needlessly and they too spoke out against the wanton slaughter east of the Mississippi, but nothing was done about it. Too few really seemed to care. Conservation of wildlife, as we know it today, was unheard of in America during the late 18th century. The country was young. It was big and new. Not even half of it had been explored. Over the vast expanse of wilderness, wild game was plentiful, and it was there for the taking.

When the American exploration of the prairies and plains began in the early 1800s, white men encountered many more buffalo than they had ever found in the east, and it was impossible for them to imagine that the killing did any harm. However, some protests were made during the first half of the 19th century. In 1820 Major Stephen Long's expedition saw many buffalo killed, and advocated a law to protect the animal from wanton slaughter.[2] George Catlin, traveling across the northern plains twelve years later, saw Indians killing buffalo for their robe. Catlin predicted, "The buffaloe's doom is sealed."[3] Plainsman Josiah Gregg echoed similar sentiments in 1845 after he crossed the southern plains:

121

The slaughter of these animals is frequently carried to an excess which shows the depravity of the human heart in very bold relief. Whether the mere pleasure of taking life is the incentive of these excesses I will not pretend to decide, but one thing is certain, that the buffalo killed yearly on these prairies far exceeds the wants of the traveler or what might be looked upon as the exigencies of rational sport.[4]

John J. Audubon, perhaps America's most gifted naturalist-artist, issued his own warning in 1843 after witnessing buffalo being killed on the plains: "Like the Great Auk, before many years the Buffalo will vanish. Surely this should not be permitted?"[5] In 1852 Joseph Leidy, a highly respected scientist, warned about the buffalo: "The time is not far distant when it will become quite extinct, unless protected by a munificent republic, as had been done by the Emperor of Russia in the case of the aurochs, or European bison."[6] Spencer F. Baird, another well-known scientist, who was to become assistant secretary of the Smithsonian Institution, went west during the early 1850s on an exploration trip into Wyoming Territory. He saw many buffalo killed. When he returned east he reported: "If it were possible to enforce game-laws, or any other laws on the prairies, it would be well to attach the most stringent penalties against the barbarous practice of killing buffalo merely for the sport, or perhaps for the tongues alone."[7]

Few people paid any attention to such warnings. It seemed so impractical to talk about needing to protect an animal that roamed in countless numbers over millions of square miles of prairie and plains. And even if protection were necessary, an army of men could not protect so many animals. During the 1850s, therefore, no measures to protect the buffalo were even considered by Congress.

After the Civil War, as immigrants headed west to settle on the plains and prairies and as people in the East were reading and hearing about the killing of buffalo, calls to protect the animal were again sounded. But these, like the earlier ones, were ignored in Congress. However, in a few western territories there was some interest to pass laws to protect wildlife, including buffalo.

In 1864 Idaho Territory, one year after becoming a territory, passed a law protecting buffalo. But it was limited. The legislation made it illegal to hunt buffalo, elk, deer, antelope, and mountain goats and sheep between February 1 and July 1. At all other times it was open season for the hunters. Besides, buffalo in Idaho Territory were few and far between in 1864. They were mostly the mountain variety in areas rarely reached by man. Slaughter had already taken its toll. But to the handful of people in the East calling for protection of the buffalo, the law was looked on as at least a start in the right direction.

Until the early 1870s, when the slaughter began in full on the southern plains, Idaho's protection law was the only one—state or territory or federal—on the books.[8] But by then some lawmakers in Washington were finally beginning to take note of the killing. Newspaper accounts told of thousands of buffalo being killed for their hides. Some Congressmen became concerned. One of them was R. C. McCormick, Territorial Delegate from Arizona, which then had few if any buffalo. In March 1871 McCormick introduced a bill (H.R. 157) in Congress aimed at reducing the slaughter. It read:

> That, excepting for the purpose of using the meat for food or preserving the skin, it shall be unlawful for any person to kill the bison, or buffalo found anywhere upon the public lands of the United States; and for the violation of this law the offender shall, upon conviction before any court of competent jurisdiction, be liable to a fine of $100 for each animal killed.[9]

But McCormick's effort failed. The bill was printed, but it was never reported back from committee to the House of Representatives for action.

That same year, 1871, the Wyoming Territorial Legislature joined Idaho and passed a law aimed at reducing the killing of game, including buffalo; in early 1872 the Montana legislature passed a similar law.

In Kansas, where perhaps the greatest slaughter of buffalo was then occurring, protection was being debated. General W. B. Hazen, commander at Fort Hays, Kansas, in buffalo country, called upon the lawmakers to take steps to prevent "this wicked waste, both of the lives of God's creatures and the valuable food they furnish." Hazen lamented that he had seen numbers of men who had killed 1,000 buffalo and sold their hides for the "paltry sum of $1 apiece, the carcasses being left to rot on the plains."[10]

The protest voice was rising. A few months later, in June 1872, a Topeka newspaper featured an article entitled "The Murder of the Buffalo." It read in part:

Few persons probably know how rapidly the American bison is disappearing from the Western plains. . . . Some idea of the extent of the ruthless slaughter may be formed from the fact that twenty-five thousand bisons were killed during the month of May south of the Kansas Pacific Railroad for the sake of their hides alone, which are sold at the paltry price of two dollars each on delivery, for shipment to the eastern market. Add to this five thousand—a small estimate—shot by tourists and killed by the Indians to supply meat to the people on the frontier, and we have a sum total of thirty thousand as the victims for a single month.

If the bison were a wild and savage animal—if to kill one required any special skill or bravery or nerve, there might be some justification for this enormous slaughter. . . . Every one remembers how Prince Alexis, under the leadership of Gen. Sheridan, participated in this "sport," to the intense gratification of his royal father and to the profit of the special correspondents. It is doubtful, however, whether even a royal precedent can justify this kind of so called hunting. However this may be in the Eastern States, the following paragraph from the letter of an army officer shows that in the Western States this kind of "sport" is estimated at its true worth, while at the same time, its reference to the number of persons who are following the Russian Princeling's example confirms the apprehension that the American bison will soon become as fabulous an animal on the dodge.

"To shoot buffalo seems a mania. Men come from London—cockneys, fops and nobles—and from all parts of the Republic, to enjoy what they call sport. Sport! when no danger is incurred and no skill required. I see no more sport in shooting a buffalo than in shooting an ox nor so much danger as there is hunting Texas cattle."[11]

By the time this article was published, the Kansas legislature already had passed a protection bill. But the governor had killed the measure. It was not politically sound for him to object to Kansan's

"Slaughter of Buffaloes on the Plains," Theodore R. Davis drawing, from *Harper's Weekly*, February 24, 1872.

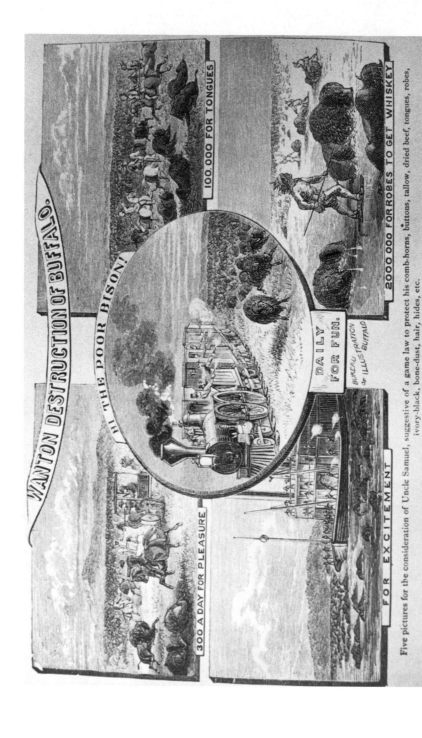

Henry Worrall drawing, from W. E. Webb's 1872 book, *Buffalo Land*.

Five pictures for the consideration of Uncle Samuel, suggestive of a game law to protect his comb-horns, buttons, tallow, dried beef, tongues, robes, ivory-black, bone-dust, hair, hides, etc.

killing buffalo. There was some criticism of the governor's veto. One widely read article, "Save the Buffalo," represented the viewpoint of at least two newspapers in Kansas and one in Colorado Territory:

> The Leavenworth *Commercial* lifts its sturdy voice in solemn protest against the ruthless slaughtering of the buffalo, in which so many "fancy hunters" are prone to indulge at every opportuity. We second the protest for the hundreth time. It is an outrage that the law should stop. A paper in Southern Kansas — the Wichita *Eagle*—says that thousands upon thousands of buffalo hides are being brought to the town of Wichita, by hunters; that in places whole acres of ground are covered with hides, spread out with the fleshy side up to dry; and that it is estimated that there is, south of the Arkansas and west of Wichita, from one to two thousand men shooting buffalo for their hides alone, and more frequently, just for "the fun of shooting." Is it not time to check this brutality? Let Kansas and Colorado enact strict laws and the evil will be abated if not eradicated. If matters are allowed to go on as now it will not be many years before such a thing as a buffalo will be an actual curiosity.—*Denver Tribune.*[12]

Even if Kansas had had a buffalo protection law, it would have been very difficult to enforce. In Idaho, Wyoming, and Montana, where protection laws were on the books, the animal was protected only on paper and not on the plains or rolling prairies or in the mountains.

In February 1872 Senator Cornelius Cole of California introduced a resolution in the United States Senate directing the Committee on Territories to investigate the expediency of passing a law to protect buffalo, elk, antelope, and other wild animals in the territories from "indiscriminate slaughter and extermination."[13] The resolution was adopted by unanimous consent. Two days later Massachusetts Senator Henry Wilson, reporting for the committee, introduced a bill (S. 655) in the Senate similar to the measure McCormick had introduced in the House almost a year earlier. But when the Wilson bill was sent back to committee, history repeated itself: the committee never reported the bill to the full Senate.[14] The committee, it appears, favored the business interests that profited from the buffalo slaughter.

Not quite two months later, McCormick renewed his appeal to protect the buffalo. On the floor of the House of Representatives he warned of the animal's possible extermination. He called the buffalo "the finest wild animal of our hemisphere." McCormick cited as evidence of the possible extermination an article that had appeared in *Harper's Weekly* showing illustrations of the slaughter. Then, in apparent hopes of strengthening his case even further, McCormick read several letters that had been sent to Henry Bergh, then president of the American Society for the Prevention of Cruelty to Animals. The letters had been written by army officers who had served on the buffalo plains of Kansas and Nebraska. These officers expressed their indignation that such cruelty to wild animals should be allowed to continue.

If McCormick had stopped at this point on the cruelty to wild animals, he might have been successful in getting some sort of protecting legislation passed. But he did not. Instead, he read a letter written by E. W. Wynkoop, who had been an Indian Agent as well as an army officer for thirteen years. Wynkoop's letter read in part:

> There is another strong reason, apart from cruelty which should compel Congress to take action; it is one of the greatest grievances the Indians have; and, to my personal knowledge, has frequently been their strongest incentive to declare war. "Little Robe," the Cheyenne chief, who recently visited Washington, at one time remarked to me after I had censured him for allowing his young men to kill a white man's ox, "You people make a big talk, and sometimes war, if an Indian kills a white man's ox to keep his wife and children from starving. What do you think my people ought to say and do when they themselves see their cattle [buffalo] killed by your race when they are not hungry?"[15]

Reading Wynkoop's letter was a mistake. The truth hurt. It did not further McCormick's cause. It reminded members of Congress that the buffalo was the Indian's commissary and that if the Indians continued to have buffalo they would be most difficult to control. Congress took no action to protect the buffalo in 1872.

Early in 1874 Representative Greenburg L. Fort of Illinois set to work to save the buffalo. He introduced two bills: one to protect the buffalo, the other to tax buffalo hides.[16] Fort hoped a hide tax would make the hide trade unprofitable and thereby reduce the killing of buffalo. The House Committee on Ways and Means voted against the hide measure, but the proposed bill to protect the buffalo, which would make it illegal for anyone other than an Indian to kill any female buffalo in any U.S. Territory, cleared the committee. Then it was sent to the House floor for debate.

During the debate, Representative Samuel S. Cox of New York said he understood it was impossible to tell the difference between male and female buffalo when they were running. Some western lawmakers, fresh from the plains, set Cox straight on the "facts of life" and showed him how he had been misinformed. Cox then objected that the bill was not fair. He said he felt it favored the Indians at the expense of the white men, and he asked whether it would be in order to strike the phrase permitting the Indians to kill female buffalo. Apparently to strengthen his position, Cox at this point reminded the House, "The Secretary of the Interior has already said that the civilization of the Indian is impossible while the buffalo remains upon the plains."[17] It was obvious by then that Cox was one of many lawmakers who leaned toward destroying the buffalo to destroy the Indian. But in spite of Cox and others in Congress who argued against the bill, it passed both House and Senate. Early in the spring of 1874 it was sent to the White House for President Ulysses S. Grant's signature, but Grant gave it a pocket veto. The unsigned bill was pigeonholed and died a simple death.

Grant's lack of action probably did not surprise Representative Fort and others favoring the legislation. They knew that Columbus Delano, Grant's Secretary of the Interior, did not want the slaughter of buffalo to stop. Only a few months earlier, in his annual report of 1873, Secretary Delano had said, "I would not seriously regret the total disappearance of the buffalo from our western prairies, in its effect upon the Indians. I would regard it rather as a means of hastening their sense of dependence upon the products of the soil and their own labors."[18] Also, it was common knowledge in Washington that the two western military chiefs, Generals Sheridan and Sherman, held the position that the only way to force the Indians to comply with military orders was to clear the plains and prairies of buffalo. President Grant himself privately supported his old comrades, Sheridan and Sherman, but publicly said as little as possible on the subject. It was also the Army's position to say as little as possible in public.

Nowhere have I found any record that the government officially helped buffalo hunters do their job. The help was unofficial. Frank H. Mayer, the last of the buffalo runners, acknowledged:

> The army officers in charge of plains operations encouraged the slaughter of buffalo in every possible way. Part of this encouragement was of a practical nature that we runners appreciated. It consisted of ammunition, free ammunition, all you could use, all you wanted, more than you needed. All you had to do to get it was apply at any frontier army post and say you were short of ammunition, and plenty would be given you. I received thousands of rounds this way.[19]

I have found only two official accounts showing that the government attempted to stop buffalo hunters from intruding upon Indian land; the accounts are in letters written by Deputy U.S. Marshal Benjamin Williams. One letter tells how on January 30, 1874, Williams and another deputy, William Talley, accompanied by eight soldiers, found six buffalo hunters in a dugout on the banks of the Cimarron River in Indian Territory. The hunters were arrested. But what happened to the hunters is not recorded. Not quite two years later, in July 1875, six other buffalo hunters were arrested by marshals Williams and Talley in Indian country. They were taken to Wichita, Kansas and charged with trespassing.[20]

Aside from these two accounts, I have found no official records indicating that the government, especially the military, stopped buffalo hunters or even tried to stop them from intruding upon Indian land, land acknowledged by numerous government treaties as belonging to the Indians. It probably is one of the few times that the U.S. Government, and especially the Army, accomplished what they wanted by officially doing almost nothing.

About one year after President Grant's pocket veto of the law that would have protected the buffalo, the state of Nebraska, in 1875, passed a law protecting the animal. But by then the Nebraska buffalo population had dwindled considerably.

In Washington new attempts to get national legislation were made during 1875 but again failed.

As buffalo hunters streamed into northwest Texas, there began to be talk in Austin, the state capital, about passing a law to protect the buffalo. Finally, someone introduced a buffalo protection bill. It was then that General Philip Sheridan, commander of the Military Division of the Missouri, was invited—he may have invited himself—to testify before a joint meeting of the Texas Senate and House. He is reported to have told those attending—I can find no official record of the meeting—that instead of stopping the hunters, the Texas lawmakers ought to give the hunters a hearty, unanimous vote of thanks. Sheridan added that the legislature should appropriate enough money to strike a medal for each hunter, a medal with a dead buffalo on one side and a discouraged Indian on the other. Then Sheridan supposedly said:

> [The buffalo hunters] have done in the last two years and will do more in the next year [1876] to settle the vexed Indian question, than the entire regular army has done in the last thirty years. They are destroying the Indian's commissary, and it is a well-known fact that an army losing its base of supplies is placed at a great disadvantage. Send them powder and lead, if you will; for the sake of a lasting peace, let them kill, skin and sell until the buffaloes are exterminated.[21]

The hunters in Texas did kill, and by 1880 the business of buffalo hunting on the southern plains was history. Ironically, that same year New Mexico passed a law protecting the buffalo, but there were few left to protect.

By 1884, the largest single group of wild buffalo left in the United States were in Yellowstone National Park in northwestern Wyoming. When in 1872 Yellowstone became the first national park, small bands of mountain buffalo roamed its more than 3,000 square miles as they had done for centuries before. Although the legislation that made Yellowstone a park had provided that there be no "wanton destruction of the fish and game [or] capture or destruction for the purposes of merchandise or profit" and also assured "the preservation from injury or spoilation of all timber, mineral deposits, natural curiosities or wonders within said park," nothing was done to make these auspicious aims possible. The park act contained no code of laws for the park, offenses were not defined, neither punishments nor law enforcement machinery was provided, and for over five years Congress appropriated not a single penny to run the park. In practice this meant that there was no money for roads, that neither the superintendent nor his aides received any salary, that the superintendent was not resident in the park but visited it only occasionally, and that the park was an attractive area for poachers and vandals—and foreign sportsmen. In 1873 a Helena, Montana newspaper lamented the practice of European nobility, especially the "British sportsman," who demoralized "guides, trappers and hunters of the plains and mountains by his lordly manner of butchering buffalo and grizzly bears. . . . Yellowstone Park bids fair to become a very tame and civilized retreat."[22]

Two years later W. E. Strong visited the park with Secretary of War W. W. Belknap and noted the deplorable work of the poachers: "During the last five years the game has been slaughtered by the thousands of hunters who killed them for their hides alone."[23] The same year, 1875, William Ludlow observed: "Hunters have for years devoted themselves to the slaughter of game, until within the limits of the park it is hardly to be found."[24] Three years later, Major James S. Brisbin, commander of nearby Fort Ellis, Montana, wrote: "Nowhere on earth are such wonders and game to be found and it will be a great shame if our authorities permit them to be destroyed." Of the game he said: "They are not slaughtered for their meat or skins but . . . simply for the pleasure of killing them." Of the wonders he said:

> Some of the most beautiful formations have already been entirely destroyed and will never be replaced during our lives or the lives of our children. This disposition to vandalism seems to possess everyone who enters the park and unless it is checked our wonderland will cease to be an object of interest.[25]

By 1882 the Northern Pacific line reached Livingston, Montana Territory, some fifty miles north of Yellowstone, making access to the park even easier. Poaching and vandalism continued. In 1884 the Wyoming territorial legislature passed an act designed "to protect and preserve the timber, game, fish and natural curiosities of the Park." The act was stringent in its provisions, but it "totally failed of its purpose. The attempt at territorial control of a national institution was in itself a blunder."[26]

Conditions became so bad in the park that on August 20, 1886, Captain Moses Harris, First United States Cavalry, relieved the civilian superintendent of his duties, and soldiers supplanted the assistant superintendents as park police. Harris soon began escorting poachers, woodcutters, and vandals out of the park whenever he or his men found them. He also posted guards at key spots around the park.

One day a patrol found thirteen buffalo heads hanging from trees in a poacher's camp, but the poacher was not caught. The following year, however, a poacher was caught in the same general area while skinning one of three buffalo he had just killed. The poacher's arrest was reported to the War Department in Washington by Captain Harris. The Captain was obeying a specific order to "protect the buffalo." But when word of the arrest reached Washington, the War Department promptly wired back, "The prisoner being detained without authority of law or military regulations, will be discharged." Surprised and discouraged, Captain Harris ordered the poacher set free, but Harris, as one version of the story goes, left the poacher in the company of one of his biggest sergeants, who proceeded to "beat the living daylights out of him." The sergeant then escorted the poacher to the park's boundary and literally kicked him over the line. That poacher supposedly never returned, nor did others who allegedly received the same unofficial treatment. Captain Harris had taken it upon himself to "protect the buffalo" as he had originally been ordered to do. As William Hornaday noted in 1889, if it had not been for Harris, most of the buffalo in Yellowstone probably would have been killed by poachers.[27]

In spite of Captain Harris' actions, some poachers continued to make excursions into Yellowstone killing and skinning buffalo and then fleeing north into Montana or in some other direction. For many poachers, Livingston, Montana was headquarters. There the poachers could sell their kills to any one of several taxidermists who had set up shops along Park Street, which ran parallel to the railroad tracks. The taxidermists would mount the buffalo heads, tan buffalo hides, and then ship them east. The buffalo meat usually found its way to the local meat markets. In the late 1880s the going price for a mounted buffalo head in New York and other eastern cities ranged from $25 to $100, depending upon the size and condition.

Some hunters, reluctant to enter Yellowstone, camped just outside the park's boundaries and waited for buffalo and other game to wander outside. When they did, the poachers killed them.

To the south, Colorado passed a law in 1889 providing severe penalties for anyone killing a buffalo before the year 1900. One rancher openly boasted that he had killed five buffalo, but when a state game warden tried to bring the man to justice, he could not find one person who would testify against the rancher.[28]

It was not until 1894 that the United States Government passed its first buffalo protection law. The law came about, in part, because of what one man, Edgar Howell, a poacher, had done. Early in the winter of 1893-94, Howell and an unidentified partner left Cooke City, Montana, with a hand sledge loaded with supplies. They quietly pushed their way into Yellowstone and set up a camp on Astringent Creek, which drains to Pelican Creek and into Yellowstone Lake. Howell and his friend were out to kill buffalo that had come down from the mountain tops to winter in the meadows. After killing the animals they would skin out the heads, hang the scalps safely in trees, and await spring. Then, as they planned, they would take their trophies out of the park on horseback and sell them over the border in Montana for between $100 and $300, then the going price.

In the days after their arrival in the park, Howell and his partner had a disagreement. The partner left camp and returned to Montana. But Howell stayed and began to hunt buffalo. One day a military scouting party checking on the buffalo east of the Yellowstone River discovered Howell's sledge trail. Although it was too old to follow, the trail aroused the soldiers' suspicions. Captain G. L. Scott, scout Felix Burgess, and a few others set out on snowshoes to check the area. What happened next is recorded in the annual report of Yellowstone National Park for 1894 by Captain George S. Anderson:

> On the morning of the 13th, very soon after starting, they came across some old snowshoe tracks which they could scarcely follow, but by continuing . . . they soon came across a cache of six bison scalps suspended above the ground, in the limbs of a tree. Securing these trophies, the party continued down Astringent Creek to its mouth and then turned down Pelican. They soon came across a newly-erected lodge, with evidences of occupation, and numerous tracks in the vicinity. Soon after this they were attracted by the sight of a man pursuing a herd of bison in the valley below them, followed by several shots from a rifle. After completing the killing, the culprit was seen to proceed with the removal of the scalps. While thus occupied with the first one my scouting patrol ran upon him and made the capture.[29]

William Hruza's meat market, Livingston, Montana, late 19th century. Note the sign, reading BUFFALO, just above the scales to the right. Buffalo meat, likely poached from Yellowstone National Park just south of the town, was a major item in Livingston's diet. *Courtesy William F. Whithorn, Pray, Montana.*

Skinning and cooking after a buffalo hunt, 1904. *Courtesy Library of Congress.*

Martin S. Garretson, well-known authority on buffalo, curator of the National Museum of Heads and Horns in the New York Zoo, president of the American Bison Society during its final four years, and one of those who helped save the buffalo. In his Bronx office, October 1931. *Courtesy Conservation Library Center, Denver Public Library.*

Edgar Howell was the man captured. As Howell, his arms, ammunition, and trophies were being taken back to Mammoth Hot Springs and the Fort Yellowstone guardhouse, the scouts met a party of conservationists, including Emerson Hough, a correspondent for *Forest and Stream,* F. Jay Haynes, a photographer, and T. E. "Billy" Hofer, a guide. Hough, quick to see a news story for the magazine in the capture of Howell, got the facts. Pictures were taken of Howell and the buffalo he had killed. After the scouting party took Howell to the guardhouse at Fort Yellowstone, plans were made to turn Howell over to civil authorities in Wyoming, where a buffalo protection law did exist, at least on the books. But the law made no provision for taking someone from the national park to Wyoming for punishment. Howell had to be released.

In the East, however, publication of Hough's story in *Forest and Stream* began to make news.[30] George Bird Grinnell, editor of the magazine, saw the story into print and then rushed to Washington with some influential friends. There they met with lawmakers and presented the facts pointing up the need for legislation to protect the buffalo at Yellowstone. Hough estimated that poaching had reduced the Yellowstone buffalo from 500 to less than 200. Thirteen days after Howell had been arrested at Yellowstone, Representative John F. Lacey of Iowa introduced a measure (H.R. 6442) providing for a jail sentence of up to two years and fines as high as $1,000 for anyone removing mineral deposits, cutting timber, or killing game, including buffalo, in Yellowstone National Park. Lacey's proposed legislation also made it illegal for any railroad company, stage line, or individual to transport any game taken from the park. Furthermore, it proposed to give Yellowstone officials the right to confiscate any equipment or vehicles used in these offenses.

As Lacey's bill was introduced, public opinion was being aroused. Newspapers had picked up the story of Edgar Howell from *Forest and Stream,* and their voices were added to the growing number calling for an adequate law to protect Yellowstone National Park, its natural resources and its game—especially the buffalo. Lacey's bill passed the House and the Senate, and on May 7, 1894 President Grover Cleveland signed the Lacey Yellowstone Protection bill into law. It was the first federal legislation protecting the buffalo. But it came ten years after man's greatest slaughter of wild game had ended.

IX

After the Kill

Nature produces nothing for nothing.
Mandan Indian 1806

One fall day in 1874 Arthur C. Bill and his cousin went shopping in Dodge City. Each bought a yoke of oxen and a wagon, camp supplies including a Dutch oven, some guns and ammunition, and one pony between them. After a good night's sleep, they got up early and headed south out of Dodge. Bill and his cousin were destined to become "bone pickers."

In 1874 only small scattered bands of buffalo could still be found on the plains between Dodge City and Camp Supply, which was south across the Kansas border in Indian Territory. Thousands, perhaps millions, of buffalo that had ranged that area only ten years earlier were nearly gone. They had been killed by white men, most of whom had taken only the animal's hide. Scattered across what seemed like an endless plain, the bones of thousands upon thousands of buffalo were the only reminders of the vast numbers of animals that once roamed those plains.

It did not take Arthur Bill and his cousin long to fill their wagons with bones that had bleached in the summer sun. And in the winter months the task of picking up buffalo bones became equally boring. Only the changing geography, the weather, the killing of a live buffalo—when one could be found—and the comfort of their nightly camp broke the bone pickers' monotony.

"We camped wherever night overtook us," recalled Arthur Bill many years later. He remembered vividly how they would turn their oxen out to graze at night, hobble their pony, make their beds on buffalo grass, and enjoy the evening meal. "We mixed our baking powder in flour and water, baking the finest, lightest bread you ever ate in our Dutch oven, using buffalo chips for fuel. We fried our buffalo steaks and bacon, made our coffee, fried our sweet potatoes and Oh Boy! didn't it taste good after being out in the wind all day!"

As soon as the two men filled their wagons with bones, they would dump them along the major north-south trail running between Dodge City and Camp Supply. To show ownership they would mark the pile with their number, one that had been assigned to them by the bone buyer in Dodge City. Then they would resume the task of picking up more buffalo bones.

Government freighters out of Dodge, who hauled goods south to Camp Supply, were glad to fill their empty wagons with the bones on their return trip. It meant a little more money in their pockets. Once the freighters reached Dodge, the bone buyer would weigh the bones and pile them along the Atchison, Topeka & Santa Fe tracks. Eventually they would be loaded aboard railroad cars and shipped to St. Louis.

Arthur Bill said the horns were used in "refining sugar . . . to make buttons, combs, knife handles, and so forth . . . their hoofs used to make glue."

"We received from $7 to $9 per ton f.o.b. for the plain bones, and $12 to $15 for the hoofs and horns," he said. "We made good gathering buffalo bones—would often come to a place where the hunters had succeeded in getting a stand on the herd and would kill them all." In such spots the two men could usually fill their wagons with bones in a very short time.[1]

Bone picking was not new on the plains of western Kansas in those days. Some trade in bones had begun several years before. Frank Root, an early Kansas newspaperman, wrote, "In the '50s and '60s [buffalo] bones were scattered promiscuously in certain localities for hundreds of miles in central and western Kansas, and between Fort Kearny and Julesburg along the Platte, as far back from the river as the eye could reach."[2] In those years small numbers of bones were picked from the plains and shipped east by wagon to be sold.

But the buffalo bone trade did not become a major industry on the plains and prairies until the railroads arrived in the late 1860s. Transporting the bones had earlier been a major problem, but almost overnight enterprising western businessmen found a market in the East for the bones which could easily be shipped by train from Kansas and Nebraska. With the railroads the bone market grew. And ships played their part, too. Large quantities of buffalo bones were also shipped by ocean-going vessels from Texas ports, bones that had been picked up on the Staked Plains of West Texas.

One early version of how the bone trade developed is found in a July 1879 Kansas newspaper article, "Buffalo Bone Business":

> About ten years ago the present manager of the St. Louis Carbon Works was traveling across the plains on the old Santa Fe trail. He particularly noticed while journeying the immense amount of buffalo bones that covered the prairies. With a Yankee's ingenuity he bethought himself that these bones might be made of practical use. He knew that in sugar refineries bones were used which were brought from North and South Carolina, and put through a steaming process. He sent to Germany and brought to America a man who was at once a chemist and practical machinist. Between them they invented and patented machinery whereby the bones could be ground dry and rapidly, and thus retain all their different valuable properties. A company was at once organized and commenced whose grinding capacity is now from thirty to fifty tons a day, and a capital of at least a hundred thousand dollars is employed in carrying on the business. After the bones are unshipped they are picked over and assorted. Of the refuse the common fertilizing phosphates are made. The other bones are ground into bone black; it is used exclusively for refining sugar.[3]

By the early 1870s the trade in buffalo bones was considerable in Kansas and Nebraska. J. A. Allen, in his 1877 report on the buffalo, noted:

> On June 2, 1875, C. F. Morse, General Superintendent of the A.T.&S.F. wrote that the "bone business is still quite heavy, and will probably last for one or two years longer." From his accompanying statements of buffalo products shipped over that road during the last three years, it appears that the shipment of bones in 1872 amounted to eleven hundred and thirty-five thousand, three hundred pounds; for 1873, twenty-seven hundred and forty-three thousand one hundred and ten pounds; for 1874, sixty-nine hundred and fourteen thousand nine hundred pounds,

or treble the amount of the previous year, and six times that of 1872.[4]

These figures are almost identical to those listed in 1877 by Colonel Richard I. Dodge. He also reported figures for the Union Pacific in Nebraska, the Kansas Pacific in Kansas, and all other railroad lines touching buffalo country:[5]

Railroad Buffalo Shipments

Product	Year	Atchison, Topeka & Santa Fe	Union Pacific, Kansas Pacific, and all other RRs	Totals
Bones,	1872	1,135,300	2,270,600	3,405,900
lbs.	1873	2,743,100	5,486,200	8,229,300
of	1874	6,914,950	13,829,900	20,744,850
		10,793,350	21,586,700	32,380,050

Many buffalo bones were picked up by homesteaders. The winter of 1873-74 was very severe on the plains from northern Texas to the Dakotas. A business panic did not help conditions. One settler recalled:

> When spring came the prospects for crops was good, but in July the grasshoppers, Rocky Mountain locusts, arrived. Since the immigration to counties in western Kansas had been made during the previous two years, most settlers were dependent upon sod corn, potatoes and garden vegetables. The locust invasion caused a panic, crops were lost, and many people were in need of assistance. Some of these people who already had been supplementing their incomes by selling buffalo hides and meat and collecting buffalo bones which everywhere dotted the prairies, turned to such work fulltime.[6]

L. C. Fouquet was luckier than most settlers in western Kansas in 1874. Having turned to farming, Fouquet planted wheat. By the time the grasshoppers came, his crop had already been harvested. For him buffalo bones were not the source of additional income. They were "a nuisance to our breaking of the sod," he wrote. But for the majority of settlers in western Kansas, this was not the case. The bones were a blessing. Picking them off the prairie, the settlers hauled them to Hutchinson or other railroad towns to sell them. "The side tracks were just lined up with long stacks of such at Hutchinson," said Fouquet. "They paid from two-and-a-half to three dollars a ton for them. And from $6.00 to $8.00 a ton for the horns."[7]

Stacks of buffalo bones were a common sight during the 1870s in railroad towns of eastern Colorado and western areas of Kansas and Nebraska. Near Granada, Colorado, for example, one stack was described as being twelve feet high, twelve feet wide, and a half-mile long.[8]

By the late '70s, however, the bone trade on the southern plains, where the slaughter was all but over, was dropping. Nearly all the live buffalo had been killed and vast numbers of bones collected and sold. On August 3, 1878, the Dodge City *Times* reported:

> A large number of teams came in from the range Wednesday loaded with bones, which were gathered for miles south. A regular business in gathering bones has long been established though not so profitable as formerly. Carcasses are not as numerous, the buffalo is becoming extinct and the long horn gives up his bones to the slaughter pen, being driven over the plains with less loss. There are thousands of buffalo and cattle that are killed and die annually; and the bones are gathered at all seasons of the year, thus affording constant employment to a large number of men and teams. The bones are shipped East by the carloads, where they are ground and used for fertilizing and manufactured into numerous useful articles.

The bone trade on the northern plains followed a pattern similar to that of the southern plains trade. During the late 1870s, before the railroad reached the Upper Missouri and before the northern slaughter got under way, some bones were picked up during the summer months and transported by wagon to Fort Benton and other river trading points. There they were sold and eventually shipped down the Missouri to Bismarck, whence they were loaded aboard railroad cars for the journey to eastern markets. But when the railroad came to the north, the bone pickers had to deliver their cargoes only to the rail-

"Buffalo Bones Were Often the Support of Settlers While Raising Their First Crop," M. S. Garretson drawing. *Courtesy Kansas State Historical Society.*

Red River carts were used to gather buffalo bones on the northern plains and in southern Canada during the late 19th century. From *John L. Stoddard's Lectures,* vol. X. *Courtesy William F. Whithorn.*

Buffalo bones stacked at the Michigan Carbon Works in Detroit, 1880s. *Courtesy Le Roy Barnett, Kalamazoo, and the Detroit Public Library.*

road. The trade in buffalo bones on the northern plains grew at a rapid rate during the 1880s.

In southern Canada much the same thing occurred. Even before the railroad arrived, many French voyageurs, trappers, and settlers had turned to bone picking. Some used high-wheeled wooden carts much like those that had been used in the Red River hunts forty or more years earlier. Often 1,200 pounds of bones could be carried by a single cart, but in southern Canada and in northern areas of the central United States bone picking was limited to late spring and summer, when snow did not cover the ground and the bones.[9]

When Julian Ralph and a friend traveled westward on the Canadian Pacific in 1888, they found buffalo bones to be a "thriving business." As Ralph recalled:

> At the outset we saw a few bison bones dotting the grass in white specks here and there, and soon we met great trains, each of many box cars, laden with nothing but these weather-whitened relics. Presently we came to stations where, beside the tracks, mounds of these bones were heaped up and rude men were swelling the heaps with wagon loads gathered far from the railroads, for a great business had grown up in collecting these trophies. For years the business of carting them away has gone on.[10]

Though the bone trade in southern Canada lasted into the 1890s, by the late 1880s the plains of Texas, Indian Territory, Kansas, and Nebraska had pretty much been picked clean. George Bock, president of the Empire Carbon Works at St. Louis, one of the largest buyers of buffalo bones during the 19th century, estimated that 70 percent of the bone shipments in the United States were processed at St. Louis. The remainder, said Bock, were shipped to Philadelphia, Baltimore, Detroit, and a few other eastern cities. When bones were delivered to Bock's firm, the price ranged from $18 to $27 a ton. Bock's records around the turn of the century showed that his plant had purchased more than one-and-a-quarter million tons of buffalo bones during a thirty-year period. If the average price paid at St. Louis was $22.50 a ton, his firm would have paid out more than $28 million for buffalo bones. It seems quite possible that the industry as a whole between 1868 and 1881 spent at least $40 million buying buffalo bones.[11]

In April 1906 Carl P. Gauthier's father homesteaded fifty miles northwest of Williston, North Dakota. Many years later Carl remembered the trip out of Williston on "The Bone Trail," a homesteaders' thoroughfare by 1906, with a few small settlements, one labeled Bone-traill (still in existence into the late 1960s). As a sixteen-year-old, Carl asked the driver of their team and wagon why it was called The Bone Trail. The driver replied:

> "Well, a lot of oldtimers made good money gathering up old buffalo bones on the prairies and hauling them to the railroads where they could be shipped to the fertilizer plants . . . . That's about done for now . . . . But I did see a couple of gondolas on the siding in town, loaded with bones."

Gauthier recalled: "I had noticed them too . . . . Undoubtedly they were the last carloads of bones shipped out of Williston."[12]

When the hide hunters killed buffalo and the bone pickers finally carried off the remains, the age-old life cycle of the plains and prairies was destroyed. Even when man began to place cattle, sheep, and other domestic animals on what had been the buffalo's range, Nature's system was not restored. Man would not let Nature have her way.

For centuries the bones of buffalo and other dead animals had bleached away in the sun, slowly decomposing, slowly returning to the earth, slowly completing Nature's complex system of recycling life's basic chemicals. Plants and animals were linked in Nature's system, each playing a definite role. Using the sun's energy, plants produced foodstuffs. These were consumed by the buffalo and other grazing animals. The foodstuffs were then passed along to other animals, to predators, for example, or returned directly to the soil to be used again and again by plants producing foodstuffs. This system of recycling had already begun to change by the time the white man arrived on the plains and prairies. The Indian, like the white man, had for centuries been a highly selective killer. He sought the fat and the productive and the young, whereas, unlike man, the predators, especially the wolf, sought the weak and the sick and the old. The wolf weeded out, cleared the vast buffalo herds of the weak and ailing, thereby keeping the buffalo a vigorous species.

This was changed by man. Man first killed the buffalo, then cleared the land of the bones. Soon cattle replaced the buffalo. Then it was only natural that the wolf would turn to cattle for survival. But man would not let wolves and other predators remove the weak and ailing from the cattle herds as the lobos had done with the buffalo. Man considered the wolf bad. The animal was not to be tolerated.

Between 1866 and 1883, the years of the Texas cattle drives to the railroads in Kansas, an estimated total of 4,707,796 head of cattle were trailed north through the southern buffalo range. Until the early 1870s, when buffalo could still be found on the southern plains in large numbers, cattle losses from wolves were not so heavy. But wolves did take advantage of cattle, especially strays. They also caused stampedes. And they were tracked down, though less vigorously than later. One of the earliest recorded wolf hunts in Kansas Territory occurred in 1859 along Walnut Creek. Frank M. Stahl and some other men set out one afternoon to kill wolves. By sunrise the following day they had killed forty, all but two of which were gray mountain wolves that had moved east from the Rockies to follow the buffalo.[13]

When Texas cattle were trailed to Colorado, Wyoming, Montana, and elsewhere in the West, and as cattlemen established ranches and grazed their animals on the open ranges, wolves continued to be a problem. So was the weather. Cattle killed by the elements or by wolves were, in the eyes of the cattlemen, "expected" losses, incidental to the business of raising cattle. This began to change as the hide hunters killed more and more buffalo. Without buffalo, wolves relied more and more upon cattle.

During the early 1870s, southern plains cattlemen were complaining about the losses from wolves. They complained even more in 1872 when the cattle business experienced its first panic. "Wolf losses" were money down the drain. As cattlemen organized cattle associations to protect their interests, they established their own "wolf control" measures. Members were assessed to provide wolf bounty money and poisons to kill wolves. By the middle 1870s it became standard practice on the open ranges of the southern plains for cowmen, when riding across their ranges, to bait every carcass they came across. A good dose of strychnine sulphate was placed inside every carcass, big or little. Then when Mr. Wolf came to dine he ate the poison and death soon followed. Unfortunately, many other forms of wildlife were poisoned, including countless numbers of kit foxes. Nature's cycle suffered further.

When the northern buffalo slaughter ended in 1884, cattlemen on the northern plains began to experience the same "wolf problem" that cowmen to the south had experienced a decade before. Many more gray wolves were seen on the Montana ranges that year. As one observer put it:

> Not being so particular as to object to beefsteak when buffalo hump was not to be had, they played havoc with the cattle herds that year. Cattlemen did not begin to pay much attention to the matter until last year [1885] when it was found that it knocked considerable from their profits to support such immense swarms of these pests. Cattle, and especially young and weak calves, dropped during the winter time, have been the food upon which they, the wolves, subsisted.[14]

As in the south, it was not long before bounties were offered in the north. One dollar on a coyote and two dollars on a wolf were the bounties in Montana Territory in the middle 1880s. As one cattleman explained it:

> In 1885 they [wolves] were on the ranges in greater swarms than ever and the damage they did counted heavily against the profits of that year. On the Chestnut range in northwestern Montana the stockmen came to the front with a handsome offer to wolf killers, which will make it a paying business for anybody to engage in.[15]

By the late 1880s the problem of wolves had spread across the border into Canada, especially on the prairie where cattle and other livestock were increasing. Ranches west of Calgary were frequently raided by wolves. They were soon outlawed and bounties were offered to kill them just as had been done in the United States.

From Montana northward into Canada bounty hunting became profitable. Many cowboys went into wolf hunting either part-time or full-time. One young cowboy named Martin took a week off and went wolf hunting in Yellowstone County, Montana during the late 1880s. A cattleman later recalled:

> To sum up he got $13 and $26 for coyote ears, and $9 and $18 for those of the wolves; after which he sold the hides for an average of about $1.50 each to a fur dealer in Billings. Total profit, $118.50 and lots of fun; cost about $5.00 for strychnine and time. Bait was had in one dead animal picked up on the range.[16]

One full-grown wolf could annually kill livestock easily worth a thousand dollars. A female wolf could do almost as much damage with a litter of half-grown pups. Together they could attack and kill many calves and colts. It was not food alone that directed the mother wolf. It was a lesson for her young in how to make a kill.

In many instances deer, antelope, and other wild game met the same fate as the wolf. When barbed wire began to fence the open ranges in the late 1870s and early '80s, ranchers and farmers resented these wild animals competing with their cattle and sheep for available forage, especially in dry years. Many deer, antelope, and proghorn were either killed or chased off the land. Those that were lucky sought out areas untouched by man.

By the time the 20th century arrived, the vast area from Texas to southern Canada had been conquered by the white man. Farmers and cattlemen had done their part, but they had changed the land. After centuries in which Nature established her system of recycling, man stopped the cycle overnight and ceased to feed the wild community. It is little wonder that Nature fought back. Grasshopper plagues, dust bowl days, no grass, and lack of water are only a few of the difficulties encountered by the white man on the plains and prairies during the last one hundred years.

Only during the last forty years or so has man really begun to understand the relationship of plants and animals to the over-all picture of the plains and prairies—the grasslands—and really begun to understand the important role played by the buffalo and all other wild animals in Nature's grand design. Man has finally begun to do something about it.

X

Battling Bulls

The buffalo bull has no more formidable enemy than another
male of his own species .

C. J. "Buffalo" Jones 1899

From late fall through winter the wild buffalo bulls usually remained to themselves. Some might wander aimlessly alone across the plains and prairies while others ran in small bunches paying absolutely no attention to the shaggy and bedraggled cows. But as the sun began to move north and temperatures began warming and the grasses grew tall, the bulls again felt their masculinity as if they had sipped a tonic. During late spring the bulls would begin to eye the cows with more than mere companionship in mind. By early July the bulls feeling strongest and bravest would begin to tussle with each other for leadership of the herd and the pick of the cows.

For wild plains buffalo such battles were instinctive. They were strenuous and sometimes bloody. George Catlin was eyewitness to a few bull battles during the early 1830s. One day while crossing the northern plains Catlin saw several thousand buffalo "in mass, eddying and wheeling about under a cloud of dust." The dust, said Catlin, was raised by the bulls as they pawed the dirt or as they were "plunging and butting at each other in the most furious manner." Catlin observed, "The males are continually following the females, and the whole mass are in constant motion; and all bellowing (or 'roaring') in deep and hollow sounds; which, mingled all together, appear, at the distance of a mile or two, like the sound of distant thunder."[1]

Surprisingly, not a great many bulls were killed during these battles in the days when vast numbers of wild buffalo roamed the plains. Perhaps it was instinct that told the males it was a sheer waste of time and energy to get involved in serious arguments since there were so

144

many bulls and so much country over which to roam. But death did come sometimes when one or more bulls became too earnest.

One spring afternoon in the 1840s, John Frémont, the explorer, saw several young bulls gang up on an old bull, apparently their leader. Frémont and his party had seen dust rising among some hills on the prairie. They went to investigate and found "a band of eighteen or twenty buffalo bulls engaged in a desperate fight." Although the bulls butted and gored one another, "and without distinction," their efforts, according to Frémont, "were evidently directed against one—a huge, gaunt old bull, very lean, while his adversaries were all fat and in good order; he appeared very weak, and had already received some wounds." While Frémont and his party watched, the young bulls knocked the old monarch to the ground several times. Although hurt badly, he kept getting to his feet to fight back. Frémont's party decided the battle was too one-sided. They attacked the bulls, but the animals "were so blind with rage, that they fought on," apparently unaware of human presence. Even though Frémont and his men fired at the animals with their guns from twenty yards away, the bulls kept fighting. When the bulls finally realized what was going on, they began to "retreat slowly along a broad ravine to the river, fighting furiously as they went." The old bull, meanwhile, slowly "hobbled off to lie down somewhere," wrote Frémont.[2]

Sometimes if a wild buffalo bull realized he was getting the worst of an encounter that had become serious, he would not hesitate to turn tail and run off to lick his wounds and restore his confidence. Apparently the old bull seen by Frémont had no choice but to stay and fight, as do buffalo confined to fenced pastures or kept in close captivity. Fatal fights frequently occur in small herds of captive buffalo, which, unlike their ancestors, do not have the vast open spaces over which to flee. Marvin R. Kaschke, manager of the National Bison Range in western Montana, figures that there is probably an average of one fatal bull fight for every year that the refuge has been in existence since 1909. "We may go several years without one and then have two or three fatalities in a season," said Kaschke, who looks after between 300 and 500 buffalo that roam the bison range of 19,000 acres.[3]

A great many battles between captive buffalo bulls were seen by Earl Drummond, who handled the animals for more than thirty years at the Wichita Mountains Wildlife Refuge in southwestern Oklahoma. After he retired in the 1950s, Drummond set down some of his observations and conclusions:

> In the breeding season, the large bulls have some battles which are as fast as wildcats fighting and a lot more wicked. They put their head to the ground and make quick jabs up with the horns striking each other in the head, neck and shoulders. The hair and wool is flying in the air while the fight lasts. If one gets a good jab at the stomach of the other the fight is ended as this is the vital spot. The hardest thing for the one that is getting whipped is to get away without exposing his side and stomach. This is where they show some speed and action.

Even in battle there is apparent respect for the opposition. Drummond noted that when two bulls were approaching each other for battle, "roaring like lions with tail up looking like a question mark, sometimes one will lie down within 100 feet of the other and roll."

> It seems to be an unwritten law they never fight him while he is down, but the others will stand and wait for him to get up. Sometimes when one gets whipped he will run and there will be two or more chase him. I have seen as many as 14 bulls in single file after one that lost the fight, and they will follow him two or three hours thinking they will catch him. If he runs they run, and if he walks they slow to a walk, and if he makes a turn they do not cut across but go almost the same route he went. If he keeps going until they get tired, they will quit him, but if he stops to fight one, they all jump on him and soon kill him.[4]

One of the best stories I have found describing a battle between two captive buffalo bulls was buried deep in the files of the *Brown County World,* a country newspaper in Kansas. It tells of a fight between two bulls kept in a fenced area at Bismarck Grove, north of Lawrence, Kansas, in 1891. The buffalo, owned by Colonel H. H. Stanton, were mostly for show. It was not unusual for people, especially on Sunday afternoons, to come from miles around to spend the day under tall cottonwood trees with a picnic lunch and perhaps a freezer of homemade ice cream to watch the buffalo. Even in Kansas in the 1890s, buffalo were scarce and a novelty.

The year of the battle, 1891, the leader of the Bismarck herd of seven cows, three calves, and two bulls was "Old One Eye." He was a wild plains bull that had been captured some years previous on the prairie in western Kansas. He had lost an eye in an earlier battle. The other bull was named "Byrd S." He was a five-year-old that had been born into the herd at the grove.

As the story goes, one day Byrd S. apparently decided to challenge the leadership of Old One Eye. The mating season—the oldtimers called it the "running" or "rutting" season—had just begun. For a few days Byrd S. only sallied up to Old One Eye and tussled with the old bull to study his weak points. Old One Eye had done that before. He knew what Byrd S. was doing and he let the youngster have his fun. One eyewitness recalled, "That 'Old One Eye' fooled young 'Byrd S.' into thinking that his once glorious strength and skill was of the past; that he had lost his cunning; that his scars had not taught him wisdom." The old bull even went so far as to imply that so strong a youngster as Byrd S. would find it as easy to kick an old bull around as he did the gentle cows. But things were to change, and Old One Eye knew it.

Late the following Saturday morning Byrd S. got up from the grassy plot where he had lain all night. The other buffalo were all standing together taking a cool drink from the large watering tank. Under the warming summer sun, Byrd S. walked slowly over to the watering tank and pushed several cows out of the way so he could get a drink. He spotted Old One Eye standing beside a young cow drinking. The sight apparently was too much for Byrd S. to take. He backed away from the tank, stamped and pawed the earth, and bellowed out what he proposed to do to Old One Eye. The old bull only looked mildly at Byrd S., who slowly backed away from the tank. Then, without warning, Byrd S. charged Old One Eye. The cows gave the old animal one look. It was all they needed to know they had better give him plenty of room. The fire in his eyes told them that it was going to be a fight to the finish. Those who saw the young bull's rush described it as "magnificent," but Old One Eye had seen the like before. Byrd S. pushed the fight, and when they came together he tried to down Old One Eye, but the old bull fought shy to save his wind and to bide his time.

As the battle became more furious, men on horseback, armed with goad sticks and pitch forks, entered the pasture and tried to separate the bulls. It was of no use. The young bull made fierce charges and gave Old One Eye thrust after thrust until his hide was gored and torn. But the old bull was cool and patient, knowing that the frantic efforts of Byrd S. would not continue much longer. Both animals backed off to rest. Perhaps a minute, maybe two minutes, passed before they started again. The two bulls pawed the earth and bellowed loudly. They were a terrible sight, each covered with dust and blood. The young bull charged Old One Eye, who took the blow well. In a flash the old bull came alive and charged Byrd S., crushing some of his ribs. Then Old One Eye, as if it were the very moment he had been waiting for, buried his horns up to his head in the side of the young bull. Byrd S. was pinned to the ground. He could not get up. He tried in vain with every breath to get to his feet but he could not.

In a short time it was over. Byrd S. died struggling bravely for another fighting chance at Old One Eye. But the old bull had not given the young bull another chance. He had not moved his horns until Byrd S. had taken his last breath. Only then did Old One Eye slowly move away, rejoining the young cow with the air of one not thinking the duel had been anything really unusual. A little while later a cowboy rode into the pasture, put a rope around Byrd S., and with some help dragged the dead buffalo outside the fence. There the animal was skinned. The cowboy had paid a shiny $20 gold piece for the remains of Byrd S.[5]

R. V. "Tex" Shrewder of Ashland, Kansas told me a story about what happened after a bull fight in a buffalo herd of which he was part owner. The herd was kept in a large pasture near Arnett, Oklahoma. About 1945, as "Tex" told the story, some young bulls decided to challenge the old bull that had ruled the herd of about 125 animals for some years. The young bulls ganged up and not only drove the old fellow from the herd but also forced him outside the pasture. The old bull went right through the barbed wire fence "just as though the fence was not there." As Shrewder recalled, the old bull traveled about twenty miles. He was so mad at the turn of events that during his journey he killed two beef cows and one cow pony, chased one cowboy up a windmill, and forced another to take refuge under a feed trough. The old bull finally attacked a farmer on a tractor, but the farmer drove off, leaving the animal standing in a field staring at the strange sounding contraption that was retreating. When the farmer reached a farm house, he called the sheriff. Soon a posse was on its way from nearby Arnett. Fifteen men arrived and set out after the old bull. They found him where the farmer had left him, and, as Shrewder said, "had a lot of fun filling him with lead."[6]

Can you imagine a buffalo bull fight in New York City? This may sound out of place, but just after the turn of the century more than 1,000 persons watched two bull buffalo, "Antonio" and "Brown Beauty," duel in the Bronx Zoo. Both bulls had been purchased from C. J. "Buffalo" Jones in 1899. For nearly two years all had been peaceful in the zoo's herd. Brown Beauty was seemingly in undisputed charge. In late August 1901, however, "grumblings and low bellowings" could be heard from the herd. Antonio began to show "a spirit of impatience and aggressiveness he never displayed before." But the animals' keepers did not worry. They felt certain the big bulls "would continue to growl menacingly, and never go beyond that." On the afternoon of September 5, 1901, things changed. What happened was witnessed by an unidentified reporter whose story was later carried by the *Topeka Daily Capital:*

It was 2 o'clock in the afternoon that the encounter began. The two bulls, massive, shaggy monsters, bellowing their defiance with all the energy that great lungs permitted, withdrew from the herd, and with tails flirting, heads bobbing and sides heaving, cantered to an elevated plain near the center of the enclosure. Then pawing the earth until they had raised clouds of dust they charged. The sound of the crash as the two big bulls came together could be heard a long distance. Their horns locked and they bent their great backs in a mighty effort to force each other back or get an opening for another move. The bellowing of the bulls as they struggled was heard a half-mile away. The other animals took up the cry and the park resounded with the growls of the beasts.

While the buffaloes were struggling, keepers Mulvhill and McEnroe got long spikes, and, jumping into the enclosure, ran to separate the combatants. The bulls were too intent upon their own affairs to pay any attention to the proddings from the keepers, but the move of the two men seemed to be considered by "Bonita," a cow, as detrimental to "Brown Beauty," and she dashed out from the herd and made for the plateau to join the strife. Both keepers had to abandon the efforts to separate the bulls and turn their attention to "Bonita." Reluctantly and with many attempts to evade them, she was forced back.

While the keepers were thus engaged, there came a sudden turn in the tide of battle between the bulls. The two big fellows had backed from each other and prepared for a charge. "Brown Beauty" was the more agile. "Antonio" was blowing hard. By a quick turn of his head "Brown Beauty" impaled his rival, one of his horns entering "Antonio's" side and sinking deep.

Slowly but steadily "Antonio" was forced back. Then, weak, and suffering, he fell. As "Brown Beauty" drew off to charge again, "Antonio" wheeled and met his enemy bravely, but the force of the other's weight knocked the wounded bull down. He rolled over, not once alone, but twice. He had been driven to the edge of a declivity, and as he rolled the second time he tumbled down in the dust and dirt below. "Brown Beauty" was king.

Out of the mouth of the victorious bull a great bellow of triumph proclaimed the end of the combat. Then he pawed the earth again and seemed to look for other rivals who might care to engage him in battle. The crowd of men and women without the enclosure cheered. They had seen a struggle of giants.

Later, keepers at the zoo determined that Antonio had suffered a wound 12 inches deep and 3½ inches wide. The bull recovered from his wound but not his lost pride.

Another unusual story about a buffalo bull who lost was told by W. E. Webb many years ago. It happened about 1871 when Webb and a small party of men were crossing central Kansas. They had just left the Saline River and were heading toward a deep canyon when their guide pointed ahead to a cloud of dust in a ravine.

"That's a buffalo," shouted the guide as the party came to a halt. "He's indulgin' in a game of bluff."

An Ernest Thompson Seton drawing was used to illustrate the official account of the Bronx Zoo battle between Antonio and Brown Beauty, from Smithsonian *Annual Report* for 1901.

"Buffalo Bulls Protecting a Herd from Wolves," William M. Cary drawing, from *Harper's Weekly,* August 5, 1871.

"Buffalo Hunt, White Wolves Attacking a Buffalo Bull," George Catlin lithograph, from his *North American Indian Portfolio.*

At first, some of the men in the party, those who were newcomers from the East, were not quite sure what was going on. But as the dust began to clear, the outline of a large buffalo bull could clearly be seen and what was going on became more evident.

"The old fellow is butting against the bank," explained the guide. The bull backed up fifteen or twenty yards from the bluff, pawed the ground for a moment, and then flung himself headfirst against the wall of earth with a tremendous force. Great clouds of dust rose into the air. For a few moments the bull continued violently hooking the soil. Then he backed up, pawed the ground, and charged the bluff again. But this time he either tripped over one of his forelegs or the blow glanced, because he fell to the ground on his forelegs, then completely flipped over. He was one surprised buffalo.

Either he felt he had been knocked down by an imaginary foe or he suddenly sensed the presence of man. He sprang to his feet and whirled his shaggy head toward Webb's party as if he were going to charge. Instead, he held his ground a moment, sniffed the air, turned, and raced away at full speed up the canyon toward the plains above.[7]

More than a century ago, when there were so many wild buffalo on the plains and prairies, the sight of a lone bull was not uncommon. Spotting a single dot off in the distance often meant it was an old buffalo bull wandering alone across the land or peacefully grazing. Many people who crossed the plains in the 1850s and '60s saw lone buffalo. Some thought they were lordly sentinels watching with fatherly care over the welfare of a herd, which, the people thought, was undoubtedly nearby. Some travelers would swear that such bulls were ready on a moment's notice to give early warnings when danger approached. Not so. Bulls found alone were often outcasts, animals driven from the herd. Others simply were stragglers, old bulls suffering from listlessness or sheer weakness and finding it difficult to keep up with their herd.

J. A. Allen, the buffalo historian, reported in 1875 that these "supposed alert protectors" were usually "the most easily approached" of any wild buffalo. "They are slower," he said, "to recognize danger when it is observed."[8]

But even though experienced hunters could easily kill the lonely old bulls, rarely was one killed by man for food. The animal's meat, what little he usually had on his bones, was very tough. Even wolves would frequently leave an old bull alone when younger buffalo or other animals could be found. But if the "pickin's were slim" and if their number was great, wolves would tackle and usually down an old bull.

George Catlin saw a wolf pack attack an old bull during the early 1830s on the northern plains. Catlin and his party chased off the wolves and moved in to see the buffalo. The old bull had put up a hard fight, but looking closely at the animal Catlin saw that

> his eyes were entirely gone, his tongue was half eaten off, and the skin and flesh of his legs torn almost literally into strings. I rode nearer to the pitiable object as he stood bleeding and trembling before me, and said to him, "Now is your time, old fellow, and you had better be off." Though blind and nearly destroyed, there seemed evidently to be a recognition of a friend in me, as he straightened up; and, trembling with excitement, dashed off at full speed upon the prairie, in a straight line.

Catlin and his party continued on their way, but in a little while when they looked back, the old buffalo was again surrounded by the wolves. "He unquestionably soon fell a victim," wrote Catlin.[9]

Many, many years ago a war party of Blackfeet were going south one day to do war. As they neared a river crossing, they thought they saw a human form high on a bluff along the river bank. A head appeared now and then as buffalo passed under the bluff on their way to water. One of the Indians, White Calf, looking through his spyglass, saw not a human being but a bear lying on the bluff waiting to jump upon a buffalo. The warriors moved a little closer, sat down, and watched. On her way to water, a cow came along a trail close to the bluff where the bear lay waiting. Behind the cow came a bull, lagging back at some distance. In a flash the bear sprang upon the cow and tried to drag her down. She cried in alarm and fought violently to shake the bear off. At her cry the bull, which apparently had not been seen by the bear, charged in. Just then, the cow swung around with the bear on her back, making it impossible for the bull to attack. A moment later, however, the bull's chance came. He caught the bear's stomach on his horns, tossed him high, caught him again, and tossed him again. When the bear struck the ground he was battered and broken in the hindquarters. His entrails were hanging out. But slowly he pulled himself from the battlefield as the bull, apparently satisfied that the bear would offer no further trouble, walked off with the cow to water.[10]

Anyone hearing this tale might immediately conclude that this wild bull acted to protect the cow. This may have been so. Journals, narratives, old newspaper stories, and the like contain several references to such acts of gallantry and devotion by wild buffalo bulls. But with most buffalo men I think that stories about bulls guarding and protecting the cows were nothing more than the creation of man's imagination. If anything is true about a wild buffalo bull's behavior, it is that it was erratic and unpredictable. Old bulls, in particular, usually ignored the cows except during the breeding season.

While telling stories about fighting bulls, I want to relate just two more about buffalo and a bear. One day in the 1870s Dr. William A. Allen and a friend were camped near a crystal clear mountain stream in the Rockies. As they drank the cool mountain water, their horses began snorting. Both men immediately investigated and discovered a small herd of about twelve buffalo, possibly mountain buffalo, grazing nearby in the center of an open meadow.

Allen was about to shoot a fat calf for meat when a full-grown buffalo bull came bounding out of the bushes on the far side of the meadow. He headed straight for the grazing herd. A moment later, a full-grown grizzly crashed through the trees into the opening. He was chasing the bull. Surprised at what was happening, Allen and his friend held their fire, remained hidden, and watched. As the bear charged out into the meadow, the buffalo cows quickly formed a circle around their calves. Then, as if on cue, the buffalo bull pivoted and charged the oncoming bear. Both came together with a dull thud. For a moment they were stunned, but in a few seconds both animals were at it again.

The bear rose up on his hind legs and prepared to do battle as only a grizzly can, but at that moment the buffalo lowered his head and shot forward. The bull's horns tore into the bear's stomach, giving the animal some deep wounds. But as the bull hit the bear, the bear grabbed the bull, digging his claws into his shoulders. As the buffalo continued forward, the backward motion forced the bear to release his hold. The grizzly fell backwards but, wild with rage, quickly regained his footing and took out after the buffalo. Again and again they clashed, and at one point the bear jumped atop the buffalo's back, fastened his claws and teeth into the bull's back, and held on. To Dr. Allen and his friend, quietly watching from a safe distance, the bear looked like a bronc rider in a rodeo, but the scene did not last long. Suddenly the buffalo did a complete somersault and threw the bear. The bear regained his footing and came back with a blow from its paw which hit the buffalo squarely on his head. It hurt. The buffalo backed off, shaking his head. Blood ran from the open wounds on both animals. Pausing, the bear and the buffalo looked at each other as if waiting for the other to make the next move. Only about twenty feet separated them.

Suddenly, another buffalo bull, a two-year-old that had been watching from the sidelines, apparently decided to test his strength with the grizzly. He pawed the ground, bowed his neck, and in a flash charged the bear. But the grizzly had played that game before. He moved aside and struck the young bull with a blow to the side of his head as the buffalo passed. Stunned, the young buffalo stopped, turned, and began walking toward the wounded bear. The bear stood his ground watching the young bull. He did not see the older buffalo shoot forward like a bolt of lightning—until it was too late. The old bull, head down, tore into the bowels of the grizzly and knocked him down. Before the bear could get up, the younger buffalo bull had also charged into the battle, striking the bear with full force.

By this time Allen and his friend were beginning to feel sorry for the bear. Allen raised his rifle, aimed, and fired. The old bull fell as he was about to charge the bear again. The other buffalo took off running into the woods. The bear immediately turned and attacked the young bull, dealing it a fatal blow, but the grizzly, too, was exhausted and fell dying. Dr. Allen walked out into the meadow, shot the bear in the brain, and then shot the dying young buffalo bull.[11]

John Palliser told of an incident related to him in 1848 by Boucharville, guide and mountain man:

> He was going round to examine his traps, and was watching a band of buffalo as they emerged from the river, and slowly ascended the bank, when he saw a bear (previously concealed in a deep rut) spring up and dash the foremost bull to the ground,

ploughing his sides with his monstrous claws and rending his heart and vitals by a succession of tremendous blows. Although, in general, the bear easily vanquishes his less formidable opponent the buffalo, I heard a very well authenticated instance related by old Provost at the Minitaree, in which both parties suffered so severely as mutually to resign the conflict, move off a little way in opposite directions, and lie down and die.[12]

A really strange fight between a buffalo bull and another animal took place in 1906. One warm July morning, Dr. R. M. Simpson of Winnipeg, Manitoba was driving a horse and buckboard on the road between Woodland and Winnipeg. About 11 o'clock, when he was about twenty miles west of Winnipeg, he saw two large animals fighting on the prairie a couple of miles ahead. As he moved closer, he saw that they were a buffalo bull and a big Shorthorn bull. The two bulls were on their knees, head to head, horns partly locked. The ground was so torn up that they must have been fighting for several hours. The two animals paid no attention to Dr. Simpson as he came closer. They continued to charge each other, fencing, working around, bellowing and slashing. Their shoulders and flanks were gored and ripped up. Both were frothing and bleeding at the nose, especially the Shorthorn; yet both seemed full of fight and energy.

For thirty minutes Dr. Simpson watched the battle. Then he had to continue on to Woodland. There he conducted his business and about 3 o'clock that afternoon started back for Winnipeg. Twenty miles out of town he found the two bulls still at it. Much of their hair was scraped off, their sides were badly ripped, and both were bleeding profusely at the nose. But now the animals were not bellowing. The rests between attacks were much longer. Their breathing was very heavy. They rested long on their knees, heads together, making occasional feeble passes at each other. It was a very hot afternoon, and they must have suffered terribly from thirst. The two bulls were too much exhausted to trouble with the doctor, who moved in close and watched them for about an hour. To him the animals looked evenly matched. They apparently had fought each other to a standstill. Neither would give in. The doctor, because of the late hour, was unable to wait to see how the battle ended. He had to get to Winnipeg. Later, he heard that the Shorthorn had been rescued by its owner and the buffalo recaptured. The buffalo had run off from the Stony Mountain buffalo herd sixteen miles north of Winnipeg.[13]

Many years ago Elmer Parker and some other buffalo men at the Wichita Mountains Wildlife Refuge in Oklahoma put a bull elk and a buffalo bull together to see what, if anything, would happen. As Parker recalled to me, "You know that elk charged the buffalo. Their heads came together. That was it. The bull elk and the buffalo backed away, looked at each other, and then walked off in opposite directions. After that they left each other alone."[14]

Among the many stories about dogs bothering buffalo, probably the best was told by R. G. Carter, an Army officer at Fort Concho in West Texas. The post was one of the newest and most remote military establishments then on the western frontier. Vast numbers of buffalo could be found near the post. On March 25, 1871, as Carter told the story, the post's headquarters, under the command of Ranald S. Mackenzie, and five companies of the Fourth Cavalry were ordered to proceed to Fort Richardson, about 230 miles northeast, and relieve the Sixth Cavalry, then under marching orders for Kansas. The troopers set out. By March 31 the column moved from a high mesa to a vast area of plains. As the soldiers emerged, they saw an almost endless prairie covered by buffalo. For a time the column pushed along beside them, but soon there were so many buffalo that the march was halted. What happened next is told in Carter's own words:

> Mackenzie, becoming a little impatient at the blockade, seized a rifle from one of the men, and dismounting, attempted, by firing at the heads of the herds, to swerve the immense throng, which were now so crowding upon the advanced company as to become positively dangerous, the horses showing great fear and becoming almost unmanageable. He fired several shots. The nearest herd swerved; but, contrary to their instincts, came roaring down beside and parallel to our mounted troopers. This was a little too much, even for well-trained, disciplined cavalry soldiers, and the men, in their intense excitement, forgetful of orders, commenced a rattling fusilade from their saddles. The buffalo veered

off, but not before several were wounded. The firing was sternly ordered to cease.

One gigantic bull, a leader, was nearest; he was badly wounded. As was the case on nearly all marches of troops changing station on the frontier, many dogs of all ages, sizes and degrees, had, under protest, accompanied the column to the Colorado River; here many of the worthless curs were left or drowned while fording; but there were several remaining, and it was these that had turned the buffalo down the column. There was among them a large, white English bulldog, belonging to the regimental band. He was a powerful brute, and had been trained to pull down beeves at the slaughter corral at Fort Concho. He was, withal, a prime favorite with the soldiers, notwithstanding his ferocity. The pack of dogs were in full cry after the stampeding herds of bellowing beasts as they rushed and tore along the column with their peculiar, rolling gait. But "King," the bulldog, singled out the immense wounded leader, who had now slackened his speed and was faltering in his tracks. He sprang at his throat with great courage, fastening upon him, and the battle commenced, with the column as silent spectators. It was a novel spectacle.

The bronzed troopers, the great, shaggy beasts thundering by; the white-topped wagon train closed up and halted; the fleeting shadows, and the almost limitless stretch of surrounding prairie and vast solitude. The bull went down upon his knees, but so great was his strength that he quickly arose and whirled the dog in great circles over his head. "King" had been taught never to let go. The entire command now watched with breathless attention the apparently unequal struggle, expecting every moment to see the dog crushed to death. Down went the bull again on his knees, this time not from any weakness, but to gore the dog; rising, he would stamp his feet in his rage, then shaking him a while, he would resume swinging and snapping him like a whip cord through the air. The foam, now bloody, flecked the long, tawny beard of the bison bull. His eyes, nearly concealed in the long, matted hair that covered his shaggy head, flashed fire, and his rage knew no bounds. The dog, which had commenced the fight a pure white, now turned to a spotted crimson from blood which had flowed from the buffalo's wounds, and still his brute instincts, tenacious courage and training led him to hold on. Had he let go for a moment, the crazed bull would have gored him to death before he could have retreated. The bull grew perceptibly weaker; he rose to his feet less often. He could no longer throw the dog in circles above his head.

The blood stained "King" to a more vivid red, and begrimed with dirt, he had lost all semblance to his former self. All were anxiously looking for the struggle to end. Impatience was already displayed upon the men's faces, when suddenly General Mackenzie shouted, "Kill the animal and put him out of his misery!" It was a merciful command. Two men stepped forward to the enormous beast, now on his knees and rocking to and fro, the dog still holding on—and placing their carbines behind the left shoulder, to reach a vital point, fired. He gave one great quiver, one last spasmodic rocking, and spread himself upon the vast prairie dead. Not till then did "King" let go!

So great had been the courage of this favorite dog in his fearful struggle, that months after when an order had been issued for all cur dogs—always an accumulative nuisance at a frontier post—to be exterminated, "King," the white bulldog belonging to the Fourth Cavalry band, was exempted by a special order.[15]

Of all the fights between a buffalo bull and another animal, perhaps the most publicized took place in Mexico in 1907. Bob Yokum, a Texan who had been transplanted to South Dakota, was in El Paso visiting two friends, Tom Powers and Billy Amonett, who ran the Coney Island saloon. One day Powers and Amonett took Yokum across the border to Juarez to see a bullfight. Instead of showing astonishment at the spectacle as a tourist might, Yokum taunted, "You think that bull can fight? Why I can get a bull that can lick bulls like yours all day long." The Mexicans, especially Felix Robert, himself a matador and manager of the Juarez bull ring, were annoyed by Yokum. They insisted that their bulls were the finest in Mexico. The animals had been bred for generations to do just one thing—fight.

Yokum was untouched. He said, "I still know of a better bull."
"What kind of bull is yours?" asked Robert.

Charles Russell painted "Death Battle of Buffalo and Grizzly Bear" (detail above) for William A. Allen's 1903 book, *Adventures with Indians and Game*.

The soldiers at Fort Union had a small herd of cattle. The cows were used for milk; the bull pulled a cart for hauling meat and wood. One day in the fall of 1847, while going after a load of firewood, driver and bull were confronted by a buffalo bull, "right in the cart track, pawing up the earth, and roaring." The driver escaped up a tree, and the two bulls fought a battle to death, the Fort Union bull victor. From John Palliser's *Solitary Rambles* (1853).

Frederic Remington illustrated the 1871 "Buffalo vs. Bulldog" story, from *Outing,* October 1887.

Scotty Philip's buffalo and a Mexican bull in the Juarez, Mexico bullring, 1907. George Philip's account mentions the buffalo's injured "left hind leg," but the photo suggests he may be favoring his *right* hind leg. Perhaps the photo has been inadvertently reversed in two generations of use or maybe George was mistaken or just possibly the buffalo insignificantly raised his right hind leg off the ground as the photographer snapped the picture. *Courtesy South Dakota State Historical Society.*

"It's a buffalo bull, of course."

Robert and his Mexican friends were doubly insulted. A common, awkward wild animal of the plains a better bull than one of their highly trained, sharp-horned prize bulls? Impossible.

"Nope," said Yokum.

The sporting conversation that followed laid the groundwork for what undoubtedly was the most unusual bull fight ever to be staged in Mexico. The wager reported was $10,000.

Yokum returned to South Dakota, where he told Scotty Philip what had happened. Philip, who had devoted much time, money, and effort to save a small number of buffalo from extinction, grinned. He thought the contest would be interesting and he agreed to furnish buffalo bulls from his herd. Yokum and Philip decided to take two bulls, an eight-year-old and a four-year-old, and rail them to Juarez. They had a boxcar fitted with heavy planking to keep the animals from breaking out. When the day came for them to head south with the buffalo, Scotty Philip could not go. A bad blizzard demanded he stay at home and watch after his cattle. But he sent his nephew George in his place. In addition to George Philip and Bob Yokum, Eb Jones, an old cowman, went along to help take care of the buffalo.

Hooked to a freight train, the boxcar with the buffalo left Pierre, South Dakota in early January 1907, during the blizzard. At Sioux City, Iowa the animals were fed and watered while the group awaited another freight train to hook on the boxcar for the trip to Omaha. In Omaha, they joined another train for the trip to El Paso, Texas via McPherson and Liberal, Kansas and Dalhart, Texas. Early the following Sunday morning, seven days after leaving South Dakota, the buffalo and their keepers arrived at El Paso, whence the boxcar with the buffalo was moved across the border to an unloading chute at the bullring.

After the noon meal, the crowd began to gather to see what the common North American buffalo could do. The ceremonies began with the parade in the bullring: the matadors followed by their banderilleros and then their picadors, all decked out in the tinseled trappings of the bullring. Then four regular bullfights were held with matadors and domestic bulls. By the time the fourth bull was meeting his death in the ring, the crowd was beginning to shout for the buffalo. It had had enough regular bullfighting.

The signal was given and from the corral came the larger buffalo bull, the eight-year-old. Almost immediately, Yokum and George Philip noticed that the buffalo's left hind leg seemed to be dislocated. Perhaps he had hurt it while kicking in the boxcar. The buffalo did not seem to mind, however, and he walked out and stood peacefully in the center of the bullring. It was as if he wondered what was going on, but he had only a few moments to wonder before a handsome, red Mexican fighting bull was released into the ring. The Mexican bull stood a moment surveying the scene, especially the large buffalo. The buffalo also saw the Mexican bull but did nothing.

Suddenly the Mexican bull charged the buffalo. The heads of both animals cracked as they came together. The sound reverberated around the ring. The Mexican bull backed away as if he thought the first blow was an accident. Again he charged the buffalo and again their heads cracked together, but this time the buffalo put a little more force in his counter-blow and knocked the red majesty back on his haunches. The Mexican bull, still unconvinced, paused. After a moment, he charged a third time, then a fourth. The buffalo showed even a greater amount of force in each counter-blow, and on the fourth charge the Mexican bull fell under the impact. It seemed to terrorize him. The poor scared creature actually tried to climb out of the bullring, but he was left inside huddling to one side. The buffalo simply remained still in the center of the ring.

Felix Robert, the bullring manager, requested the right to turn in another bull with the buffalo. Yokum did not care. "Sure," he said, for he knew what would happen. The second Mexican bull entered and the same thing happened. After three charges at the buffalo, the second Mexican bull also tried to get out of the ring, but he too was left inside and huddled near the other Mexican bull.

When a third Mexican bull was released into the ring, the script was repeated. But when a fourth bull appeared, it was too much for the buffalo. He got really mad. His tail went up, and he decided to tackle all four Mexican bulls. But his injured left leg slowed him down as he took out after them. Before he reached them, the chute gates were opened and the bulls fled the ring, leaving only the buffalo inside to cool off. The contest was over and the common buffalo from north of the border had "done his stuff."[16]

XI

How They Ran

The trampling bison hurl along
A black and bounding fiery mass
That withers, as with flame, the grass.
Oh, terrible—ten thousand strong.
    Sam L. Simpson 19th century

The sun was just peeking over the horizon when a buffalo hunter named Dodge and his two sons drove their four-wheeled freight wagon out of Great Bend, Kansas. It was the summer of 1871. The day before, they had laid in a good supply of food and ammunition. Now they were getting an early start in search of buffalo.

The early morning was pleasant. The cool air off the plains, brushing past their faces, gave a new freshness to everything. The three men in their wagon forded the Arkansas River, then, crossing the sand dunes which glistened in the warming sun, they made their way south toward higher ground. Soon they reached the open prairie and began to watch for buffalo. The first day they saw no buffalo nor even any fresh signs of buffalo. They did spot a coyote in the lowlands and a pair of antelope and many jackrabbits on the upper prairie, but all showed the hunters their tails. Anyway, it mattered little. The Dodges wanted buffalo.

It was dusk when the three men stopped to make camp near Rattlesnake Creek. They unhitched their team and picketed the horses a little way from camp. As darkness came, the three hunters prepared to make the evening meal. Dodge and his two sons speculated, dreaming perhaps, about how many buffalo they would find the next day. Suddenly, one of the men stopped talking and held up his hand as he looked off into the gray distance.

"Listen," he said.

Heads cocked intently, the trio listened.

161

At first it sounded like thunder, away off in the distance. But there were no clouds in the sky and the first stars of the evening shone brightly. Gradually the earth seemed to tremble and the thundering sound intensified. In a matter of seconds the three men, each almost at the same instant, realized what they were hearing. It was an approaching herd of stampeding buffalo.

The horses, picketed nearby, sensed the situation. They began snorting and pulling at their ropes. The three men ran to the horses, untied them, and led them to the wagon, where they were tied securely to the wheels.

From the southeast the roar of the stampeding buffalo grew louder and louder. Dodge and his sons quickly gathered behind their wagon. Shortly, they saw the approaching line of huge dark forms heading straight for their camp. The Dodges had not yet had time to build their campfire. It might have caused the buffalo to skirt them. There was nothing now but the darkness of their own camp. The three men thought of the worst, of being trampled to death. But they were not ready to give up. With their buffalo rifles, they began firing into the oncoming animals. The Dodges did not have the fine-grained, quick-acting gunpowder they normally used. At Great Bend, heavy blasting powder was all they had been able to buy. It suited their predicament perfectly. With each shot the powder made a loud report and gave off a great flash of light, unlike the finer powder.

The three men kept firing into the oncoming mass of buffalo as rapidly as they could reload their rifles. The loud explosions and the great flashes of light had the right effect on the buffalo. When the first animals neared the camp, they swerved to the right and to the left, thundering past the wagon on both sides so closely that if the men had wanted they could have touched the shaggy creatures. But this was furthest from their thoughts. The thundering hoofs, the click and clatter of clashing horns, and the animals' bellowing almost drowned out the loud shots. For several hours the buffalo kept coming in a seemingly endless procession. The tiring hunters kept firing, ever thankful that they had a good supply of powder for their rifles.

A little before dawn the numbers of buffalo began to dwindle. As the first faint glow of morning spread across the prairie, only a few stragglers could be seen and they were passing far out from camp. In the dull light the Dodges saw the outlines of dead buffalo around the camp. Some had been dropped in their tracks by the hunters' shots. Others, wounded and downed, had been overrun and trampled to death by the herd. Exhausted, the three men smiled at one another and sank to the ground to rest, not saying a word. They had their buffalo and the stampede was history.[1]

Many such accounts of buffalo stampedes are recorded by hunters, soldiers, and early settlers on the plains and prairies.[2] Nearly all who saw such sights described them as spectacular. As one old frontiersman put it, a buffalo stampede is "indescribably grand, yet so awful in its results."[3] A herd of wild plains buffalo would sometimes stampede on the slightest provocation. "Buffalo" Jones maintained they would stampede at "nothing."[4] The barking of a prairie dog, the cry of a wolf or coyote, a flash of lightning, or a clap of thunder might set off a stampede. If a single buffalo snorted and started to run off on his own, others might join him and set off a chain reaction. In a matter of seconds a peaceful herd of grazing buffalo could become a surging mass pushing across the land for ten or twenty or more miles.

Once running, the buffalo might trample everything in their path, including other buffalo too slow to keep up with the others. "Buffalo" Jones and other oldtimers said it was impossible to turn a stampeding herd, but there are many factual accounts of stampeding herds being turned.[5] Even the largest herds were "easily turned from their course by a single man who may intercept them," wrote one observer.[6] Another noted that even "the noise of the dogs and children" in an Indian camp would turn a stampeding herd.[7]

A horse once supposedly caused a stampeding buffalo herd to swerve. In April 1846 George Andrew Gordon and four friends were hunting buffalo on the plains of northwest Texas. Gordon became separated from his companions. To make matters worse, in trying to find them he lost all but one of his rifle bullets. Thinking perhaps he could see his friends from higher ground, Gordon began riding toward the crest of a nearby hill. Suddenly he heard a low murmuring sound "as of wind in the tops of pine trees." But there were no pine trees within miles, only a small cluster of other trees, probably live oak or cottonwood, on the crest of the hill he was approaching. As the sound grew louder, there was no doubt in Gordon's mind that he was hearing an approaching herd of buffalo. "The roar became deafening, the ground trembled, and my horse shook with fear," Gordon recalled. As he scanned the countryside, the only refuge he could see was that small grove of trees on the hill. A moment later he saw the approaching buffalo. This is how he described what happened next:

> Toward the grove I put my horse at full speed. The grove was free from underbrush, the trees about a foot in diameter, straight and without limbs for twelve or fifteen feet except one that had three branches a few feet above the back of my horse. I stood up in my saddle with great difficulty on account of my heavy gun and clambered into the forks of the tree. I had run right into danger, for the center of the mass struck the little grove.

Gordon held tight to the branches with all his strength as the oncoming herd of buffalo headed straight for the grove of trees. Gordon was certain the end was near, but suddenly when the leading buffalo in the center of the mass were about a hundred feet from the trees, they saw or smelt Gordon's horse and tried to turn or stop. "In the twinkling of an eye," recalled Gordon, those animals "were overwhelmed by the pressure behind. I have never seen two railroad trains come together, but one who has seen the cars piled up after a wreck can imagine how the buffalo were heaped up in an immense pile by the pressure from behind." The buffalo to the rear kept charging forward. They kept coming, but as they reached the heap of downed buffalo just in front of Gordon, they veered to one side or the other. Gordon watched from his perch in the tree. His horse "stood shaking with fear."

> I could now enjoy a spectacle which I fancied neither white man nor Indian had ever before seen. The front rank as they passed was as straight as a regiment of soldiers on dress parade. The regularity of their movements was admirable. It appeared as though they had been trained to keep step. If one had slackened in the least his speed, he would have been run over.

After about fifty-five minutes of what Gordon described as "alternate terror and pleasure," the buffalo passed. He just sat in the tree for several minutes wondering whether he had had a bad dream. Then he climbed down from the tree to his horse, which was still shaking.[8]

"Breaking a Herd," Theodore R. Davis drawing illustrating how stampeding buffalo were turned by an army contingent while crossing western Kansas "by the Smoky Hill route in 1865," from *Harper's,* January 1869.

Frederic Remington drawing to illustrate Theodore Roosevelt's account of an 1877 hunt in Texas by his brother, Elliott, and his cousin, John Roosevelt. The two hunters were caught on the Staked Plains in front of a stampeding buffalo herd, and "their only hope for life was to split the herd." (*St. Nicholas,* December 1889)

Hunting buffalo seems to be pictured as an idyllic pastime in this illustration from the 1870s in Frank Fossett's *Colorado, Its Gold and Silver Mines.*

"Sport on the Plains," from *Harper's Weekly,* March 21, 1874.

A similar account is related by O. W. Williams. In the summer of 1878 Williams headed a surveying party working on the Staked Plains of Texas, in Lamb County—"Up to that time the map of that county in the General Land Office was only a blank sheet."

Although there was a gracious promise in this warm red soil of bountiful harvests in the long years of the future, yet I believe that I have never looked upon any other country as destitute of the graces which go to make up what we call scenery as this plain which then lay before our eyes. The blue sky met the horizon of brown grass, then in the sere of a long drought, in a circle unmarked by tree, stone or hill, save a little undulation in the southwest where the sand dunes under the eternal southeast wind had risen in small waves. The margin of sky and earth danced up and down in heat waves just enough to throw doubt upon the wavering outlines of the dunes. But away from that margin there was no movement; even the birds of the air were not to be seen. Nor was there sound. There was no hum of bees—no chirping of crickets—no swish of a bird's wing—no rustling of leaves.

It might have been a dead world. But we could not consider it a desert, in the modern use of the word, so long as that spring kept up its flow and that grass sward lay so solidly on the dark red earth. With better grace we might have called it a desert (in the ancient sense of a country deserted by its former inhabitants), for of life and motion there was little evidence. Yet it was only a few hours until we saw animal life passing over it in great mass and swift motion, with the sound of falling water.

We ran our line of meander down the creek on the north side—two chainmen and two flagmen afoot—while I carried the transit from station to station on a gentle horse—wise to frontier life, as we soon learned. We had come some six or eight miles down the watercourse when, as I was setting up my instrument, the flagman asked if I had not heard a peculiar sound. I stopped my work to listen and caught a faint throbbing sound of somewhat irregular cadence such as I had heard two years before, when twenty miles away from Niagara Falls. It came from the north, and looking in that direction, we could make out what seemed to be a low-lying cloud sweeping down on us quite rapidly.

It was late July, so it could hardly be a norther. There was nothing in its appearance to lead us to suspect it might be a rain cloud. We were for a moment at a loss to account for it. Then we caught sight of dark objects showing up on the ground-front of the grayish white cloud and then dropping back from sight everywhere along its width from the eastern sky to the western horizon. Almost at once the cry went up, "Buffaloes! A stampede!" Immediately, we began to prepare to meet the storm. Looking back, we saw our wagon and our ambulance (passenger wagon) about half a mile away, apparently being lined up to meet the charge with the smallest possible front, but too far away for us to join them in time. So we prepared our little party to face the stampede on our own ground.

We stood in single file, facing the oncoming herd. Our transit was set up in the middle of the file, with the last man holding the reins of my saddle horse. With the only rifle in the party, naturally I was at the head of the file, in order to split the pass-ing animals by the firing of the gun—if they did not divide to either side on catching sight of us. There was no greater danger at the head of the file than at its foot; for once broken at the head, immediately the whole line would go down. It seemed to me inevitable that the mere sight of us would divide the herd; however, I might have been wrong on that point.

We were not long in getting set for the rush of the buffaloes. But we were barely ready, when they were on us with a swirl of dust and a thunder of hoofs—yet so far as I could judge, they were absolutely mute. The front line was thickly packed shoulder to shoulder, and the eyes of the animals were cast back as if trying to see something behind them. When I realized this atti-tude of the buffaloes, I began firing the gun—although they must have been one hundred feet away—because it began to look as if they might run over us without seeing us.

The shot had no effect; I do not think the sound of the gun was heard by the animals. I concluded that we could not split them by sight; so I commenced to shoot as rapidly as possible, but without effect until they were a distance of about thirty or thirty-five feet. At this point I saw some of the animals in front of me begin to push their neighbors to one side or the other to make an opening in that crowded front line. That opening must have been about twelve feet wide when the front line passed us, although it seemed to me that I could touch a buffalo on either side with the point of my gun.

Behind this dense line there was no regular formation; the animals came on in loose order, gradually thinning out to the rear. As soon as I felt safe in doing so, I turned back to see in what shape the onrush had left our party. During the terrific uproar of the passing multitude, I had dimly made out sounds which might have come from the men or the horse behind me, and when I turned I greatly feared to find that some calamity had befallen us. But beyond a horse that was trembling, a party of four men exceedingly dust laden and full of strange oaths, there was nothing to show that we had been in any danger.

After the big bodies had passed us, they plunged through the small tule-bordered creek. The water and mud lay spotted on the brownish-yellow grass for some forty yards out, while the banks and bed of the creek were streaked with the marks of the huge briskets and cloven feet. The water for a few minutes ceased to run. We saw no animals mired or bogged down; but if the stream had been wider, we were certain, some of the weaker lagging ones would never have gotten out to hard ground. We had seen along the larger streams numbers of places in boggy ground where the carcasses of these beasts were locked up thickly in dried mud, and we could now understand the reason.

As soon as the last of the animals had passed us, our vision opened to the west. We could see that our main party had weathered the storm and was beginning to get ready to follow our march. We learned later that the herd was not so dense and heavily packed with buffaloes as at our point of contact; but the estimate made by members of our main party as to the number of animals in the stampede ranged far higher than ours: it was set by some of them at fifty thousand. It was impossible for us to make any reliable estimate, however, because neither party could determine the end of the flanks on the east and the west. But the herd was almost surely the last great herd of the

Southern buffaloes after they had been cut off from any migration to the north, and after five years of the Sharps rifle in the hands of the professional hunters.[9]

Some early observers considered that any herd of running buffalo was a stampeding herd, running in panic from some real or imaginary danger. But this was not always the reason. One buffalo hunter, James McNaney, saw a herd of buffalo running in a column on the northern plains during the winter of 1882-83. McNaney and some fellow buffalo hunters were camped on the plains of southeastern Montana, about seventy-five miles southeast of Miles City. It was a bright morning as McNaney and his friends watched the herd pass near their camp. McNaney described their gait as a "long lope." He said the animals were four to ten abreast. The calves were running with their mothers. The animals galloped past the hunters' camp over a course no wider than a village street. They stretched out in a long column clear across the prairie. From all appearances, McNaney said, the run was organized. It certainly was no stampede.[10]

Other early observers maintained that buffalo, more often than not, would run into the wind when alarmed or stampeding. "Buffalo" Jones and other early plainsmen noted this during the 1870s. One of them, John R. Cook, who hunted buffalo on the southern plains during the '70s, reported that even unmolested buffalo traveled into the wind. In July 1875 Cook and some hunting companions found large buffalo herds in the Wolf Creek country north of the Canadian River in what is today Oklahoma. As Cook described it:

> The wind, what there was at a certain time, came from the southwest for seven consecutive days, and every buffalo was either traveling or headed that way; and on the eighth day the wind changed to the east about midnight, and blew pretty strong all the next day. All day long the buffaloes moved eastward. That night there was heavy thunder and sharp lightning in the south; and just before daylight the wind whipped to the south and rain began to fall. As soon as it was light we noticed the buffaloes were headed south, and moving en masse.

That was not the only time John Cook noticed unmolested buffalo moving into the wind. In November 1875 Cook and some other hunters were on a tributary of the Red River of Texas when a "norther" rolled in, dropping temperatures considerably. "For three days," said Cook, the buffalo "were headed north and northwest" into the cold wind.[11]

Cook's experiences were proof enough for him that buffalo "always" traveled into the wind. But it is difficult to conclude that they "always" did, though there is some logic in thinking that wild buffalo might travel much of the time into the wind. Catching the scent of fresh grass or water on the wind would have been sufficient reason for the animals to move into the wind. Also, when Indians or white men hunted buffalo, they usually approached the herds from the downwind side so the buffalo could not scent them. Attacked from downwind, buffalo naturally would run away in the opposite direction into the wind. Running into the wind enabled them to smell danger far ahead.

Though the actions of wild plains buffalo were often unpredictable, when they stampeded they nearly always followed the leader. This trait often caused the death of many buffalo. Lewis and Clark observed this in 1805 as their party headed west near present Great Falls, Montana. Their journals recorded that the buffalo "go in large herds to water about the falls, and as all the passages to the river near that place are narrow and steep, the foremost are pressed into the river by the impatience of those behind."[12]

Another early observer on the northern plains, George Catlin, noted that stampeding buffalo "run chiefly by the nose, and follow in the tracks of each other, seemingly heedless of what is about them, and of course, easily disposed to rush in a mass, and the whole tribe or gang to pass in the tracks of those that have first led the way."[13]

A later observer, Joel Allen, writing in the 1870s, compared a herd of wild stampeding buffalo to a panic-stricken crowd rushing pell-mell from a public building when someone yells, "Fire." Allen thought that a herd of stampeding buffalo possessed "the sheep-like propensity of blindly following its leaders" when running from some "real or fancied danger."[14]

The Indians on the northern plains knew how the buffalo would react and capitalized on it to stampede buffalo over precipices. Even the white man used this technique as late as the 1870s to kill buffalo. In the summer of 1873 the *Denver News* reported:

Out on the plains, about two hundred miles from Denver near the Kansas-Colorado border, is a vertical bluff seventy-five feet high. A party of hunters recently stampeded a herd of buffaloes right to the brink of the precipice. The foremost brutes, appreciating their critical situation, attempted to avert the calamity, but the frightened hundreds behind crowded forward with characteristic persistency. The front rank, with legs stretched toward each cardinal point of the compass, bellowed in concert and descended to their fate. Before the pressure from behind could be stopped, the next rank followed, imitating the gestures and the bellowing of the first. For thirty seconds it rained buffaloes and the white sand at the foot of the bluff was incarnadine with the life-blood of wild meat, and not until the tails of fifty or seventy-five of the herd had waved adieu to this wicked world did the movement cease.

Colonel Richard I. Dodge witnessed a herd of perhaps 4,000 buffalo attempting to get across the South Platte River near Plum Creek late in the summer of 1867. There had been high water on the river, but it had subsided to a depth of only one or two feet. However, the river bed channels had filled or were filling with loose quicksand so that when the first buffalo hit the water they were hopelessly stuck.

"Those immediately behind, urged on by the horns and pressure of those yet further in the rear," wrote Dodge, "trampled over their struggling companions to be themselves engulfed in the devouring sand." Dodge noted that the pushing and shoving continued "until the bed of the river, nearly half a mile broad, was covered with dead or dying buffaloes." Only a small number of the animals actually crossed the river, and those were soon driven back by hunters on the other side. More than 2,000 buffalo died that day in the South Platte River.[15]

About four years later, Colonel Dodge saw buffalo reacting strangely along the Santa Fe line in Kansas. The winter had been so severe that the ponds and smaller streams had frozen solid, forcing buffalo from a wide area to head for the major rivers, such as the Arkansas, for water. To get to the Arkansas, the buffalo had to cross the Santa Fe tracks, which that winter stretched only as far west as Newton, Kansas. Colonel Dodge thus described what he witnessed and could not explain:

If a herd was on the north side of the track, it would stand stupidly gazing, and without a symptom of alarm, although the locomotive passed within a hundred yards. But if the herd was on the south side of the track, even a mile or two away, the passage of a train set the whole herd in the wildest commotion.

The buffalo would make for the track at full speed. If the train happened not to be in their way they would cross the track and stop satisfied.

But if the train was in the way, each individual buffalo went at it with the desperation of despair, plunging against or between locomotive and cars, just as blind madness chanced to take them. Numbers were killed, but numbers still pressed on to stop and stare as soon as the obstacle was passed.[16]

Did the plains buffalo become a faster animal during the 1870s and early '80s? George Bird Grinnell thought it did:

During the last days of the buffalo, a remarkable change took place in its form, and this change is worthy of consideration by naturalists, for it is an example of specialization—of development in one particular direction—which was due to a change in the environment of the species, and is interesting because it was brought about in a very few years, and indicates how rapidly, under favoring conditions, such specialization may take place.

This change was noticed and commented on by hunters who followed the northern buffalo, as well as by those who assisted in the extermination of the southern herd. The southern hunters, however, averred that the "regular" buffalo had disappeared—gone off somewhere—and that their place had been taken by what they called the southern buffalo, a race said to have come up from Mexico, and characterized by longer legs and a longer, lighter body than the buffalo of earlier years, and which was also peculiar in that the animals never became fat. Intelligent hunters of the northern herd, however, recognized the true state of the case, which was that the buffalo, during the last years of their existence, were so constantly pursued and driven from place to place that they never had time to lay on fat as in earlier years, and that, as a consequence of this continual running, the animal's form changed, and instead of a fat, short-backed, short-legged animal, it became a long-legged, light-bodied beast, formed for running.

This specialization in the direction of speed at first proceeded very slowly, but at last, as the dangers to which the animals were subjected became more and more pressing, it took place rapidly, and as a consequence the last buffalo killed on the plains were extremely long-legged and rangy, and were very different in appearance—as they were in their habits—from the animals of twenty years ago [1872].[17]

So far as I can learn, only Grinnell has set forth this thesis. Nowhere can I find that anyone has proved or disproved Grinnell's conclusion. Unfortunately, trying to do so today would be next to impossible. Those persons who were present on the plains when the change supposedly occurred are dead. Records of the animals were not kept, and the facts are simply not available. There are many stories and tales about the speed of buffalo, but none mentions any changes in the buffalo's speed over the years.

Colonel Richard Dodge noted: "A buffalo can run only about two-thirds as fast as a good horse; but what he lacks in speed he makes up in bottom or endurance, in tenacity of purpose, and in most extraordinary vitality. If a herd is not overtaken in 500 or 600 yards, the chase had better be abandoned, if any regard is to be had for the horse."[18]

Other men on the plains noted the same thing. Another military man, Colonel R. B. Marcy, reported: "The buffalo has immense powers of endurance, and will run for many miles without any apparent effort or diminution in speed." Marcy once recalled: "The first buffalo I ever saw I followed about ten miles and when I left him he seemed to run faster than when the chase commenced."[19]

While mentioning the subject of chasing buffalo, I want to tell the story of Ben Trott. About 1850, Ben and some young friends left the East and headed for the gold fields of California. All were young and adventuresome and eager to make their fortune in the West. The young men took it easy on their journey, especially Ben, who had a fine horse which he prized highly. However, one day while crossing the southern plains near Sulphur Springs in what is now south central

Oklahoma, Ben and the other young men came upon a large herd of buffalo. They started out after the animals, hopeful of separating a bunch of cows and calves from the main herd and killing some of the calves for meat. Ben, wanting to show what his horse could do, raced ahead of the others. A few seconds later, as his horse was out in front of the others, the animal stumbled and fell. Ben went sprawling on the prairie, but he got to his feet quickly, unhurt. His horse also regained his footing and took off *toward* the buffalo as fast as his legs could carry him. Jarred and mortified by the fall, Ben broke into tears. He just stood there crying as he watched his prize horse disappear in the distance with the running buffalo. Ben never did see that horse again.[20]

A running herd of wild buffalo had a strange effect on man and beast. Ben Trott's horse was apparently drawn to the running buffalo as if they were magnetic. And this is not an isolated incident. Colonel R. B. Marcy told a similar story. Marcy was commanding a group of soldiers on their way to Utah when they came upon a buffalo herd on the plains. One of the officers, already tasting buffalo steak, jumped from his horse and ran forward to shoot a shaggy. He forgot to tie his horse to something, and a moment later the animal, like Trott's horse, was streaking off across the plains with saddle, bridle, and the other accoutrements of an officer's horse to join the running buffalo. Marcy later recalled that the horse went with the buffalo "far over the prairies out of sight, and has not, I believe, been heard from since."[21]

Some fared better with their horses, of course. Yale paleontologist Othniel C. Marsh tells of a horse that saved his life:

> In October, 1872, I was exploring the chalk cliffs of Western Kansas for fossils. . . . [T]he buffalo were there in countless numbers, and herds of thousands were daily in sight from the bluffs on which we were at work. . . .
>
> Our first camp was on Smoky Hill River, and we then moved South, near Chalk Creek, a tributary of the Smoky, about thirty miles South-east of Fort Wallace. Our tents were pitched near a dismal waterhole, nothing more nor less than a large buffalo wallow; and as we found later, to our consternation, at least one aged bull had here been recently mired and found his last resting place. The water itself was mostly covered with a thick green

slime, and was scarcely fit for man or beast. It was, however, all
we could find in reach of the rich fossil deposits we were working, and we made our coffee from it, over a fire of "Buffalo
chips," and were contented if not happy.

One afternoon, when returning from a long fossil hunt, the
guide [Ned Lane], Lieutenant Pope, and myself were riding
slowly abreast, discussing the day's fossil hunt and the prospect
for the morrow. . . .

As we rode to the crest of a high ridge, the guide . . . suddenly called out,—"Great God, look at the buffalo!" and we saw
a sight that I shall never forget, and one that no mortal eye will
ever see again. The broad valley before us, perhaps six or eight
miles wide, was black with buffalo, the herd extending a dozen
miles, up and down the valley, and quietly grazing, showing that
no Indians were near. The animals were headed to the South,
and slowly moving up the valley in the direction of the great table
land where pasturage for the night was to be found. The sight
was so wonderful that we sat on our horses for some time,
watching the countless throng, and endeavoring to make some
estimate of how many buffalo were in sight before us. The
lowest estimate was that of the guide, who placed the number at
50,000. I thought there were more; and our military comrade,
with his mathematics fresh from West Point, made a rapid calcu-
lation of the square miles covered, and the number of animals
to the mile, making his total nearly one hundred thousand. . . .

While wrapped in the wonderful prospect before us, the guide
quietly remarked,—"We must have one of those fellows for
supper." . . .

The three agreed that since Marsh was mounted on the best horse,
"a fleet Indian pony," he should be the one to shoot the buffalo.

My hunting weapons consisted of a cavalry carbine and a pair of
navy revolvers, not too many for an Indian country, and I hoped
soon to bring down a young buffalo that would give our camp
the wished-for meal. I rode slowly down toward the herd, avoid-
ing a few old bulls outside the main body, so as not to disturb
them. The wind was in my favor, and I soon was near the herd.
Selecting my animal, I promptly gave chase, hoping to get in

my shot before the herd started, which I knew would soon be the case. The animal selected, a young cow, proved especially fleet, and it was some minutes before I was along side, ready to shoot, in the exact manner my first guide, Buffalo Bill, had taught me long before. While the chase was still going on, I heard one or two shots behind me in the distance, and concluded that my comrades were firing at some straggling animal, but I had no time to look around, as my pony, knowing what was wanted, made a direct chase for the buffalo selected, and soon carried me, where a shot from my carbine, brought the animal to the ground. I then had a chance to look back.

To my amazement, I saw that the main herd, alarmed by the shots of my comrades, had started and was moving rapidly southward. I saw also what I had not before surmised; that in my eagerness, I had pushed well into the herd without noticing it, and as the great mass of animals in the rear started, they began to lap around me, and I would soon be enclosed in the rapidly moving throng liable at any moment to be trampled to death if my pony should fail me. My only chance of escape was evidently to keep moving with the buffalo and press towards the edge of the herd, and thinking thus to cut my way out, I began shooting at the animals nearest to me, to open the way. Each shot gave me some gain, as those near pushed away, and when one went down, others stumbled over him. The whole mighty herd was now at full speed, the earth seemed fairly to shake under the moving mass, which with tongues out, and flaming eyes and nostrils, were hurrying onward, pressed by those behind, up the broad valley, which narrowed as it approached the higher land in the distance. My horse was greatly excited by his surroundings, and at first seemed to think I wanted some particular animal, and was thus inclined to make chase after it, but he soon came to understand the serious problem before him, and acted accordingly.

A new danger suddenly confronted me. The prairie bottom had hitherto been so even that my only thought was of the buffalo around me and the danger of being overwhelmed by them, if my pony could not keep up the race. The new terror was a large prairie dog village, extending for half a mile or more up the valley. As the herd dashed into it, some of the animals stepped into the deep burrows, and near where I was riding, I saw quite a number come to earth and now and then a comrade from behind fall over them. My trained buffalo horse here showed his wonderful sagacity. While running at full speed along with the herd, he kept his head down, and whenever a dangerous dog hole was in his path, he either stepped short or leaped over it and thus brought me through this new danger in safety. The race had now been kept up for several miles, and my carbine ammunition was nearly exhausted; while my pony, after his long day's work and rapid run, showed unmistakable signs of fatigue. My only hope was that he could hold out until we reached rougher ground, where the herd might divide. This came sooner than I expected.

As the valley narrowed, the side ravines came closer together at the bottom, and our course soon led us among them. The smaller gullies were leaped with ease by the buffalo close around me, and my pony held his own with the best of them. As the ravines became deeper, longer leaps were necessary, and my brave steed refused none of them. Soon the ravines became too wide for a single leap, and the buffalo plunged into them and scrambled up the opposite bank. My pony did the same, and several times I could have touched with my extended hands the buffalo on either side of me as we clambered together up the yielding sides of narrow canyons we were crossing. This was hard work for all, and the buffalo showed the greater signs of fatigue, but no intentions of stopping in their mad career, except those that were disabled and went down in the fierce struggle to keep out of the way of those behind them.

As the valley narrowed, I saw ahead, perhaps a mile distant, a low butte, a little to the left of the course we were taking. This gave me new courage, for if I could only reach it, it would afford shelter, as the herd must pass on either side of it. Drawing a revolver, I began to shoot at the nearest buffalo on my left, and this caused them to draw away as far as the others would let them, and when one went down, I gained so much ground. They were now really more afraid of me and my steed, than we were of them, and for this reason did not charge, as a single wounded buffalo might have done. Continuing my shooting more rapidly as we approached the butte, I gradually swung to the left, and when we came to it, I pulled my pony sharp around it, and let the great herd pass on. . . .

We could not move until the herd had gone by, and it was more than an hour before the last of them left us alone. . . . The yellow chalk butte that sheltered us, now glistening in the last rays of the setting sun, was a characteristic feature of that region, being mushroom in shape, with the sides worn smooth and deep where the buffalo for ages had rubbed against, to rid themselves of the gnats that annoyed them.

The danger was now over, and the pangs of hunger reminded me of the supper I had promised to secure for my comrades. One of the last stragglers of the herd in the twilight was a young heifer, and a shot brought her to my feet. To draw my hunting knife and remove the tongue and hump steaks, sufficient for our small party, was the work of a few minutes; and thus laden, I was ready to start for camp, some half a dozen miles to the eastward. Meanwhile, I had not forgotten my faithful Indian pony. He had saved my life, and I did all I could for him.[22]

Did the wild plains buffalo trot? "Buffalo" Jones thought not. "When you read of buffaloes 'trotting off,'" said Jones, "you may rest assured that the author of the statement knows nothing of the habits of the animal." Jones asserted that buffalo "invariably walk or gallop."[23] However, Martin Garretson, former secretary of the American Bison Society, disagreed. Buffalo, said Garretson, have three gaits. He described them as the gallop or long lope, the pace, and *the trot*.[24]

Both Jones and Garretson were wrong. The buffalo has four distinct gaits: the walk, trot, bound, and gallop. Each is determined by the sequence in which the buffalo lifts and lowers its hoofs. When a buffalo walks, it begins by lifting its left hind hoof; next the front left hoof is raised, followed by the right rear and the right front in a one-two-three-four order. To trot the animal first moves its left front hoof, then left hind hoof, followed by the right front and right rear. The bound gait can be seen when the buffalo suddenly springs forward in flight; as it bounds forward, the animal uses both front hoofs, first the left and then the right, followed by the left and right hind hoofs. Its fourth gait, the gallop, is similar to the bound gait, with the same sequence, but it appears smoother.

To see a small herd of buffalo take off running across the rolling country in the middle of the Wichita Mountains in southwestern Oklahoma or down the slopes of the National Bison Range in western Montana is strange but beautiful. At first, the animals appear to be clumsy in their movements, leaning first to one side for a time, and then to the other. Yet they constantly gain much ground.

E. N. Andrews, whose account of hunting buffalo by train in western Kansas is detailed in Chapter VI, described the buffalo's gait as "remarkable." Andrews wrote: "One who has seen it can never forget it; while it cannot be described to those who have not seen it. It may be said, however, that the general movements, including the slowness or apparent awkwardness, the contour of the form, very much resemble those of the elephant."[25]

Earl Drummond, who probably knew more about the buffalo in his day than anyone else, noted that when buffalo get older, "their hind legs get more bent in the hock joints, the feet set forward more, and their hind parts droop down lower than the younger animals."

> They have very short front legs, and their front quarters are deep up and down so as to give them plenty of lung space as they have large lungs, which gives them plenty of wind for long runs; and they are cut high in the flanks, which helps them to run fast and a long ways.
>
> Some oldtimers said the buffalo could eat breakfast in Texas, dinner in Oklahoma, and supper in Kansas—their meals probably would be many hours apart, but they can out wind a good horse and out run him after one mile.
>
> [Buffalo] run with their tongues out and their heads down, and make a noise puffing the wind in and out of their mouths.
>
> When they start to run and make a quick get-away, they spank themselves with their tails by slapping the tail up and down. The quicker the get-away the faster they spank. When they get up speed they don't spank with their tails very much, but if they slow down and start again they repeat the spanking. The calves do this more than the older buffaloes.

Drummond noted that buffalo "can run at almost full speed and pivot the front feet and hit the ground with the hind feet and be headed in the opposite direction and not lose but very little of their speed." However, Drummond said, running buffalo do have one problem. When a bunch of bull buffalo are running, the bull in the lead "will turn his head from one side to the other thinking he will bluff the others so they will stop or turn off to one side." They do this, said Drummond, because they know it is very dangerous to have a buffalo bull behind. "They have to keep on guard all the time," he wrote. "Another bull will not pass up a chance to take a prod at the other fellow if he thinks he can whip him."[26]

XII

Trails and Travels

Travel the buffalo trail, it's the best route.
Frontier advice 19th century

There is a certain fascination about trails, whether they are in the mountains, the woods, or out on the open prairie. To see a new trail is to want to see where it leads, even if it is nothing more than a cattle trail winding through the bluestem of the Kansas Flint Hills. To most outdoorsmen, trails offer the same kind of irresistible charm that side roads offer the Sunday driver or that rivers provided for the early explorers.

Long before the white man arrived in North America, Indians were using animal trails. When buffalo ranged east of the Mississippi into the Appalachians, buffalo trails were numerous, and the Indians made good use of them. When white men arrived and later moved westward across the Appalachians, they followed animal trails, especially those of the buffalo. The routes of many early roads and trails established by the white men were over old buffalo trails.

When white men began to explore the plains and prairies, they frequently followed buffalo trails. During the early 1800s, Lewis and Clark, traveling along the upper Missouri River, saw hundreds, perhaps thousands, of buffalo trails. Their journals reported that because buffalo "have wonderful sagacity in the choice of their routes, the coincidence of a buffaloe with an Indian road, was the strongest assurance that it was the best."[1]

When John Bradbury, an English naturalist, traveled along the Niobrara River in what is now northern Nebraska, he also saw many buffalo trails. These he described as "excellent roads." In Bradbury's opinion, "No engineer could have laid them out more judiciously."[2]

When there were buffalo, there were buffalo trails. On the open

180

plains and rolling prairies, there were so many buffalo that their trails were almost everywhere. Roaming the same range year after year, the buffalo would wear deep ruts into the earth, more often than not by their going to water in single file. Usually once a day they would seek water at a creek, pond, or river. On especially hot days, when insects would torment the almost naked buffalo, they might travel to water in the morning, then stand in the cool water free of insects until late afternoon. About dusk they would move back to the prairie or plains to graze upon the sweet grasses until the next morning, when they would repeat the process. All the time they would be wearing the trails deeper and deeper into the earth.

When Zebulon Pike and his party crossed the plains in 1806, they found what appeared to be a road between the Solomon and Republican Rivers in what is today north central Kansas. Pike's road was in reality a buffalo trail worn bare of grass by thousands, perhaps millions, of buffalo over many centuries. It was a trail to water. Sixty-five years later, Charles S. Scott, an early Kansas settler, found the remains of the old trail on the level prairie southwest of Mankato, Kansas. It was twenty to thirty feet wide and hollowed out in the center. The trail was depressed below the general surface of the land by a few inches, but where it crossed a ravine the banks on both sides had been worn down three or four feet. The trail ended at the Republican River, at the head of Beaver Island, where a rocky ford provided for a good crossing of the stream. Today, however, only very faint traces of that old buffalo trail are visible. During the early 1900s a farmer plowed the land level so that he could plant crops. Hardly any portion of the trail remains today as a reminder of those days when millions of buffalo roamed the Kansas prairie and plains.[3]

One early traveler crossing what is today Oklahoma reported finding the ground in many areas "intersected in every direction by buffalo trails worn deep." He advised anyone traveling that way to follow the buffalo trails. They make "tolerable crossings over the bad creeks."[4]

When white men began bringing wagons onto the plains and prairies, they were not always able to follow buffalo trails. Most were too rough. Colonel Richard Dodge observed, "The buffalo is extremely careful in his choice of grades by which to pass from one creek to another; so much so indeed that, though a well-defined buf-

falo trail may not be a good wagon road, one may rest assured that
it is the best route to be had." Dodge noted, as did other observant
pioneers on the plains, that a buffalo "seems to have a natural anti-
pathy to the exertion of going up or down steep places" unless he is
being chased.[5] In this regard the buffalo seems to be like human beings. More often than not, they take the line of least resistance in
walking.

Yet there is at least one unique fact about the old buffalo trails.
"Buffalo" Jones pointed it out in the late 1890s. A buffalo trail, said
Jones, "cannot be found anywhere that is longer than four hundred
yards without a change of direction, but the general course of the herd
would be comparatively straight for a distance of thirty or forty
miles." Jones said the reason buffalo trails were crooked was that
the buffalo "never pursue a straight course in traveling; their eyes are
so placed in the head that it is impossible for them to see directly in
front." Also, buffalo cannot look backward "on account of their
immense shaggy shoulders." Jones concluded: "They are compelled to
keep one side or the other turned in the general direction in which
they are going. Not being good travelers sideways, they look ahead
with one eye and to the rear with the other, deflecting to the right
and then to the left for a distance of two or three hundred yards."[6]

Some oldtimers would swear on a stack of bibles that buffalo mi-
grated annually from Texas to Canada and back, just as some birds
fly south for the winter and north for the summer. This belief was
prevalent among many early Jesuit explorers east of the Missouri
River.[7] But such migrations never occurred.

At ten to twelve miles a day, the buffalo would have taken at least
six months to complete a one-way trip from Texas to northern Montana and southern Canada on the open plains. They could not possibly have accomplished the round trip in any one summer.[8]

Furthermore, there are numerous accounts of buffalo wintering in
the cold north—Montana and southern Canada—which in itself dis-
proves the migration theory.[9]

Yet there is no doubt that buffalo on the plains and prairies wandered considerably. One veteran buffalo hunter, Frank Mayer, contended that the animal wandered in a small circle. A large herd would
thus clear the prairie of grass in one area and then move on to find
more grazing. They "follered the feed," as Mayer put it.[10]

One of the first white men to assert that buffalo did not migrate was George Catlin. He did so in the early 1830s. But even though his conclusion was based upon more in-the-field experience and observation than that of any other writer of his day, his reports on the buffalo, among other things, were downgraded by some. As Frank Gilbert Roe said, "Catlin had one fatal disqualification—he sympathized with the Indians! To those propagandists and 'educators' of public opinion who coveted the Indian lands, it was of the first importance to decry Catlin as half liar, half fool. This embraced everything he said, and naturally included buffalo."[11] If Catlin had been believed, perhaps the myth of buffalo migrations might never have existed.

Ernest Thompson Seton, writing in 1929, concluded that buffalo were migratory in theory. Seton pointed out that wild plains buffalo traveled 300 to 400 miles northward in the spring and about the same distance south in the autumn. But he admitted that it was hard to tell whether there was any regularity in such movements. Natural events such as bad weather and incidents caused by both Indians and white men often caused changes in the animal's movements. These things, said Seton, obscured whatever regularity might have existed in the buffalo's movements.[12]

Cowman Ed Lemmon, writing earlier in the *Belle Fourche Bee,* prefigured Seton. Prior to the railroads, which "somewhat retarded migration," Lemmon said, buffalo movement

> was on an average of about 300 miles for each migration, say those of Texas migrated in the spring to the Indian Territory and Kansas and back in the fall. Those of Kansas to the Plattes and those of the Plattes to the Dakotas, and those of the Dakotas to northern Montana, and those of Montana to Canada. . . . And another peculiarity was the fact that after reaching northern Montana climatic conditions had a bearing on their color and compact build and they were universally known as bison and no longer buffalo. . . . They seem to move in more compact bodies migrating north than south, for their return was more leisurely.[13]

Roe, writing in 1951, after more than fifteen years of research and study on the wild buffalo, concluded that regular buffalo migrations did not "as historical facts, invariably take place. These wanderings were utterly erratic and unpredictable and might occur regardless of time, place, or season, with any number in any direction, in any manner, under any conditions, and for any reasons—which is to say, for no reason at all."[14]

XIII

Cows and Calves

She brings forth in the spring, and rarely more than one.
George Vasey 1851

It is early afternoon in southwestern Oklahoma. The hot July sun beats down on the prairie, which is slowly turning brown for lack of rain. Off in the distance, perhaps a quarter mile away, a small herd of buffalo is grazing near a grove of oak trees. As you watch, you see a cow slowly leave the herd and move off in a very deliberate manner toward the grove of oaks. There in the shade of the trees the cow lies down to give birth. Mission accomplished, she slowly gets to her feet, turns, and starts to clean the little calf, which is beginning to squirm on the ground. The Wichita Mountains buffalo herd has increased by one.

It takes a few minutes for a buffalo calf to be born. Sometimes a cow will drop her calf while standing, but she usually lies down. W. H. Blackburne, former head keeper at the National Zoo in Washington, D.C., timed a buffalo birth many years ago. The cow had cleaned her calf and had begun nursing the little buffalo twenty-six minutes after she had stopped eating hay to lie down.[1]

When a buffalo is born, it does not look much like a buffalo. There are no signs of the familiar hump. Frequently a young buffalo can pass for an ordinary domestic calf, although its tail is sometimes shorter. But unlike the calves of domestic cattle and of nearly all other bovines, a buffalo calf will lose its birth color within a few weeks. A newborn buffalo calf is usually soft brick red, but once in a while one may turn up cinnamon-colored with a tinge of yellow. As the weeks pass, the color gets darker until finally the brown, almost black, coloring of the adult buffalo replaces the birth color and remains for life.

185

For the first three or four days, the young buffalo usually views the world from ground level. A calf spends most of that time lying down or nursing. Sometimes, however, a newborn calf will walk around an hour or two after birth. An instance has been reported in which a calf followed its mother around within forty-five minutes after birth. That calf "was particularly strong, and ran bucking around its mother when it was only a little more than a half hour old."[2]

As a buffalo calf begins to grow, its hump develops rapidly. Soon there is no question in anyone's mind that the little fellow is a buffalo. During the first few months, a calf can be quite playful. Elmer Parker told me, "Playin' is quite common in buffalo up to about a year old. They're like a bunch of kids. They'll play anytime they feel like it. I've seen a calf run around for forty-five minutes trying to get old buffs to play. Sometimes the older ones will. But then as a calf grows up, it becomes shy like an older buffalo."[3]

For any buffalo calf, long spells of nursing are unusual, almost unheard of. A calf must attack the problem frequently to get enough milk because the supply is limited at any one time. A buffalo cow's storage capacity for milk is much less than a domestic cow's. Several visits are usually required before a buffalo calf can get its fill. Many buffalo men have told how they have watched young calves trying to get their fill but finding that the supply has run dry for the time being. One buffalo man said, "Sometimes you see a little feller who looks downright disgusted with his Ma when the spout runs dry." However, L. Roy Houck, president of The National Buffalo Association in 1973 and the owner of 2,500 buffalo on his South Dakota ranch, concludes that buffalo milk "is very rich and high in protein, and a buffalo calf only needs to consume a small quantity of it at a time for his existence."[4]

Does buffalo milk taste good? Is it as good as or better than domestic cow milk? Although I have never tasted buffalo milk, I have my doubts, based on conversations with several buffalo men who admitted to me they had tasted it. One of them, who had spent many years at the National Bison Range in Montana, said he once tasted buffalo milk but did not like it. He said other buffalo men had sampled it and felt the same way. Until a few years ago, it was an unofficial ritual for any new buffalo man at the National Bison Range to taste buffalo milk. It was all in good fun, something like a fraternity initiation. But the practice, I am told, no longer exists.

Several years ago some buffalo men down in Oklahoma were sitting around one spring day talking buffalo. One of them suggested they ought to milk a buffalo. The others agreed. It so happened that a buffalo cow was penned up in a nearby corral. The men ran the animal into a squeeze chute and pushed the sides together to keep the cow from moving. Then one of them reached in and began to milk. "The old cow kicked and fussed, but she got milked," he said. The quantity was small, but there was enough for each man to have a thimble full. Some liked it; most did not. Some of the milk was saved and sent to a nearby town to be tested. When the report came back a few days later, the analysis showed that the buffalo milk tested richer than that of a Jersey cow. The report, however, did not change the buffalo men's drinking habits. They stayed with "store bought milk" from domestic cows. Tasting buffalo milk once had been enough for them.

When a buffalo cow gives birth, usually about nine-and-a-half months after mating, she normally has only one calf. Sometimes twins are born, and they are not so rare among buffalo as some oldtimers have led us to believe. When I questioned every known private and government buffalo herd in North America and Hawaii,[5] one of the questions dealt with buffalo twins. Out of 207 replies, twelve reported the birth of twins and two of these reported more than one set in their herds during the preceding twenty years.

The National Bison Range in Montana reported two sets. One pair was born about 1940, the other in 1967. A Gillette, Wyoming ranch reported that during a three-year period, 1966 to 1969, three sets of buffalo twins were born into a herd of 1,700 buffalo, owned by the Durham Meat Company of San Jose, California. Another set of buffalo twins was reported by Joseph Armington of Anderson, Indiana, but they did not survive. Armington wrote:

> In April, 1964, I was preparing to sell some buffalo. I drove them all into an enclosure. I noticed that one cow kept trying to get out. The next evening when I went to the farm, that cow had jumped out and had twin heifer calves. I didn't think she was due for about another month. The calves looked normal. Whether they were born alive, I don't know. I imagine the ex-

citement of being driven into the enclosure caused the premature births.

What may have been the only buffalo twins born in a zoo arrived at the San Francisco Zoo in 1968; both animals were healthy. As far as I can determine, the only known buffalo twins born in Canada was a pair born to a cow owned by Fred Burton of Claresholm, Alberta. Canadian officials have no records of twins in any of the government herds, but this does not mean there have not been any. Some of the larger Canadian government herds are not watched carefully enough for this kind of check.

It would appear overwhelmingly true that over a century ago, when there were so many buffalo roaming the plains and prairies from Texas northward into Canada, buffalo twins were even more numerous than today. A cow with a pair of calves could easily have been lost among a herd of 5,000 or 10,000 wild buffalo. Also, during those years, most white men on the buffalo range took little interest in the habits and social life of the buffalo. They were interested only in tracking down the animal, getting a good stand, and then getting the meat or hides or both.

The earliest account of buffalo twins on the plains dates back to about 1840, when Dick Wootton, trapper and plainsman, found orphan twin buffalo calves somewhere in what is today southeastern Colorado. He took them to Bent's Fort, where he finally, after much effort, convinced a domestic milk cow to provide nourishment for the little buffalo.[6]

Even today, milk cows are sometimes used to raise orphaned or unwanted buffalo calves. Lee Giles, who ran fifty head of buffalo on his ranch south of Greeley, Kansas, had a buffalo cow that gave birth to twins. She would accept only one of the calves. Giles had to put the other buffalo calf with a domestic cow.[7]

One evening in late fall, after spending most of the day observing a herd of buffalo in Oklahoma, I sat with a buffalo man in front of a crackling fire and we talked buffalo.

"You know," said my friend, "watching buffs so much I sometimes get the feelin' that the cow, not the bull, really wears the pants in the buffalo family."

"Why do you think so?" I asked.

"Well, it's kinda like some human families I know," he replied. "The woman makes the husband feel like a king. She lets him have his fun, lets him go out with the boys and she'll do things to please him, but all the time she's quietly runnin' things, doin' just what she wants. She bears the young, watches out after the young, and makes the important decisions."

Whether a buffalo cow really wears the pants in a buffalo family is a question that naturalists, zoologists, or other animal authorities will have to answer, if they can. Chances are they will conclude it all depends upon which buffalo are studied, and undoubtedly it does. Each animal, like every human being, has a different personality. But, like humans, buffalo do form habits. The habits of buffalo are similar to those of domestic cattle: eating, sleeping, watering, and so forth. C. J. "Buffalo" Jones, however, once observed that buffalo were more clannish than domestic cattle.

"There is no animal in the world more clannish than the buffalo," he said. "Each small group is of the same strain of blood."

Jones also observed, "The resemblance of each individual of a family is very striking, while the difference between families is as apparent to the practiced eye as is the Caucasian from the Mongolian race of people."

This difference is particularly noticeable today in large captive buffalo herds in which different strains exist and comparisons can be made.

But even though a family of buffalo may be close, there comes a time, as it does for human beings, when the young must leave the family group. Jones pointed out that the male buffalo calf "follows the mother until two years old, when he is driven out of the herd, and the parental tie is then entirely broken."

"But the female fares better," he said, "as she is permitted to stay with her mother's family for life, unless by some accident she becomes separated from the group."[8]

Even today, among captive buffalo, much the same thing holds true. Julian Howard, long-time manager of the Wichita Mountains Wildlife Refuge in Oklahoma, and others at the refuge note that buffalo bulls stay pretty close to their mothers until they get to be about two years old. Then they gradually leave.[9]

C. J. Henry, who for many years managed the National Bison

Range in Montana and then retired to a nice home down the road, pointed out that only during the rutting season would an adult bull be seen with a cow and calf. Otherwise you rarely see bulls with cows unless a group of bulls happen to wander through a group of cows and calves. "Except for the breeding season, I don't think you will normally see adult bulls in company with anything except other adult bulls, or occasionally solo," said Henry.[10]

The late Earl Drummond, after many, many years watching and working with the Wichita Mountains buffalo, concluded that a buffalo cow "is not a very regular breeder." Drummond, who spent most of his adult life handling buffalo, said that most cows had a calf every two years, though he did demur:

> But then there was "Texas." We called her "Texas" on account of her long horns, which in the years I knew her raised one heifer calf every year for eight years and a bull calf the ninth year. There was no calf the tenth year, but then she raised a heifer calf the eleventh year. That was the last. She died a short time later.
>
> Cows very seldom breed until they reach two years of age, and they raise their first calf when they are three years old. Buffalo cows continue to grow until they reach the age of six or seven years, and one cow in the Wichita herd raised a calf at the age twenty-eight and lived to the age of thirty-one years. Others have raised calves at twenty-six years, and many of them live to reach thirty years of age.

Bulls, Drummond said, continue to grow to the age of "nine or ten years." He remembered one old bull, "General Lawton," that was the star attraction for many years at the Wichita Mountains Wildlife Refuge. That old bull "lived to be twenty-six years old and sired a calf when he was twenty-three years old," said Drummond.[11]

Dr. Warren D. Thomas of the Henry Doorly Zoo at Omaha, Nebraska had a bull that sired calves out of five different females every year until his death at the age of twenty-three years.[12] But what may well be the record for buffalo belongs to a cow purchased in 1922 by the Little Buffalo Ranch near Gillette, Wyoming. The animal came from the Scotty Philip Ranch in South Dakota and lived to be forty-one years old; when she died in 1963 she had a record of calving "almost every year."[13]

Buffalo, like most animals, can communicate with one another. Such communication is especially noticeable between a cow and her calf, but it is limited. Charles Aubrey, an old buffalo hunter and Montana trader, remembered watching buffalo cross wide Montana rivers in the late 1870s and early 1880s:

> In calling her young, the buffalo mother does not make the bawling cry of the domestic cow, but gives out a peculiar muttering, grunting sound, the calf answering in much the same manner. . . .
>
> If all was quiet, and there was no reason to hurry, the mother in swimming kept her calf on the upstream side. When the calf grew tired, if small, it might climb on its mother's back. She would ferry her young across swift-rushing rivers this way.[14]

More recently, in 1961, Penelope Jane Marjoribanks Edgerton studied the kinship of buffalo cows and calves in Canada. She spent many weeks observing the captive buffalo at Waterton Lakes National Park and the wild buffalo of Wood Buffalo National Park. Afterward, she concluded not only that cows and calves communicated with one another, but also that they could do so over a long distance. "The moo and the bleat are used over long distances while the grunt is used for communication at close range," she said.[15]

Probably the longest running debate among buffalo men concerns the intelligence of the buffalo. A century or more ago when wild buffalo still roamed the plains, some plainsmen considered the buffalo downright stupid. Others thought the animal highly intelligent. Even today there are men who firmly believe either extreme. Most modern buffalo men, however, think the animal's intelligence lies somewhere in the middle. There is little dispute among buffalo men when you ask them whether a buffalo bull or a cow is smarter. They usually answer that the cow is smarter. Then they usually recall an experience or two to substantiate their opinion. And the feeling that buffalo cows are smarter than bulls is nothing new. George Bird Grinnell felt this way about buffalo cows more than seventy-five years ago:

> The cows were much more alert and watchful than the bulls. They were always the first to detect danger and to move away

from it, while the bulls were dull and slow, and often did not start to run until the herd at large was in full flight. Moreover, the cows and younger animals of the herd were much swifter than the bulls, and so pressed constantly to the front, while the bulls brought up the rear. The disposition of the males had nothing to do with any desire to protect the herd, but resulted from the fact that they were slower than the others.[16]

But what Grinnell and others have described or felt as a higher intelligence for buffalo cows may, in fact, be instinct. In the female of the species the instinct for protecting and raising the young is generally stronger than in the male. In most stories citing the cows' superior intelligence, the presence of young buffalo is mentioned.

Most modern buffalo men feel that a buffalo cow takes pretty good care of her young. Elmer Parker remembered only one occasion in his twenty years of handling buffalo when a cow "pushed her calf off." Said Parker:

> For a day or so the calf and cow were close together, but then I began seeing them apart. I didn't pay too much attention to them, but about three weeks later we found the calf dead in Lake Rush. The calf was blind. It had been since birth. There was skin over both eyes. The little feller apparently wandered into the lake and drowned. The mother, after being around the calf for a couple of days, apparently sensed that something was wrong with her little one and just left him alone.[17]

Though there are many accounts by early plainsmen telling how buffalo cows took good care of their young, Colonel Richard Dodge contended that it was the bull, not the cow, that protected the calf. "The cow seems to possess scarcely a trace of maternal instinct and, when frightened, will abandon and run away from her calf without the slightest hesitation," he said. Dodge asserted that it was the bull's "duty" to protect the calves. To back up his opinion, he would tell a story told him by an Army surgeon friend about an experience one evening as he was returning to camp after hunting on the plains of western Kansas, probably sometime in the late 1860s. As the doctor rode across open country, he noticed a group of six to eight buffalo in the distance. As he moved closer, he saw they were all bulls. They were standing in a close circle, their heads outward from the center. Twelve to fifteen yards out from the buffalo were at least a dozen large grey wolves, licking their chops, sitting in a half-circle watching the buffalo. After a while, the buffalo broke up, and, still keeping in a compact group, started to trot off for the main herd about a half-mile away. The doctor then saw the reason for the buffalo's strategy. It was a small calf, so new to the world that it could hardly walk. After the bulls moved about seventy-five yards, the little calf tired and lay down. Immediately, the bulls formed a circle around the calf. The wolves, which had trotted along on each side, sat down and licked their chops some more. Unfortunately for history, the doctor did not continue to watch. It was nearly dusk and there was some distance to travel. But, as he later told Colonel Dodge, he felt certain the bulls carried the little calf to the safety of the herd.[18]

A somewhat similar tale was told many years ago by Doc Barton, said by some to have been the first Texas cattleman to drive a herd of longhorns to Dodge City, Kansas. He credited both bulls and cows with protecting a calf. Barton, who had the reputation for "spinning a windy," recalled seeing a band of big gray wolves try to grab a buffalo calf on the plains west of Dodge City. The youngster let out a loud bawl. In a flash the calf's mother, father, aunts, uncles, and cousins rushed to the rescue. They forced the wolves back. Then the bulls and cows formed a cordon around the little calf. Barton said the wolves sat down around the buffalo and began to howl. Soon their yelps brought reinforcements. In a little while, Barton claimed, there must have been 200 or 300 wolves on the scene, circling the buffalo, which held tight. When a nervous cow advanced to smash a wolf, the pack snagged her flanks or cut the tendons at her heels and dragged her out of the guarded circle. Behind her the ranks would close, and again the front would be impenetrable.

Doc Barton maintained that the buffalo held ground for forty-eight hours, the wolves yelping continuously day and night. Finally, one big wolf, apparently the leader, became impatient. He realized that one old bull was leading the standoff. Yelping defiantly, the huge wolf ventured near the old bull and barked in his face. The bull could see the glitter in the wolf's eye and smell his foul breath. It was too much. With a sudden lurch, the old bull crushed the wolf to the ground, ripped him up with his horns, and pounded him with his feet. It was a superb attack, quick as a flash. But the wolf pack also moved with lightning speed. They leaped on the big bull from both sides, tore at his neck, flanks, and heels. The other buffalo only watched. They made no effort to save him. The strategy the old bull had successfully used failed him now. The remaining buffalo only tightened their circle as the wolves swarmed about feasting on the carcass of the old bull. Some of the wolves even devoured the remains of their fallen leader. Finally, stomachs full, the wolves went slinking off across the prairie. The buffalo had saved the calf, but the cost to the herd had been high.[19]

Although the struggle to survive was a continuing effort for wild buffalo, they did enjoy life to the extent that any animal "enjoys" living. One of the most memorable stories about the lighter side of the wild buffalo's life, particularly that of the cows, was told many years ago by Charlie Norris, a cowboy who worked on the southern plains during the 1880s. One day Charlie rode up on a bunch of buffalo, mostly cows, at a creek. Several of the older cows were jumping off a steep bank into the creek. They would then swim to the bank, climb out at a low place, and repeat the performance. For some time Charlie sat still and watched. He said the animals acted very much like a bunch of young children down at an old swimming hole on a hot summer day.[20]

The same is true today for captive buffalo. They may play even more than their wild ancestors of a century ago, since survival is not so difficult as it once was. Elmer Parker told of a time when some men were baling hay on the Wichita Mountains Wildlife Refuge:

> Some buffalo came up to where the men were workin'. One animal seemed to want to play. And everytime a new bale of hay was tossed to the ground, the buffalo—a cow—would roll and butt it around on the ground, just like it was a ball. Once she must have rolled one bale a good fifty or sixty yards before stoppin'. It's not uncommon to see buffalo—young and old—playin', although the real old ones don't play as much as the younger ones."[21]

C. J. Henry had much the same experience on the National Bison Range in Montana. One day some workers cut a field of hay. They baled it and trucked off all but a few bales. That evening some of the refuge's buffalo discovered the remaining bales. They hooked them with their horns, rolled them around on the ground, and stomped on them. They literally tore the bales apart. One buffalo then hooked a bale with his horns and tossed it into the air. "They sure had a lot of fun," said Henry.[22]

Victor "Babe" May, foreman at the National Bison Range, told of seeing some buffs "playing sort of a king of the mountain game." In the refuge exhibition pasture, the ground is rolling and rises to some knobs and small bluffs. As May watched one day, he saw the buffalo in the exhibition pasture—bulls, cows, and calves—"racing to the top of the knob." Once there, the animals jumped off, but, as May recalled, "they didn't push each other off when they got to the top"; The younger and smaller buffalo jumped off first and let the big ones be kings.[23]

There is a belief among some buffalo men that the weather is responsible for such antics. Elmer Parker asserted that buffalo were particularly active when the weather changed. "Then you'll see them buffs runnin,' buckin,' and playin' for no apparent reason," he said.[24]

The late Earl Drummond agreed, adding, "They're good weather forecasters. They rest before it rains or snows. They all lie down and get a good rest. Then when the moisture comes they're all on their feet. By then they are rested and can and do stay on their feet through the wet weather, whether it's rain or snow."[25]

XIV

Heads and Horns

A buffalo head and horns on a sportsman's wall!! I'd rather see the critter livin', breathin', kickin' and free.
Anonymous 19th century

Remove a pair of horns from the head of a dead buffalo bull and place them alongside a good pair of Texas longhorns. The buffalo's horns look rather puny. They just do not compare with longhorns in size, length, and over-all beauty. But place a good mounted head of a large buffalo bull beside a mounted longhorn's head—even if the longhorns are five or six feet from tip to tip—and the sheer bigness of the buffalo's dark shaggy head puts the monarch of the plains into immediate competition with the longhorn for honors. This may be one reason why so many mounted buffalo heads, instead of just horns, hung in saloons and other joints in the West, and even in the East, during the second half of the 1800s. Of course, the buffalo was considered wild game by the sportsmen of the day. The longhorn was not.

Some of the sportsmen of that era, although not by design, helped to preserve something of the gallant buffalo that once roamed the plains by the millions. One such sportsman was Albert Friedrich. In 1881 Friedrich, a young Texan, went to work as a bartender in the old Southern Hotel on the Main Plaza in San Antonio, a hangout for cowmen whenever they hit town. For about six years Friedrich dispensed drinks and listened to and talked with dry cowboys and cattle barons who dropped by the hotel to clear their throats of the south Texas trail dust. Off duty, Friedrich hunted.

In 1887 Friedrich decided to go into business for himself. He opened a saloon just across Dolorosa Street from the Southern Hotel. He called it the Buckhorn and put the heads and horns of buck deer and other game animals on the walls and ceiling. As Albert Friedrich's business grew, so did his horn collection. It became the subject of many conversations in and out of the saloon. From all parts of Texas, Friedrich's customers and friends began to bring unusual specimens of horns and heads. Sometimes he paid cash for the trophies. Sometimes he paid in trade. His horn collection continued to grow— and it included buffalo horns.

In 1889 a cowboy getting ready to join a trail drive rode up to the Buckhorn with five antelope horns and two sets of buffalo horns tied to a pack horse. The cowboy said he had shot the buffalo on the ranch where he worked in Runnels County near Ballinger. He said that there were large numbers of buffalo in that area and that it was no trick at all to shoot one at any time. It is doubtful there were many wild buffalo left in Texas by 1889, but the fact remains that the cowboy did provide Friedrich two more buffalo horns for his collection.[1]

Today, more than eighty years later, that collection is still preserved and intact. The Buckhorn saloon is no more, but its horn collection is owned by the Lone Star Brewery in San Antonio and is on public display at the brewery. It is indeed impressive.

Wild buffalo heads and horns were also preserved by the Union Pacific, which crossed Nebraska, and by the Kansas Pacific, later part of the Union Pacific system. Both lines crossed the heart of buffalo country and adopted the buffalo head as a symbol of their western rail travel. Almost overnight, buffalo heads appeared on the walls of the larger Union Pacific depots and in most of the nation's large hotels beside "elaborate and beautifully painted signs of the great overland pioneer road."[2] Modern advertising agencies probably could not have found a more appropriate symbol to publicize rail travel.

Though sportsmen and railroads had played significant roles in the wanton slaughter of the buffalo, they also did a little to preserve something of the animal. Unfortunately, few such trophies exist today. Even in the Union Pacific Museum at Omaha, Nebraska, there is only one mounted buffalo head, and it is not one of the original ones from the 19th century.

Most mounted heads of wild buffalo killed between 1860 and the late 1880s have not stood the test of time. Taxidermy then was not the art it is today. Until about 1890, buffalo and other animals, big and little, were literally stuffed with straw, cotton, newspapers, sawdust, anything the taxidermist found handy. Even after new methods

of preservation were developed during the late 1890s, not all trophies were properly mounted. As one observer noted in 1905:

> Between the trained scientific taxidermist and the "animal-
> stuffer" there is a wide and deep gulf into which the latter should
> be thrown. The skin of a beast or of a bird rammed full of any
> sort of rubbish that may be at hand makes a sorry figure of the
> creature at best, but in many instances the "animal-stuffers"
> have afflicted the people with actual atrocities in their manipula-
> tion of buffalo skins, the distorted objects they have produced
> being but little better than caricatures of the fine animals they
> purport to represent.[3]

Joe P. Jonas, Sr., of the Denver taxidermy firm of Jonas Brothers,
confirmed that there were very few buffalo heads still around that had
been mounted in 1890 or before. Jonas knew of one owned by Henry
Zietz of Denver which was mounted in 1890. It hung with other
buffalo, deer, moose, and elk heads in Zietz' Buckhorn Pioneer Lodge
on Osage Street in Denver. Zietz said his father killed the animal
about seventy-five miles east of Denver in 1890. "This is an unusually
large buffalo and is probably the reason my father had the head
mounted," said Zietz.[4]

Another specimen, vintage 1890, was found near Denver at the
Buffalo Bill Museum atop Lookout Mountain. It was a fully stuffed
buffalo kept inside a glass case. Although not in perfect condition,
it had stood the test of time remarkably well when I saw it in 1969.

When new taxidermy methods were being developed in the East
during the middle 1880s, William T. Hornaday, chief taxidermist at
the Smithsonian, read a magazine article describing the near extermination of buffalo in the West. Wondering what buffalo specimens
the museum had, he checked and was shocked to find a sorry array of
only one mounted buffalo, a couple of mounted heads, two skeletons,
and some fragmentary skulls. All were in poor shape. Hornaday immediately informed Spencer F. Baird, secretary of the Smithsonian,
who in turn was alarmed and directed Hornaday to find better specimens as soon as possible while a few wild buffalo still roamed scattered parts of the West. Thus William Hornaday made plans for a
buffalo hunt that would last many months.

Wild buffalo were nearly gone in Montana, but Hornaday heard that a few still existed in remote areas. He led an expedition to Miles City near where the animals had been seen. But there he was told the buffalo had been killed. Discouraged, and about ready to return east, Hornaday happened to meet Henry R. Phillips, a rancher, who told Hornaday there were still a few live buffalo on his Montana cattle range. "One of our men killed a cow on Sand Creek; and about thirty-five head have been seen," reported Phillips. Excited by the news, Hornaday led his party to Phillips' ranch. Two days after setting up camp, they captured a young buffalo bull calf alive; a few days later they killed two old bulls. But the animals had already begun to shed their winter coats; their skins were unfit for mounting.

Hornaday decided to return to Washington and wait until fall, when the buffalo would again have their heavy winter robes, perfect for mounting. But before leaving Montana, Hornaday pleaded with everyone he saw, including cowboys and hunters, not to kill the buffalo. None was killed.

When Hornaday returned the following September, the buffalo were still roaming near the Phillips ranch. Hornaday carried back to Washington twenty-four fresh skins, sixteen skeletons, and fifty-one buffalo skulls. The biggest buffalo taken, an old bull, stood 5 feet 8 inches high at the shoulder. It was fully two inches taller than any of the other bulls killed. The animal weighed about 1,600 pounds and was what old buffalo hunters sometimes referred to as a stub-horn. Part of the outside around his horn tips had either worn away or been broken off.

Back at the Smithsonian, after months of work mounting six of the buffalo, Hornaday unveiled his display in the fall of 1887. The principal figure was the old stub-horn bull.[5] For sixty-nine years Hornaday's buffalo display remained at the Smithsonian. When it was dismantled in 1956, the condition of the buffalo was excellent. However, when Hornaday mounted his buffalo in 1887, he had perhaps not expected them to last. Smithsonian officials learned this when workmen, dismantling the display, discovered a sealed metal box buried in the dirt floor of the huge glass display case. The box contained a copy of an article from *Cosmopolitan* magazine, October 1887. The article, entitled "The Passing of the Buffalo," had been written by Hornaday. Across the top of the first page, in Hornaday's own handwriting, was the following message:

My Illustrious Successor,

Dear Sir:—Enclosed please find a brief and truthful account of the capture of the specimens which compose this group. The Old Bull, the young cow and the yearling calf were killed by yours truly. When I am dust and ashes I beg you to protect these specimens from deterioration and destruction. Of course, they are crude productions in comparison with what you produce, but you must remember that at this time (A. D. 1888, March 7.) the American School of Taxidermy has only just been recognized. Therefore give the devil his due, and revile not.

[signed] W. T. Hornaday.
Chief Taxidermist, U.S. National Museum

Not long after the display was dismantled, the buffalo were shipped back to Montana and presented to the Montana State Historical Society in Helena. There the animals, including the old stub-horn bull, were placed on public display. They had returned to the state where they had been killed three-quarters of a century before.[6]

Hornaday's old bull, the stub-horn, had nine rings on each horn. Some oldtimers maintained that rings could be used as a measurement for age in buffalo bulls. These believers thought that when a buffalo bull reached the age of three, one ring would appear at the base of the horns. Then each year after that another ring would develop. If that is true, Hornaday's stub-horn would have been eleven or twelve years old when killed. But I have yet to meet a buffalo man who thinks you can "always" tell a buffalo's age by this method. Most think it is an old story started by some buffalo hunter back in the 1870s. Most buffalo men say the general condition and appearance of the whole animal are a far better index to a buffalo bull's age than just the horns.

A few oldtimers in Kansas still remember the last years of Old Barney, a buffalo bull belonging to Colonel H. H. Stanton, who for many years ran the Union Pacific Hotel at Topeka. In 1879, Stanton paid $8 for Barney, then an orphaned three-day-old buffalo bull calf at Wallace, Kansas. Stanton took the young animal back to Topeka by train and there raised him on a bottle. Barney was kept in an open field across the street from the hotel. He grew rapidly, and within a

Proud display at the general offices of the Kansas Pacific Railway Co., Kansas City, Missouri, c. 1869. R. Beinecke photo. *Courtesy Fred Mazzulla, Denver.*

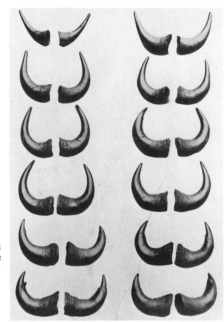

Development of buffalo horns
from 2 years to 20 years of age
(read left to right, top to bottom).

H. S. Poley photo, late 1880s.
*Courtesy Denver Public Library
Western Collection.*

William T. Hornaday's Smithsonian display, 1887-1956. *Courtesy Smithsonian Institution.*

Two Society employees paused to have their picture taken, c. 1920, as they moved Old Barney from the State Capitol building (in background) to the new Society building across the street. *Courtesy Kansas State Historical Society.*

Skull and world-record horns of Old Tex, Yellowstone National Park.

Buffo. *Courtesy Woolaroc Museum, Bartlesville, Oklahoma.*

year he was getting big and becoming something of a problem. He became so unruly that Union Pacific officials strongly suggested to Stanton that Barney be moved. He might hurt someone, they said. Colonel Stanton found a new home for Barney at Bismarck Grove, near Lawrence. It was a 240-acre piece of bottom land covered with cottonwood trees just northeast of Lawrence and only twenty-five miles from Topeka. It was owned by the Union Pacific Railroad. There a corral was built and Barney was moved to his new home. Stanton then bought two buffalo cows for $25 each and moved them to the grove to keep Barney happy. For seven years he was. But about 1886 Barney died from some unknown cause.

Whether Colonel Stanton or the Union Pacific footed the bill is unknown, but Old Barney, as the animal had become known, was stuffed. About 1890 he was placed in a glass showcase in the office of the Kansas State Department of Agriculture at the statehouse in Topeka. There, for many years, Old Barney remained. He was a good conversation piece and a tourist attraction. About 1920, Old Barney, his glass case and all, was moved across the street to the newly completed Kansas State Historical Society building. There, in a fourth-floor museum, Old Barney became even more of an attraction. But gradually time caught up with him, and he began to come apart at the seams. His condition became so bad that during the 1930s he was moved to a storeroom. It was there that Old Barney literally fell to pieces. From what persons at the museum remember, the end came one day during the early 1940s when a trash man was called to cart Old Barney off to the city dump.

Nowhere does there appear to be a record of Old Barney's horn spread, but if the word of oldtimers around Topeka means anything, it was "pretty good from tip to tip." But Old Barney did not have a very long beard. Like many mounted buffalo, he met his fate during the summer months after shedding much of his hair, including that from his beard. Also, bulls would often lose hair from their beards during the battles of the rutting season.

Another reason for short beards on buffalo may be their grazing habits. Edwin Carter, an early Rocky Mountain naturalist from Breckenridge, Colorado, thought that the normal process of grazing kept the beards short since they rubbed along the ground much of the time. Carter said the longest wild buffalo beard he ever saw measured only twelve inches. The buffalo bulls with the longest beards seem to be those kept in close captivity and fed daily in a trough. These animals do not graze like their wild cousins of a century ago or other captive buffalo in large herds. Therefore, they do not wear away much of their beards.

An impressive mounted buffalo, complete from beard to tail, is Old Charlie. He stands today for all to see in the Panhandle-Plains Historical Museum at Canyon, Texas. His story began about 1922 when Charles Goodnight sold two buffalo calves to West Texas State College at Canyon. The male was named Charles, after Mr. Goodnight, and the little heifer Mary Ann, after Goodnight's wife. For many years these buffalo were mascots for the college. One night in the early 1930s the bull, by then known simply as Old Charlie, broke out of his enclosure. Boone McClure of the Panhandle-Plains Museum remembered what happened:

> In those days I was going to school and working as a soda jerk in a downtown Canyon drug store. One night some students ran into the drug store out of breath wanting a drink of water. They said Old Charlie had gotten loose and they were chasing him. Later that night, some of the students came back by the drug store. They told me Old Charlie had died, apparently as a result of injuries he received during that chase.[7]

The year Old Charlie died, the Wyoming State Game Commission was having trouble with an old buffalo bull in its herd at Afton, Wyoming. The winter of 1936, the bull really caused havoc. It killed several elk and a couple of horses in the winter feed yard. When spring came, officials had made up their minds to remove the trouble maker from the herd. The bull was killed, and taxidermist Bob White got the job of mounting the animal. Today, that fully mounted bull stands, as big as life, in Slim Lawrence's Jackson Hole Museum at Jackson, Wyoming, causing little trouble for anyone.

What may be the most unique stuffed buffalo in the world is "Buffo." He greets visitors at Woolaroc, a 3,500-acre wildlife preserve established by Oklahoma oilman Frank Phillips in 1925. It is located fourteen miles southwest of Bartlesville, Oklahoma, on state highway 123. Buffo is mounted in a life-like diorama. Visitors, especially children, can feed the stuffed animal any type of litter including paper, cellophane, and empty soft drink cans and cups. The buffalo acts as a trash container. And he also talks. By recording, Buffo not only thanks visitors for feeding him, but the slow, deep voice coming from his stuffed frame tells "Buffo's Ode to All Buffalo," a capsule history of the buffalo in America. He also sings a rendition of "Home on the Range."

Buffo once was a live member of the Woolaroc buffalo herd. He weighed about 1,500 pounds and stood about five feet high at his hump. His hide was prepared for exhibition by Jonas Brothers, Denver, Colorado, and then Clinton Reser, a New Strawn, Kansas taxidermist, mounted it over a rigid fiber form identical in size to Buffo's original dimensions. Litter placed in Buffo's mouth is carried in a long vacuum pipe through the body and into an adjacent room where the trash drops into large containers. Although the vacuum is maintained at a level sufficient to transport empty beverage cans and other items of similar size and weight, it is not strong enough to lodge a child's hand, for example, in the mouth opening.

Buffalo bull horns are short, very thick at the base, but they taper rapidly to a sharp point on top. In old bulls, the point usually becomes worn on the lower side, and the end is often shortened by wear. Sometimes horns on old bulls become splintered. Earl Drummond once noted:

> The horns of the bulls come out from the head two or three inches then turn straight up and grow eight or ten inches high. When the bulls get to be four or five years old their horns then begin to turn in at the point, but they don't seem to grow much longer. Instead, they turn in at the point until they get in about a quarter circle and are almost always uniform.[8]

A young bull who has either one or both horns growing nearly straight out of his head is called a spike-horn. His horns look like spikes. Russ Greenwood of Sundre, Alberta is the only man I know of who has ever changed the shape of a spike-horn on a live and kicking buffalo. When he was seventy-six years old, Greenwood told me that a few years earlier he had the job of training two buffalo to pull a two-wheeled cart for the Calgary Stampede. One of the buffalo had a perfectly natural left horn, but the right one came straight out from the head like a spike. Greenwood decided to see whether he could reshape it. He built a wooden frame to go around the animal's head like a yoke and he attached springs and weights to the frame and the horn. Time did the rest. Greenwood was able to curve the horn to match the shape of the other. It was a slow process. "It took three months, but it worked," he said.[9]

Charles Goodnight thought the size and shape of an animal's horns were an indication of strength:

> Take your cattle of small horns, take the animal with the crinkly horn, it is likely to be the cull of the herd. Nothing denotes strength in a buffalo or in cattle like a powerful horn. The same is true in the nails of a man or woman. The daintily curved finger-nails of the society lady come from doing nothing. I have seen a blacksmith who could hold between thumb and first finger what with me required a vise. And the result of such exercise is the strong, flat nails of the hand that is powerful and beautiful from work.[10]

If we assume Goodnight's rule of thumb to be correct, then a buffalo bull killed in the middle 1920s by Samuel T. Woodring, chief park ranger at Yellowstone National Park, Wyoming, was mighty strong. In 1902, twenty-one buffalo were purchased for the park. Three of the animals, all bulls, came from Goodnight's private buffalo herd in Texas. The rest came from a private herd owned by Michel Pablo, a well-known rancher in western Montana. Shortly after the new buffalo arrived at Yellowstone, one of Pablo's cows gave birth to a bull calf. Over the years the animal grew very large. The bull became known around Yellowstone as Old Tex.

About 1925 or 1926, the bull became quite dangerous and Ranger Woodring decided to kill him. Old Tex was then almost twenty-five years old. The deed done, the skull and horns were saved and placed on display in the Fishing Bridge Museum at Yellowstone. They remained there for many years. In 1951 Edmund B. Rogers, then Yellowstone's park superintendent, measured the horns and skull for the Boone and Crockett Club, an internationally known sportsmen's organization. He found the horns to be a record. The greatest distance between the horns—from curved part to curved part—was a little more than 35 inches. The base of the animal's right horn was 16 inches around, the left horn 15 inches; from horn tip to horn tip it was 27 inches. It was and still is the world's record.

The skull and horns were kept on display at the Fishing Bridge Museum until the museum was remodeled in the middle 1960s, when they were taken to park headquarters at Mammoth. In June 1969 I found them hanging over the fireplace in the curator's office at the Mammoth Museum. They were a mighty impressive sight.

Down in Oklahoma, Edwin Dummond, a second generation buffalo man, has a pair of buffalo horns that are not as impressive as those of Old Tex, but the story is as colorful. The horns were given Drummond by his father, who had been hired as a ranger not long after the Wichita Mountains Wildlife Refuge was started as a National Forest back in 1907. In that year fifteen buffalo were taken from the Bronx Park Zoo in New York City and shipped by train to Oklahoma to start the Wichita Mountains herd. In all, there were six bulls and nine cows. One of the cows became known as Topsy. Edwin Drummond's father, Earl, used to describe Topsy as the meanest buffalo he ever saw.

In the early days of the herd, the buffalo were kept behind fences and not on an open range as they are today. Earl was responsible for feeding them. But everytime he tossed hay over the fence, Topsy would come up and butt the fence at the spot where Earl was standing. This became annoying, especially after Drummond felt the points of her sharp horns a few times. He then decided to cure the cow, if he could, from butting him through the fence.When feeding time came one day, Drummond tossed some hay over the fence. As usual, Topsy came to the fence. With a pitchfork in his left hand and a handful of hay in his right, Drummond made a motion as if tossing the hay over the fence. As expected, Topsy butted the fence. At that moment, Drummond let go with the pitchfork; not hard, but with enough force to prick her tough hide. The pitchfork did the job. After that, Topsy left Drummond alone at feeding time.

Many years later, Topsy died at the ripe old age of thirty-two. Drummond saved her skull and horns. After Drummond's death in 1968, they were passed on to his son Edwin. As far as I can determine, the skull and horns are the only remains of the original fifteen buffalo used to start the herd in the Wichita Mountains of Oklahoma. Today, that herd numbers more than 1,000 buffalo.[11]

XV

White Buffalo

They probably occur in the proportion of not more than one in millions.

J. A. Allen 1877

Many years ago, long before white men made the plains their home, a Cheyenne war party was traveling north one day across a wide expanse of rolling hills on the northern plains. The sun was warm. There was a gentle breeze blowing up from the south carrying the sweet smell of the tall grasses that covered the black earth for as far as the eye could see. But the Indians' thoughts were not of the grass or the land. They were moving north to do battle with the Crow Indians, their enemy. They were thinking of the victory they felt certain would be theirs.

As the scouts reached the top of a small hill, they suddenly stopped. Ahead were buffalo. Spread out across the wide valley before them were several thousand buffalo. Some were grazing, others lying on the ground enjoying the sun. But what caught the Indians' attention was a white buffalo, a cow, lying peacefully in the middle of the herd. Almost hypnotized by what they saw, the Indians gazed in wonder at the white buffalo. It was the first white one they had ever seen. They had no idea such an animal existed.

The white cow and the other buffalo lying near the center of the herd rose to their feet and slowly began moving across the valley. Other buffalo that had been grazing nearby joined them. But none of the buffalo moved very close to the white cow. They gave her plenty of room as if they had great respect for her.

On that summer day, a long time ago, the Cheyenne Indians came to believe that a white buffalo was something special, a chief among other buffalo. From that day on, so the legend goes, the Cheyenne

Indians worshipped white buffalo; they believed them to be "good medicine."

Though Cheyenne legend does not record whether that first white buffalo was killed, others discovered later by the Cheyenne and other tribes were. When a white buffalo was slaughtered, great care was taken so as not to anger the gods who the Indians knew watched over it.

Cheyenne women, the workers of the tribe, would not dress the robe if someone else was around to do the job. A captive Kiowa or Pawnee—someone not bound by Cheyenne customs—would be called upon to dress the robe. In that way the gods would not be angry at the Cheyenne. Should no outsider be available, however, the medicine man would select one Cheyenne woman for the task, and before she could even touch the dead animal, she would have to go through a ceremony that would protect her and the tribe. The woman would be ritualistically painted to ward off evil spirits, and the medicine man would come forward to offer a prayer. This was to assure everyone, especially the woman, that the spirits would not harm anyone while the robe was being dressed.

The Mandan Indians followed a similar procedure. Some of their medicine men used white robes in ceremonies to cure illness. They also considered a white buffalo to be good medicine, as did the Arapahoe and Pawnee tribes. The Pawnees would often keep white robes as part of their medicine bundles or would wrap the bundles in a white robe.

Some Plains Indians believed that a white buffalo was the property of the Sun God. The Blackfeet hung a white buffalo robe outside in the sunlight near the medicine man's tipi as an offering to the Sun God.

Indian chiefs sometimes carried a white buffalo robe into battle, believing that it would shield them from any harm. One Cheyenne chief, Roman Nose, reportedly was carrying a white robe when he led a charge against Colonel George "Sandy" Forsyth's band of scouts in the battle of the Arickaree near the Kansas-Nebraska border in 1868. But Roman Nose, like others, found the white robe of little protection. He was killed in that battle.[1]

A buffalo's robe did not have to be completely white—a true albino—to be considered powerful in the eyes of most Plains Indians. Any buffalo with unusual coloring was often held in great reverence.

Even buffalo with only small patches of white or cream color were thought to be good medicine.

During the early 1870s, J. A. Allen bought one of these pied buffalo from some hunters at Fort Hays, Kansas. The animal, a cow, had a white face. From the horns to the muzzle, the hair was pure white, but the rest of the body was of normal color. Allen had the head mounted and gave it to the Museum of Comparative Zoology at Harvard, where it remained on display for many years.[2]

There were plenty of pied buffalo on the plains when wild buffalo abounded. A former agent of the American Fur Company reported around 1850 that he had seen hundreds.[3] Edwin James, who recorded Major Stephen H. Long's expedition to the Rockies in 1819, also reported the presence of pied buffalo: "A trader of the Missouri informed us that he had seen a grayish-white bison, and that another, a yearling calf, was distinguished by several white spots on the side, and by a white frontal mark and white fore feet." He also told of seeing in an Indian village a buffalo head with a white star on its forehead. The Indian who owned the head believed that buffalo kept returning to the area "to seek their white-faced companion." The Indian valued the buffalo head highly and would not sell it.[4]

During the early 1800s, few Indians would part with white or partly white buffalo robes or those of light coloring. But as the Indian of the plains had more and more contact with the white man, customs changed. By the 1870s, many Indians thought nothing of selling or trading a white buffalo robe for something from the white man's world. High prices may have had their influence. On the Upper Missouri about 1879, George Bird Grinnell saw a pied buffalo skin that Indians had sold. It was white on the head, legs, and belly, with a wide band of white bordering the normal dark brown coloring— "beautiful," Grinnell said. "If I recollect aright, this particular hide was sold on the river to an Englishman for $500."[5]

On the other hand, of course, robes were sometimes given as gifts out of simple friendship. James W. Schultz tells of such an incident and his own subsequent role. In April 1881 Blackfeet Spotted Eagle and his son presented a pied buffalo skin to their friend and Schultz' fellow trader, Joseph Kipp:

The robe was perfectly tanned, as soft as velvet; and on its flesh side the old man had painted some pictographs of enemies he had killed, enemy horses he had stolen, and encounters with grizzly bears. Said Kipp when we had extended and pinned it to a wall of the trade room: "It is the only pure-white-spotted robe that I have ever seen. And it is worth something. Anyhow a hundred dollars."

Word that we had a white-spotted buffalo robe spread all up and down the river. On a day in August, when Kipp was in Fort Benton, the *Red Cloud,* St. Louis bound, tied up at our landing, and Captain Williams and his passengers came in to see the robe.

"How much?" asked one of them.

"A hundred dollars," I answered.

"There you are," he said, laying two fifty-dollar bills upon the counter.

I learned afterward that he was a Montreal man, and have often wondered if he had been generous enough to give it to some museum. When he had gone with it, I began to feel very uneasy about selling it. Kipp had said that it was worth a hundred dollars, and I had been paid that price for it. Still, perhaps I should not have let it go.

I was right in my misgivings. When Kipp returned, he stopped just within the doorway of the trade room, stared at the bare wall, roared: "Where's that spotted robe?" I meekly answered that I had sold it for a hundred dollars. "Oh, my God," he groaned. "I had a hunch somehow that you were likely to do it, so I hurried back. Why, Charlie Conrad, you know, told me that I ought to get at least five hundred for that robe."[6]

The earliest historical account I can find of a white buffalo is dated October 1754, when Antony Henday, a trader among the Blackfeet, saw a white buffalo skin in a Blackfeet village somewhere in the lower Battle River country of Canada. The skin was used as a seat covering by a Blackfeet chief.[7]

Another early account is dated 1800, when Alexander Henry wrote in his journal that the Crees had seen in the middle of a buffalo herd a calf as white as snow. Four years later, Henry reported that he had bought a white buffalo skin, probably the first time a white man was able to buy a white buffalo robe from Indians. "The hair was long, soft, and perfectly white, resembling a sheep's fleece," wrote Henry. The Indian he bought it from apparently did not value it highly.[8]

Perhaps the first white man to kill a white buffalo was William Craig. He was of that hearty breed of mountain men who crossed the plains to trap and trade in the Rockies. Sometime during the early 1830s on the plains just east of the central Rockies, Craig shot and killed the animal. Many years later, Craig told a Montana newspaper editor that the buffalo was not pure white, but of a light cream color. In Craig's mind, however, it still qualified as a white buffalo. Craig gave the skin to Sir William Drummond Stewart, a captain in the British Army, whom he met while Stewart was on a pleasure and hunting trip in the American West. Stewart reportedly carried the robe back to England with him.[9]

As tales of life on the frontier were carried east and retold during the middle 1800s, stories about white buffalo ranked high with yarns about fighting with Indians, the hardships of living in the West, the vastness of the plains, and prospecting for gold. In 1868 Theodore R. Davis, an eastern writer and illustrator, came west to see what the frontier was really like. And he saw a white buffalo. But, he wrote, "being mounted upon a pony tired from much travel and somewhat long run, I failed to secure a position sufficiently near the the animal to make a sure shot, but a white buffalo it certainly was."[10]

In 1870 a white buffalo was killed on the plains. It became known as the Morgan White Buffalo and may have been the first white buffalo killed in Kansas. One day near the Kansas-Colorado border James Morgan and his brother John were hunting buffalo when they spotted a white one in a large herd and gave chase. James finally shot the animal, but it was John who recognized the potential value of the white oddity. Carefully, John Morgan skinned the buffalo and took the hide to Denver, where it was mounted. For several years the Morgan White Buffalo was exhibited in various towns and cities. In 1875 the Morgans grew tired of showing the trophy and placed it in storage in Kansas City, where it remained for several months until some Kansas businessmen decided that the buffalo would be a good attraction at the 1876 Centennial Exhibition in Philadelphia. They got in touch with John Morgan, who then lived at Strong City, Kansas. Morgan gave them permission to ship the buffalo east.

At the Philadelphia exhibition the white buffalo received much attention. Newspaper articles were written about it, and many of them found their way back to Kansas. Some Kansans then typically decided that if folks in the East thought the white buffalo was something special, it must really be. Thus, when the trophy was returned to Kansas after the exhibition, Morgan agreed to let officials display it at the statehouse in Topeka, where it was placed in a glass case near Old Barney. A Kansas lawmaker, Hill P. Wilson, impressed by the trophy and the publicity it had received, felt the state should buy it. Wilson introduced a bill in the Kansas Legislature, but when the measure reached the Ways and Means Committee, members decided to offer only $50 for the white buffalo.

Morgan laughed. "No," he asserted, "I wouldn't even consider thinking about selling the buffalo for that. Why, P. T. Barnum says he wants it for his museum. He's already offered $1,500 for the specimen." Whether Barnum had actually made an offer or Morgan was only trying to get a better offer is unknown. The fact is that the trophy remained in the state agricultural office at Topeka for several years, during which time Morgan's health failed and he died. His brother James had died earlier.

The Morgan White Buffalo became just another object gathering dust and taking up room in the state agricultural office. Few people paid any attention to it. Thus no one objected in 1898 when someone confiscated the glass case that had protected the trophy. The Morgan White Buffalo was left in the open on top of the glass case containing Old Barney. When a Topeka newspaper reporter happened by the office and saw the animal in its decaying condition, he wrote an article condemning the negligence of authorities in allowing the relic to be destroyed. "Mice and moth lunch off the hide continually," the reporter noted. A few days later, apparently as a result of the newspaper story, a woman walked into the office and claimed the white buffalo. She said her husband had killed the animal many years before. Officials, taking the woman at her word, were glad to let her cart off the old buffalo—mice, moths, and all. It had become an eyesore.

Officials forgot about the trophy until early in 1903 when a daughter of James Morgan visited Topeka and asked to see the white buffalo her father had killed. The people at the state agricultural office tried to explain. When Morgan's daughter asked for the name of the woman who had taken the trophy, no one had it. No one had written it down. The daughter threatened an investigation, but the white buffalo was gone. So far as I can determine, it has never been found.[11]

Another white buffalo was killed in Kansas in October 1871. James Caspion and another hunter, Sam Tillman, started out on horseback looking for buffalo in far western Kansas. A third man followed in a wagon to pick up the kill. To cover more ground, Caspion and Tillman separated but kept in sight of each other over the rolling plains. In the late afternoon Caspion reached the top of a long ridge and saw some buffalo grazing peacefully in a wide valley, perhaps twenty-five miles across. As his eyes scanned the herd, they suddenly stopped focused on one spot. On the outside of the herd was a milk white buffalo feeding on the tall grass about a mile from Caspion, its whiteness contrasting vividly with the dun tints of hundreds, perhaps thousands, of buffalo nearby.

Caspion signaled Tillman to join him, dismounted, and moved to the crest of the ridge to see the best way to move up on the white buffalo. As he reached the ridge, he glanced back to see where Tillman was. At that moment he forgot about the white buffalo. Tillman was riding for his life in the opposite direction, a party of about fifty Cheyenne warriors after him. The Indians' chase was a short one. As Caspion watched, helpless to do anything, the Indians shot Tillman's horse, then closed in on the white man. Moments later, with Tillman's scalp borne aloft on a lance, the Indians turned and started for Caspion, who was about a mile away. Caspion jumped atop his horse and quickly scanned the countryside. He could not ride to the right or left without giving the Indians the advantage of being able to cut him off. So he headed his horse straight for the buffalo herd.

As Caspion bore down upon the buffalo, they stampeded. In a few moments the rider and horse were swallowed up by the mass of buffalo, and they were in the rush sweeping across the prairie. The last thing Caspion saw over his shoulder, before the dust shut everything from view, was the party of Cheyenne warriors coming over the crest of the hill he had just vacated. Crowded, jostled, holding on for dear life, Caspion gave the horse his head. The animal held his own among the pushing and shoving buffalo. The dust was so thick Caspion could see nothing. He held his head low. The dust was choking. By nightfall Caspion could only tell by the feel of the ground that the buffalo had moved out of the valley and were now in rough and hilly country but were at least twenty-five miles from where the stampede had started.

Suddenly, the leading buffalo, unable to scale some steep bluffs, divided. Most turned to the left. Some followed a valley, which went to the right. Caspion's horse was forced to turn with the buffalo heading into the valley. After a narrow opening, the valley widened and the buffalo began to thin out. The dust was not so thick. Caspion realized it was getting dark. A bright moon was shining. It made the ride seem even more unreal.

Then Caspion saw the white buffalo ahead of him, running with other buffalo about a hundred yards away. They came to a place where a deep ravine, worn by water, cut close against the side of a bluff. The buffalo nearest the bluff kept their footing, but those near the edge were crowded off into space. As his horse passed the spot safely, Caspion could hear the dull thuds as the buffalo hit the floor of the ravine twenty feet below. The painful bellowing of buffalo in the ravine could be heard above the thundering hoofs of the lucky buffalo that had passed the spot safely. When the valley widened again, Caspion slowly reined to a stop outside the dwindling flow of buffalo. The horse was tired, almost ruined. Caspion dismounted, tied his horse, and lay down. Exhausted, he fell asleep, paying little attention to the fading roar of buffalo.

As the faint rays of morning light crept across the valley floor, Caspion awoke. Only a handful of buffalo remained in the valley. They were a mile or more away, grazing peacefully. Caspion untied his horse, mounted, and slowly walked the horse back over the trail he had been forced to follow the night before. When he came to the ravine, he saw scores of buffalo lying at the bottom. Many were dead. Others were too badly hurt to move. Among them was the white buffalo. The animal, a young bull, was leaning against a bank. It had landed on other buffalo when it fell and had broken its leg. Wasting no time, Caspion climbed into the ravine, shot the white buffalo, and skined him on the spot. For five years Caspion kept that white robe, believing like the Indians that its possession would bring him good fortune. For a time it did. But about 1876 Caspion sold the robe for $100 while on a spree one night at Fort Lyon. Shortly afterward, Caspion was killed by Comanches in New Mexico.[12]

Not all the white buffalo reported on the plains were actually white —or even buffalo. In 1876, for example, there were stories of a white buffalo running with a large herd of buffalo along the Bow River in Canada. When a prairie fire stampeded the herd, many of the animals, including the "white buffalo," rushed blindly over a cutbank and plunged to their death on the river's rocky banks. The "white buffalo" turned out to be a white Texas longhorn that had taken up with the herd.[13]

Another "white buffalo" yarn concerned a settler from the East who had built a sod house on the northwestern Kansas prairie and started farming. Like many other settlers, he largely depended for his survival on what the land produced, especially wild game. One day, when his meat supply got low, he decided to go get a buffalo. He took his gun and horse and soon found a small herd grazing near a stream. On the edge of the herd was a buffalo at least half of whose body was white. The settler gave chase. The buffalo took off into the wind, the settler following, determined to "get the white one" if he had to chase the animal all the way to Texas. He got off a good shot and the "white buffalo" fell dead on the prairie. The settler smiled, but as he neared the fallen animal the smile changed into a frown. If ever a Kansas settler was disappointed, this settler was. The dead animal was a buffalo all right, but, as the settler recalled years later, "You know that ornery bufferlo had been over by them chalk cliffs, and all those white markings came from rubbing against them blamed cliffs." Nevertheless, the settler cherished his not-so-white buffalo, and as the story goes, he had the animal's head mounted and kept it for many years under his bed—one of those beds built high off the floor—along with his high silk hat.[14]

Aside from chalk that might shade a brown buffalo white, dried mud sometimes had the same effect. When a buffalo rolled in a muddy wallow in an area of light-colored earth, the clay that stuck to the animal's robe would bake in the sun to be "very nearly white."[15]

Then there were the "white buffalo" that were not all buffalo. In the spring of 1886, C. J. "Buffalo" Jones was returning to his Kansas ranch from Texas. At a point between the two Canadian rivers in West Texas, Jones and his party came upon what appeared to be a mixed herd of buffalo and domestic cattle. Wanting a closer look, Jones followed a nearby ravine until he was almost within touching range. One of the herd was a milk cow, all white, but there were three white buffalo: a three-year-old, a two-year-old, and a yearling. They were not as white as the driven snow, recalled Jones, but they were white enough to be called "white buffalo."

Although his horse was used up, Jones was determined to capture at least one of the white buffalo. He returned to his horse and headed straight for the herd. The white cow spotted him, rolled her long tail up into the air, and led the buffalo south at a fast gait. Jones' horse could not keep up the pace and finally gave out. Years later Jones concluded that those "white buffalo" were actually prototypes of "cattalo," or crosses between buffalo and domestic cattle.[16]

David Morrow might be considered typical of many easterners who came to the Kansas buffalo range about 1870. He was single and somewhat rebellious, and there was a streak of adventure in him. Like many others, he had joined the western migration in hopes of finding himself and making a fortune. But unlike most, Morrow was from a prominent family. He was an uncle of Dwight Morrow, later U.S. Ambassador to Mexico and France. When David reached Kansas, he became a buffalo hunter. Between hunts he did odd jobs and loafed around Hays City enjoying everything that colorful trade center for buffalo hunters and soldiers had to offer. When things were quiet, Morrow used to spin "windies"—tall tales—either to impress or to scare newcomers.

Morrow had trained a couple of prairie dogs he had caught on the edge of town. There were plenty of them around Hays City. He had them in his pockets one day when a train pulled in from the East. Morrow walked along the station's platform talking with passengers, getting the latest news from the East, and spinning his tales about how lovable his pets were. To his surprise, a passenger offered Morrow $5 for the little critters. Morrow gladly parted with them and that day entered the prairie dog business.

To replenish his supply, Morrow hauled some barrels of water out of Hays City to a nearby prairie dog town and poured the water down the animals' holes. As the little dogs floated to the surface, he picked them up. But the method was slow. So Morrow built a better prairie dog trap. The trap consisted of a barrel full of sand, open at one end. Morrow would place the barrel over a prairie dog's hole, open end down. As the sand flowed into the hole, prairie dogs would surface through the sand into the barrel and find themselves trapped. The sand meantime would fill the hole and cut off return.

Morrow set many of these traps, and at his convenience returned with a sack and picked up the merchandise. Within a short time his prairie dog business was booming. But as with any good thing, competition was just around the corner. Other young men entered the field, and soon the prairie dog market at Hays City was glutted. The going price for prairie dogs dropped from $5 a pair to $1, and then to 50 cents, and finally a quarter. At that price, trapping prairie dogs was hardly worth the effort. Morrow decided to return to hunting bigger game—buffalo—but by then the nickname Prairie Dog Dave had become permanently attached to him.

As Morrow began hunting buffalo again late in 1872, he heard other buffalo hunters talking about a white buffalo supposedly seen on the plains southwest of Hays City toward Dodge City. Most people, including many old buffalo hunters, had some doubts that such an animal really existed, but reports persisted. Prairie Dog Dave decided to hunt the white buffalo. He bought a gallon of formaldehyde, took his gun, some ammunition and supplies, and set out in his wagon to search for the animal. Just how long he was out on the plains or where he went, no one is sure. But around January 1, 1873, Dave Morrow drove his wagon into Dodge City, pulled up in front of Robert Wright's general store, and tied his team to the hitching post. He was smiling. In the back of the wagon was the carcass of the white buffalo. Wright paid Prairie Dog Dave $1,000 cash on the spot for the animal's remains and had the white robe and head mounted and shipped to Kansas City, where it was put on display. Later it was shown in Topeka at the statehouse. It finally ended up in the Hubbell Museum at New York City, where it was destroyed by fire a few years later.[17]

In October 1876, almost four years after Dave Morrow killed his white buffalo near Dodge City, another white one was killed on the plains of West Texas. J. Wright Mooar was camped near where Snyder, Texas stands today. He had spent the day scouting the countryside for buffalo. It was almost sunset, and he was about to head back to camp when he saw a small herd. The setting sun flashed on a white object in the midst of the herd. Mooar instantly know what it was—a white buffalo.

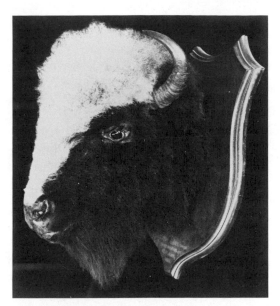

Pied buffalo cow, killed near Fort Hays, Kansas, 1870, given by J. A. Allen to the Museum of Comparative Zoology at Harvard University. *Courtesy Museum.*

"Prairie Dog" Dave Morrow, Dodge City, 13 years after he killed his white buffalo. *Courtesy Kansas State Historical Society.*

J. Wright Mooar (right) with his white buffalo hide. *Courtesy J. Wright Mooar Collection.*

Statue and historical marker commemorating J. Wright Mooar, "champion hunter of buffalo," and his killing of a white buffalo. Courthouse Square, Snyder, Texas. *Courtesy Snyder Chamber of Commerce.*

Big Medicine, the white buffalo at National Bison Range, Montana. *Courtesy Montana Historical Society.*

Big Medicine. E. P. Haddon photo. *Courtesy Bureau of Sport Fisheries and Wildlife, Washington, D.C.*

Big Medicine, mounted, Montana Historical Society museum. *Courtesy Society.*

Pure albino bull, son of Big Medicine, National Zoo, Washington, D.C. *Courtesy Kansas State Historical Society.*

Mooar made a wide circle around the herd so as not to disturb the animals, and then he headed as fast as he could for camp, about a mile away. At camp he stayed only long enough to get Dan Dowd, another hunter. Then they both headed back to the buffalo herd. The two men tied their horses some distance from the buffalo and followed a creek bed on foot until they reached the prairie near the animals. From there they crawled through the tall grass until they were close to the buffalo. Mooar pointed ahead. There, directly to the front, was the white buffalo, a cow, about four years old.

"Take a look," said Mooar in a whisper. "There is the gamiest animal on earth."

Dowd said nothing, but just stared at the white animal.

Mooar, putting his rifle to his shoulder, was ready for action. He aimed and fired. The white cow fell. The rest of the herd stampeded, not away from the hunters and the gun sound, but toward them. As Mooar recalled many years later, he had to shoot three bulls to keep from being run over, but he had his white buffalo, his first.

He had seen another almost four years earlier in Kansas and had tried to kill it. "I took four shots at her and hit her every time, but I didn't kill her. I was six or seven hundred yards away, and it was after sundown. The next day another man run her down and killed her."[18] The white buffalo Mooar had failed to kill was on Three Mile Ridge, west of Dodge City, Kansas, the same general area where Prairie Dog Dave killed his white buffalo. Looking back and comparing the stories, one could not unreasonably conclude that the white buffalo Mooar shot at was the very animal killed by Dave Morrow.

The most famous white buffalo of all time—the one that received the most publicity—was born in captivity at the National Bison Range in western Montana. The year was 1933, the month May. Cy Young, then refuge manager, was checking the pastures, as he did almost every day, when he discovered that a cow had given birth to a white buffalo calf. The little fellow was only a few hours old when Young found him near his mother. Young checked the animal over as best he could from a distance and found him normal in every way except in his coloring. The herd, including the calf's mother, apparently had accepted the bright-looking little buffalo even though he stood out like a snow-covered mountain top in July.

The calf grew large and strong. The buffalo men at the refuge soon discovered that he had blue eyes. As he grew larger, they noticed a woolly knot of brown hair appearing between his horns. It was the only "normal" coloring on the animal's whole body. By then the refuge people were calling the buffalo Whitey. By the time Whitey was two years old he was becoming well known. He was a tourist attraction at the National Bison Range, and he also had become known as Big Medicine.

In his second year he was moved to an enclosure with some suitable females in hopes of extending the albino strain. As Cy Young told me, "Theoretically when calves would be born to his granddaughters, sired by himself, some of them would be albino." In 1936 refuge workers were delighted when three calves were born to Whitey's harem, two of them heifers. The plan then called for the two heifers to be bred to the white bull. The breeding experiment, however, was discontinued on orders from Washington. Cy Young said that the experiment was discontinued supposedly because "We are not in the business of raising freaks."[19]

Washington, however, could not control nature, which took its course. In May 1937 the white bull's mother, bred by Big Medicine, her own son, gave birth to a pure albino calf. The calf was completely white, a true albino with white hoofs and pink eyes, but the little calf was totally blind at birth. In spite of the handicap, the blind white calf grew and became a strong animal. When he reached the age of six months, he was shipped to the National Zoo in Washington, D.C., for public display. Carefully watched and cared for, he remained there for twelve years, until 1949, when some baling wire got into his hay. He ate the wire and died of indigestion.[20]

Back in Montana the calf's father, Big Medicine, had developed into an extraordinarily fine animal. Allowed to roam the range with the refuge herd much of the year, Big Medicine spent his summers with the small herd in the exhibition pasture, where thousands of tourists viewed him. In 1943 the exhibition pasture was reseeded, so Big Medicine was left to roam the range during the rutting season. That summer Big Medicine was one of the main contenders for herd leadership. Refuge workers, who by then were calling him Old Whitey, saw the animal in the thick of battle with other bulls on several occasions. It was impossible for him to come through such battles unscathed, and by August of that year Big Medicine had a number of fighting scars and wounds, the most serious a swelling over his right eye, but it resulted in no permanent damage to his eyesight.

The following summer Big Medicine was returned to the exhibition pasture, where he was lord of the herd. He jealously protected his harem to the point that no one was safe near him. In 1947 his supremacy came to an end. A nine-year-old bull named Yellowstone, a new arrival in the exhibition herd, soon took over. Big Medicine was forced to accept a second-rate position and remained aloof from the cows during the rutting season. Refuge officials then decided to restrict Big Medicine to the exhibition pasture all year. It would protect him more from the perils of winter. Big Medicine had become almost a living legend and was quite a tourist attraction.

Although he was gradually declining in vigor and alertness, Big Medicine was still an impressive sight in the exhibition pasture—and he proved he still had the stuff buffalo are made of. At the age of twenty, during the summer of 1953, Big Medicine regained leadership of the exhibition herd by vanquishing a five-year-old bull. During the next few years, Big Medicine's age began to show. He preferred to remain solitary most of the time. No longer did he demonstrate any interest in the cows. His teeth became so worn that feeding him became a problem. He had to be put on a special diet of steamed barley soaked in molasses, special protein pellets, and tender third cuttings of alfalfa to pamper his aging teeth. For a time, things looked better, but by January 1959, Old Whitey's eyes began to look unhealthy. A veterinarian diagnosed that cancerous growths resulting from old age were developing and nothing could be done about them. Another veterinarian confirmed the diagnosis. Big Medicine's life was near an end.

As usual, thousands of visitors saw Big Medicine that summer of 1959 at the National Bison Range. But he paid little, if any, attention to them. He was growing weaker and thinner. Finally, on August 25, Big Medicine was found lying in the pasture near the refuge headquarters, unable to rise. A few hours later he was gone. Big Medicine had lived a life of twenty-six years and three months. During his prime he had approached 1,900 pounds in weight, but when he died he had shriveled to about 1,200 pounds.[21]

Bob Scriver of Browning, Montana undertook the task of mounting the remains of Big Medicine for posterity. It took Scriver about two years. Although the old bull had died in August when his coat was not in its prime, the patient work of Scriver produced a most dominating figure of the white buffalo. In July 1961 Montana Governor Donald G. Nutter dedicated the still imposing figure of Big Medicine in ceremonies at the Montana State Historical Society Museum in Helena. Today, Big Medicine remains a popular attraction, viewed by thousands of people annually.[22]

For many persons who had seen or known of him, Big Medicine's death meant the end of an era when white buffalo were more than just legend. Old Whitey had been living proof that white buffalo did exist, that they were not just make-believe, fairy-tale animals. For many oldtimers, that white buffalo had been the living symbol of the good old days when the West was still wild, the grass free, and civilization only a word in a dictionary.

Though Big Medicine was by far the best known white buffalo to have ever lived, there have been others since. In 1928 the U.S. Government took twenty buffalo from the National Bison Range in Montana and shipped them north. Washington had decided to try to establish a herd in the Territory of Alaska. The buffalo undoubtedly carried with them the inherited albino trait, the same trait that appeared in Big Medicine, because several white or partly white buffalo were seen in the Alaskan herd established at Delta Junction.

Records in the files of the Alaska Department of Fish and Game show that one white buffalo was born in 1938, another in 1939. They were seen together on several occasions, but by 1941 both animals had disappeared. In 1949 another white buffalo was seen in the Alaskan herd but he was struck and killed by a truck shortly thereafter. In 1958 two more were seen and in 1961 three more. All were completely white except for one which had a brown patch on the top of his head like Big Medicine. But by the end of summer in 1961, none of the calves could be found.

The Alaskan buffalo ran wild, unlike those at the National Bison Range in Montana and in other herds owned by the U.S. Government. Thus, it was hard for Alaskan Fish and Game people to keep track of the buffalo. Also, because white buffalo are usually not so healthy as normal buffalo, Alaskan wildlife specialists have concluded that their white buffalo either have died from the elements or were killed by wolves or bears. None of the white buffalo born in Alaska is known to have survived to reach the age of three years in the wild and virgin country across which they roamed.[23] The twelfth white buffalo known to have been born in Alaska was spotted near Chitina in the fall of 1973. Government officials set out to capture it, with the intention of giving it to the Children's Zoo in Anchorage, where its survival chances would be greater than in the wilderness; but their capture attempt was unsuccessful. Private parties took up the search, but the animal eluded them too.[24]

It should be noted that the non-white buffalo thrived in Alaska. They even bothered farmers and motorists—and flyers. By 1940 farmers in the Big Delta area complained that the animals were a hazard to their crops. And in 1941 bush pilot Vern Bookwalter was ferrying a Loening aircraft up to Fairbanks. When he came in to land at Delta Junction, he was dismayed to see buffalo in possession of the runway. Some of the animals wandered away, but two of them stood their ground for awhile. Finally, after Bookwalter circled and buzzed them a few times, these two drifted far enough apart so that he could land.

"I was glad," Vern said, "that the one people called Big Joe was not there. He had a reputation of wrecking cars along the Richardson Trail, and he might have got the notion that he didn't like planes either."[25]

XVI

Those Who Saved the Buffalo

It would be a real misfortune to permit the species to become extinct.

Theodore Roosevelt 1907

Since the Ice Age, the greatest natural disturbance to strike the North American continent was caused by the white man. It began when he first set foot on these shores, and it continues today. Within three centuries this man has reduced the rich wildlife resources of this continent to a fraction of their former abundance. In some instances entire species have been exterminated. Others, like the buffalo, were slaughtered until they were nearly gone. Yet, it was this man, with help from the Indian, who saved the buffalo. A handful of men, working independently of each other, hung up their rifles to preserve the buffalo. Though some of these men had profit in mind, most also wanted to preserve the animal. Many were branded as dreamers, impractical men, by the "practical" materialists. Some were described as warped sentimentalists living in the past. In truth, some were dreamers, some were not practical, and some were living in the glories of the past. But all had one thing in common. Each was an individualist. Each lived by his own thoughts and convictions and generally rejected the increasing conformities of society.

According to oldtimers' stories, one of these men was a Pend d'Oreille Indian named Samuel Walking Coyote. He lived with the Flathead Indians on their reservation in what is today western Montana. Although some writers have discredited the story of Samuel Walking Coyote, it persists. What follows is the version I like best.

In the summer of 1872, Walking Coyote left his Flathead wife and his adopted home and traveled east across the Rockies to the Blackfeet reservation, where he spent the winter of 1872-73 hunting buf-

222

falo with the Blackfeet Indians along the headwaters of the Milk River. Walking Coyote fell in love with a beautiful Blackfeet woman and married her. For a while, Walking Coyote thought very little about his other wife back on the Flathead reservation, but when spring came he began to think of her and his adopted home. Walking Coyote decided to take his new wife and return to his old home.

As he made plans for the journey, Walking Coyote began to worry. It was against Flathead law to marry out of his adopted tribe, as he had done. He also knew that the Jesuit Fathers at St. Ignatius Mission on the reservation did not approve of more than one wife for any man. It was not until eight orphaned buffalo calves wandered into his hunting camp several days later that Walking Coyote thought of a possible solution to his problem. The little calves—some motherless from the hunt, others separated from their mothers in the chase—attached themselves to the Indians' horses. Walking Coyote thought the little buffalo might make an excellent peace offering to the Jesuit Fathers and to the Flatheads. Perhaps, thought Walking Coyote, the gift of the buffalo would smooth over any ruffled feelings.

With his eight buffalo calves and his Blackfeet wife, Walking Coyote began the long journey toward the Flathead reservation in the spring of 1873. It was rough going through the Rockies. Two of the calves died, but six survived the trip. However, Walking Coyote's homecoming was anything but happy. When he and his wife and the six calves reached the Flathead Mission, the Indian police reportedly met them and gave the couple a thrashing they never forgot. Then the police ejected Walking Coyote from the tribe. This was the punishment for having two wives. This so angered Walking Coyote that he decided to keep the buffalo calves for himself. He was not about to give the Jesuit Fathers anything, especially the buffalo calves, after what they had directed the Indian police to do. Walking Coyote, with his Blackfeet wife, took the calves and left the Mission, but they did not go far. The Flathead Valley was his adopted home. He wanted to stay. So Walking Coyote and his wife made their home in the valley and raised the buffalo calves on its good grass.

According to stories by oldtimers around the reservation, the buffalo became unusually tame. When the heifers reached four years of age in 1877, each gave birth to a calf. Others were born in the years that followed. By 1884, Walking Coyote had thirteen buffalo and they were becoming a problem. They were hard to control. Walking Coyote had come to love the buffalo, but they taxed his meager resources and at times they annoyed the neighbors. He reluctantly decided to put them up for sale.

D. McDonald, a trader on the nearby reservation, began to negotiate with Walking Coyote. Meantime, Charles A. Allard, a successful cattle rancher on the Flathead, heard that the buffalo were for sale. For some time, he had considered raising buffalo. He talked with another rancher, a boyhood friend, Michel Pablo, and convinced him that they should buy Walking Coyote's herd. Before McDonald could act, Allard and Pablo made the deal with Walking Coyote. Walking Coyote refused to accept a check and insisted on cash. Ranchers Allard and Pablo got the cash and met Walking Coyote beside a stream for the payoff. They counted out $2,000 in gold, Walking Coyote's price for the thirteen buffalo. The animals changed hands.

Walking Coyote took the money and went south to Missoula, where he went on a colossal binge. Soon thereafter, he was found dead under a bridge, probably the Higgins Bridge, at Missoula. Walking Coyote apparently died from natural causes. He was broke.

In spite of the tragic ending to Walking Coyote's life, history has not forgotten his contribution. His buffalo calves became the nucleus of a herd whose descendants today populate many private and government buffalo herds throughout North America. But Walking Coyote was not solely responsible for saving the buffalo from extinction. No one man was.

Under the care of Allard and Pablo, Walking Coyote's buffalo prospered, and the herd increased. In 1893 the two ranchers bought twenty-six more buffalo and eighteen cattalo from C. J. "Buffalo" Jones in Kansas. Pablo and Allard thought there might be some profit in raising buffalo, and both were very optimistic.

In 1895 Allard suffered a knee injury. It did not heal and became so bad that a doctor in Montana sent Allard to Chicago in hopes that doctors there could help him. They operated on Allard and sent him home. Within weeks, Charles Allard was dead at forty-three.

When Allard died, the buffalo herd he and Pablo owned numbered about 300. These were equally divided between Pablo and Allard's estate. Mrs. Allard sold her share almost immediately to Charles

Konrad of Kalispell, Montana. The animals that went to Allard's daughters and son Charles were sold to Howard Eaton. Eaton later sold fifteen of the buffalo to the federal government, which placed them in Yellowstone National Park. The remaining animals, owned by Allard's son Joseph, were sold to Judge Woodrow of Missoula, Montana. These animals eventually were sold to the Miller Brothers' 101 Ranch in Oklahoma.

Michel Pablo did not sell his buffalo until 1906, when he learned that the Flathead reservation was about to be opened to home-steaders. Since his buffalo grazed that land, Pablo asked the federal government for other land. He heard nothing from Washington. After much waiting, Pablo offered to sell his buffalo to the government. President Theodore Roosevelt and Secretary of the Interior E. A. Hitchcock favored buying the buffalo and asked Congress to appro-priate the money. Congress refused. Disgusted, Pablo approached the Canadian government for grazing land. Instead of offering land, they offered to buy his entire herd. He agreed. The final negotiated price was $200 a buffalo. In the fall of 1906 Michel Pablo signed with Canadian officials a contract that would ultimately remove the buffalo from the United States.

The contract called for Pablo to deliver the buffalo to the rail-head at Ravalli, Montana. Getting them there was no easy matter. The twenty or more cowboys he hired to move the herd had their hands full. The buffalo were on the range in the Little Bitterroot Hills and along the Flathead River thirty to forty miles from Ravalli. The cowboys soon found they could handle only a few head at any one time. Even then the animals would scatter to the four winds fre-quently and delay the operation even more.

It took Pablo six years, not the anticipated two years, to move all the buffalo to the railroad. After two years of trying to herd them overland, which was not always successful, Pablo's men began to crate the buffalo and carry them by wagon to Ravalli. From there the animals were shipped by train to Elk Island Park, near Wainwright, Alberta. The last shipment of seven buffalo was made in June 1912. The final count on delivery was more than 700. (1)

In the early 1870s, as Walking Coyote was taking his buffalo calves over the mountains to the Flathead reservation, other Indians far to the north in Canada were capturing five buffalo calves somewhere west of Winnipeg. Returning east they passed through Winnipeg, where James McKay, better known to the Indians as "Tonka Jim," bought the calves. McKay, who at one point was provincial secretary of Manitoba, let the calves run about on the outskirts of Winnipeg until about 1877, when he sold five of them to Colonel Sam I. Bedson for $1,000. Bedson took the animals to his home at Stony Mountain, near Winnipeg.

Under Bedson's watchful eyes the buffalo thrived. By 1888, when he had nearly eighty full-breed buffalo and thirteen half-breeds, or cattalo, Bedson decided to sell his buffalo. Sir Donald A. Smith, later Lord Strathcona, bought a few of the animals which he gave to the Canadian government. Eighty-six other buffalo, including the cattalo, were sold to C. J. "Buffalo" Jones in Kansas for $50,000.[2]

It was a cold and snowy December day in 1888 when Jones prepared to load his buffalo aboard a train at Stony Mountain and head south for Kansas. He cut thirty-three buffalo from the eighty-six he had bought and started driving them to the railroad. But several old bulls broke and ran, returning to the main herd. When Jones finally reached the railroad stock pens, he had only a few buffalo. And his trouble was not over. The buffalo began to fight among themselves. However, he succeeded in getting them loaded and started south, promising to return to get the others in the spring. Near St. Paul, Minnesota, three of the mixed-blood buffalo were killed by others in the close confines of the railroad cars. When Jones and his buffalo reached Kansas City, as if he had not had enough trouble, thirteen buffalo broke loose as they were being unloaded. They stampeded through Kansas City and caused quite an uproar. After several hours of freedom, the animals were recaptured near the junction of the Missouri and Kansas Rivers, where they had taken refuge.

Although his troubles might seem to indicate otherwise, handling buffalo was not a new experience for Jones. He already had a rather large buffalo herd on his ranch near Garden City, Kansas. Between 1886 and 1889, Jones had captured a few wild buffalo calves among remnants of the wild buffalo that roamed the Texas Panhandle and northeastern New Mexico. With these calves and a few grown buffalo purchased from ranchers in Kansas and Nebraska, Jones had a herd of about fifty buffalo. By the time he completed the job of trans-

Charles Russell was in on the Pablo herd roundup in October 1908. In a letter to
Philip R. Goodwin he described the frustration of trying to catch the buffalo, and
he illustrated one sentence of the letter for Goodwin:

The first day they got 300 in the whings [wing corral] but they broke
back an all the riders on earth couldent hold them   they onely got in
with about 120
    I wish you could have seen them take the river   they hit the water on
a ded run . . . .
    we all went to bed that night sadisfide with a 120 in the trap but woke
up with one cow   the rest had climed the cliff an got away
    the next day they onely got 6 an a snow storm struck us an the round-
up was called off till next summer [*Good Medicine*, 138-39; date for
Russell's roundup participation furnished by George D. Coder, Lake-
wood, Ohio]

N. A. Forsyth recorded on film the roundup of the Michel Pablo herd for shipment to Canada. This 1908 shot illustrates the trouble the cowboys had handling the buffalo. *Courtesy Montana Historical Society.*

Trailing Pablo's buffalo to Ravalli, the shipping point, proved too difficult; specially built wagons were the solution. Forsyth photo, 1908. *Courtesy Montana Historical Society.*

Buffalo bull emerges
from wagon into
stockpens at
Ravalli, Montana.
Forsyth photo.
*Courtesy Montana
Historical Society.*

Spectators came from miles around to watch the Pablo buffalo being loaded into
railroad cars at Ravalli. On May 29, 1907 the *Daily Missoulian* reported: "The
first bull came up the chute. . . . The lariat slipped around the snubbing post and
the bull entered the car under pretty good headway. He hit the back side of the
car with such an impact that it shook the very rails and rattled the spectators on
the roof in more ways than one. Some of them didn't wait to climb down the
ladders; they just jumped." Forsyth photo. *Courtesy Montana Historical Society.*

This Pablo bull challenged his captivity but died in the process. Forsyth photo. *Courtesy Montana Historical Society.*

The buffalo bull who broke his neck trying to escape from a cattle car at Ravalli was given to the Flathead Reservation Indians. Forsyth photo, 1907. *Courtesy Montana Historical Society.*

porting all of the Bedson herd to Kansas, Jones' herd numbered nearly 150 animals. The herd continued to increase rapidly. To maintain it, Jones began selling a few buffalo to zoos and parks and to persons who, like himself, wanted to raise buffalo. He also began experimenting in the crossbreeding of buffalo with domestic cattle, but that story comes in Chapter XVIII.

Everything went fine for Jones and his buffalo until the early 1890s. A try at establishing a buffalo ranch near McCook, Nebraska failed. Other investments failed, and by 1895 Jones was in serious financial trouble. He was forced to sell his buffalo. Some were purchased by Michel Pablo in Montana and eventually were sold with Pablo's herd to the Canadian government. Still others were sold to a California rancher and shipped to the west coast.

For the next few years, Jones stayed out of the buffalo business. For a time he was sergeant-at-arms in the Kansas Legislature. By 1900, he had begun working to establish wildlife refuges for buffalo and other wild animals. When he heard that officials at Yellowstone National Park were going to establish a captive buffalo herd in the park, Jones asked for the job to build the herd. He received a commission in 1902 as the park's game warden. Jones helped to supervise the construction of a buffalo compound about a mile south of Fort Yellowstone. A partly tree-covered area crossed by a small stream was enclosed with Page woven-wire. Then pure-bred buffalo were purchased from private owners. Fifteen cows came from the Allard herd in western Montana, and three bulls were selected by Jones from Charles Goodnight's herd in Texas and shipped to the park. By late October 1902, the eighteen buffalo were in the enclosure at Yellowstone.

From what is known, Jones did a fine job in getting the new herd established at Yellowstone. But Major John Pitcher, who took over as park superintendent in 1901, later recalled that after the herd was established, Jones' "usefulness in the park ended absolutely."[3] Though Jones was an excellent animal man and tolerated even the most difficult of buffalo, he was not tolerant of human beings who drank whisky, used tobacco, or played poker. For example, Major Pitcher placed Jones in charge of the park's scouts in early 1903 to assist Jones in capturing wild buffalo calves in the park, but soon Jones was lecturing the scouts on their vices, calling them "habitual drunkards and gamblers," and openly questioning their honesty. Because of this, one scout resigned and the rest refused to cooperate with Jones. Major Pitcher then removed Jones as their supervisor.

At that point the relations between Major Pitcher and Jones became worse. When President Theodore Roosevelt visited the park and set out on a camping trip with Major Pitcher and a handful of men mainly to keep camp, Jones took his hounds and joined the party uninvited. Jones hoped to tree a mountain lion and entertain the President. Before nightfall the President asked that the hounds and their keeper be sent back. Jones and his dogs returned to Mammoth. Jones was so mad that he began to write critical letters and to generally criticize the park's administration.[4] He tasted success when the secretary of the interior directed that Major Pitcher "should be no longer burdened with the details of the preservation of game in the reservation" and that the game warden, Jones, should be held accountable for the care, preservation, and management of the Yellowstone game.[5] Jones thought he had been successful until he found that his new responsibility was meaningless without the park personnel's cooperation. He resigned on September 15, 1905 and returned to Kansas. For a time he worked for wildlife conservation by starting a buffalo ranch north of the Grand Canyon in Arizona. Then he traveled to Africa twice, lecturing upon his return. But his health soon began to fail. Finally, at the age of seventy-six, Jones died at a daughter's home in Topeka, Kansas, in October 1919.

Although Jones' reputation as an authority on buffalo was in part valid, Jones was, as the late J. Frank Dobie called him, "a poser."[6] Charles Goodnight, a well-known Texas rancher, likewise did not have a very high regard for Jones, once labeling Jones' so-called statements of authority on the buffalo as nothing but "hot air." Goodnight himself was a rather quiet man, the exact opposite of Jones, who did have the reputation of being somewhat "windy." When Goodnight heard that Jones had claimed he helped guard Billy the Kid at Mesilla in 1882, he noted that it was quite possible Jones did guard Billy the Kid—in his grave—since Billy had been killed in July of 1881. Goodnight then asked John W. Poe to set down on paper what actually had happened. Poe, a member of the posse that hunted down the outlaw, gave his account. Much to Goodnight's delight, Jones was not even mentioned by Poe.[7]

Regardless of Jones' negative qualities, he was among those who helped save the buffalo and perpetuate the species, for his contribution, like that of Goodnight's, is an acknowledged fact of history.

Goodnight's efforts began around 1866. In the spring of that year, four years before the hide hunters began killing buffalo in large numbers on the southern plains, Goodnight decided to capture some buffalo calves. He had learned by experience that if he chased a bunch of buffalo cows with calves, eventually the calves would grow tired and fall behind their mothers. Then, if he changed course and headed away from the older buffalo, the calves would follow his horse and not the buffalo. In this way Goodnight captured three buffalo calves on the plains of Texas in 1866. The calves simply followed Goodnight's horse back to his Elm Creek ranch, where he put the little buffalo with some domestic cows, who accepted and nursed them as their own.

Later, Goodnight got two more buffalo calves to follow him home. Still later, he found another. It was only about one day old. Goodnight chased off its mother, but she returned and attacked. He had to kill her. More than half a century later, he said he still regretted having killed that cow.

Goodnight took the six buffalo calves to a friend in Park County, Texas. For one-half of the eventual profit, Goodnight's friend agreed to raise the animals. But the friend grew tired of the buffalo and sold them. Goodnight, busy with other things, had little time for buffalo, but the seed had been planted in his mind.

Ten years later, about 1876, after Goodnight had established his ranch in the Texas Panhandle, his wife told him one day that she was concerned about the slaughter of so many buffalo. At that time the killing of buffalo for their hides was at its peak on the southern plains. Goodnight gave it some thought, and, while working cattle a few days later, he ran into a bunch of wild buffalo. He roped one young heifer. After tying the animal down, Goodnight took off on his horse after the fleeing buffalo. Soon he spotted them and took out after a young bull calf. After much trouble, he cornered the animal and tied him up. Later that day, Goodnight had his cook take a wagon out on the plains to pick up both calves. They were brought to the ranch house and placed with domestic milk cows. Although the cows resented the strange calves at first, they finally came around and provided the young buffalo with more than adequate nourishment.

Goodnight got other calves captured by friends and neighbors, and by 1887 he had a herd of thirteen buffalo. By 1910 the number had increased to 125 buffalo. Over the years he sold several to zoos and parks, including Yellowstone, and he used others for breeding purposes. Like "Buffalo" Jones, Goodnight experimented in crossbreeding. And like others, Goodnight envisioned a renewed market potential in buffalo meat and hide. But unlike most others, Goodnight saw *too* much; he attributed panacea to buffalo products. Take the case of his buffalo wool and medicinal tallow:

> Buffalo tallow, Goodnight claimed, was medicinal. In 1916 he sent batches of it to E. C. Seymour in New York, president of the American Bison Society (so much that Seymour complained, "What in thunder will I do with this buffalo fat?"). When Seymour sent some to Dr. Robert T. Morris, Madison Avenue physician, to have its supposed medicinal benefits assayed, Morris answered that he couldn't experiment until he knew what it was. "Tell me what is in the bottle and I can charge somebody twenty-five dollars for finding out if I have hurt him by using it." Another doctor commented that Buffalo Balm's viscosity would be good for automobile gears. On the other hand, Goodnight received a request for another twenty pounds of the stuff from Dr. Tillotson of Los Angeles, who said he had used it on rectal cases and the privates with "great benefits." Goodnight wrote Seymour later in the year of other qualities of tallow: it was a fine preservative—mixed with mince and unrefrigerated, it remained edible for over a year. . . .
>
> Charles wrote Seymour again in 1916 about their "discovery" of buffalo soap: "It is not soap at all but so made. . . . Now we take this soap and mix it with water and it makes a compound with far reaching qualities. It has no taste and smell of soap and has qualities unknown to us. . . . I am satisfied it will relieve rheumatism. By all means have it tried on infant paralysis. Try it for tuberculosis. I do believe it will work. It is harmless. We do not know what we have found. Help me hunt it out. I believe it stands a fair chance to become the discovery of the age." . . .
>
> In addition to its medicinal qualities, buffalo soap, Goodnight

claimed, could be used to clean oil paints, to clean "silver and book of all description," as a "disinfectant," or to "kill insects." . . .

Sometime later Goodnight clipped "superior" wool from the buffalo and made it into blankets "superior" to the ordinary sheep's wool product. He saw curative power in these blankets, too, for they sparked in the dark, not like phosphorous, but like electricity, for "some claim they can feel the shock." A beast with electric wool, Goodnight felt, was worthy of experiment by Edison or endowment by Carnegie. But these gentlemen proved unin-. terested in beastiology.[8]

When Goodnight died in 1929, the Texas Legislature authorized the Texas Game and Fish Commission to purchase Goodnight's buffalo, provided a suitable place could be found for them. Unfortunately, the legislation was vague, and it provided no funds to establish an "appropriate place" for the animals. Thus the state of Texas did not take over Goodnight's buffalo herd. When it appeared the herd might have to be sold, the Great Southern Life Insurance Company formed a syndicate headed by A. C. Nicholson of Dallas. They continued to raise the buffalo, maintaining the herd at a good level and selling only the surplus animals.[9]

Although not so well known as Charles Goodnight, Pete Dupree was another man who helped preserve the buffalo. Dupree was the son of Frederick Dupree, an early-day French trapper who settled among the Sioux Indians, married into the tribe, and built a home on the Dakota Sioux reservation along the Cheyenne River.

About 1881, Pete Dupree captured five buffalo calves on the prairie about a hundred miles west of Fort Bennett in the Dakotas, took them home by wagon, and turned them loose to graze with the Dupree cattle. Two months later, two of the calves died. The three others reached adulthood and raised young. The herd prospered, although one buffalo was killed by an Indian who thought it was a wild buffalo.

By 1888 Pete Dupree had nine pure-blood buffalo, four of which were bulls, and seven animals of mixed blood. For ten years the herd grew. Then, in 1898 Peter Dupree died. His estate, including the buffalo, was divided among his heirs. D. F. "Dug" Carlin, who had married one of Pete's sisters, administered the estate. He decided to dispose of the buffalo, and they were sold to James Philip, an immigrant from Scotland, who had picked up the nickname of "Scotty."

In late 1901 Scotty Philip took the buffalo from the Dupree ranch to a pasture near Fort Pierre. By then there were nearly fifty buffalo in the herd, and it continued to increase naturally and otherwise: from time to time, ranchers in the area would find a few scattered wild buffalo in their cattle herds at roundup time and give them to Philip. By 1904 Scotty Philip's herd numbered more than eighty buffalo.

Unlike most buffalo men of his day, Philip did not like buffalo of mixed blood. He wanted only "pure" buffalo, no crossbreeds. Philip was once quoted as saying that mixed-blooded buffalo were "not worth a damn." He removed all the cattalo from his herd and either sold or butchered them. This left nothing but full-blooded buffalo in his herd.

All went well for Philip and his buffalo until July 1911. On a Sunday evening, at fifty-three, James "Scotty" Philip died unexpectedly. Two of his sons, Stanley and Roderick, took over the buffalo business. The Philip herd then numbered about 400 buffalo, but financial problems developed and some of the buffalo had to be sold for working capital. In the fall of 1914 the sons sold thirty-six buffalo —six bulls, eighteen cows, and twelve calves—to the state of South Dakota. Only a few months earlier the state legislature had set aside more than 60,000 acres in Custer County for a state game preserve. Philip's thirty-six buffalo were the nucleus for a herd of nearly 2,000 head today in Custer State Park, South Dakota.

In the early 1920s more buffalo in the Philip herd had to be sold. William Randolph Hearst bought 100 and had them shipped to his animal preserve at San Luis Obispo, California. Another 100 were sold to the Miller Brothers' 101 Ranch in Oklahoma. The Miller Brothers had a small herd, but the animals had not been healthy. The Philip buffalo were purchased to strengthen the 101 Ranch herd. Oklahoma oilman Waite Phillips also bought 100 of the Philip buffalo for a ranch he owned near Cimarron in northeastern New Mexico. Their descendants are viewed annually by hundreds of Boy Scouts today on the ranch, which is now a Boy Scout camp.

Other buffalo from the Philip herd in South Dakota were sold to other private persons and to parks. By November 1925, the Philip herd near Fort Pierre, South Dakota numbered only about 250 buffalo. Philip's sons then decided to throw part of the herd open to sportsmen and charge $100 a head for each buffalo killed. Just how many animals died in this way is unknown, but a few persons around Fort Pierre who had known Scotty Philip wondered at the time just what he would have said about the killings. Philip had spent much of his life building the herd to save, not kill, the buffalo.[10]

What was left of the Philip buffalo herd was sold in the late 1920s to Henry O'Neil of Rapid City, South Dakota, who maintained it for many years.

While the efforts of Philip, Goodnight, Jones, and the others already mentioned are generally known to students of the animal's history, the contribution made by two brothers, Allen and Miner McCoy, is not. They also helped to preserve the buffalo in a rather unusual way.

The McCoy brothers and J. W. Summers settled in what is today the Oklahoma Panhandle when it was only a neutral strip of no-man's land between Texas and Kansas. They operated a cattle ranch. In 1883, not far from the present-day Beaver, Oklahoma, the McCoy brothers caught a pair of buffalo calves, a bull and a heifer. The young buffalo were among the last wild buffalo on the southern plains. They had bunched together on the plains between Beaver Creek and South Canadian River, west of the head of Coldwater Creek.

About two years later, the McCoys, after raising the young buffalo on milk cows, decided to make a gift of the animals to the city of Keokuk, Iowa, Allen M. McCoy's home. McCoy offered the animals on the condition that the McCoy brothers would receive any offspring from the original pair. The city agreed, and the two buffalo were shipped to Keokuk and placed in Rand Park. By 1896 the two buffalo had produced four offspring. Allen McCoy asked for the offspring. The Keokuk city attorney, A. Hollingsworth, acknowledged the claim, but told McCoy the city also had a claim. The city wanted to be paid for the care and feeding of the offspring. A settlement was finally reached with the McCoys taking one pair of buffalo and giving the other pair to the city as payment.

The city of Keokuk soon sold two of their four remaining buffalo to the Page Woven Wire Fence Company of Adrian, Michigan. Shipped to Michigan in large crates, the two buffalo were used to demonstrate the holding capability of a new kind of wire that had been created by J. Wallace Page. The company claimed their wire would hold the largest and fiercest animals, including domestic bulls, wild horses, elk, deer, longhorn cattle—and buffalo. Between 1896 and 1904, the thirty-acre exhibition enclosure built by Page at Adrian, Michigan held many animals, including the buffalo. It was a convincing demonstration of the Page wire. And during the late summer and early fall months the buffalo were taken to state and county fairs throughout the midwest and enclosed in corrals of Page fence — "a fence that would hold a buffalo would hold any farm animal," said one advertisement.

When William Hornaday, whose brother lived at Keokuk, Iowa, learned of Page fence wire, he used it to contain buffalo and other animals in the National Zoological Park in Washington, D.C. and later at the New York Zoological Park.

The New York Zoological Society purchased three buffalo cows and a bull for $1,200 from the Page Wire Fence Company in October 1904. The zoo bought them to increase their small herd started late in the 19th century. Three years later fifteen buffalo from the New York zoo were shipped to Oklahoma. They were the nucleus of the buffalo herd established in the Wichita Mountains. Several of these buffalo were descendants of the original pair of buffalo captured by the McCoy brothers in 1883.[11]

Allard, Dupree, Goodnight, Jones, the McCoys, Pablo, Philip, Walking Coyote and others on their own initiative helped to save the buffalo from extinction. And each of these men, in one way or another, profited from the buffalo.

By 1902 there were more than 700 buffalo in private herds in the United States. However, it became evident that the future of such private herds was in doubt. Owners died and herds were sold. Some owners tired of raising buffalo, partly because of the drain on their pocketbooks. Others, to make ends meet, had to sell buffalo continually. There was no guarantee that private persons would continue to preserve the buffalo on their own.

During the summer of 1904, naturalist Ernest Harold Baynes visited the Blue Mountain Forest and Game Preserve in New Hampshire, owned by Austin Corbin, a wealthy man who had developed New York's Manhattan Beach and Coney Island. The preserve had all kinds of big game, including 160 buffalo, some of which had originally come from "Buffalo" Jones' ranch in Kansas.

Charles Goodnight and mounted buffalo. *Courtesy Panhandle-Plains Historical Society.*

Buffalo Bill Cody at Pawnee Bill's Oklahoma Ranch.

Black Dog, c. 1916, 10 years of age, at the Wichita National Forest and Game Preserve, now the Wichita Mountains Wildlife Refuge near Cache, Oklahoma. Black Dog weighed 2,800 pounds and was considered the largest living buffalo in 1916. As a calf in 1907 he was one of the 15 buffalo from the New York Zoo used to start the Wichita Mountains herd which now numbers about 1,000. *Courtesy Wichita Mountains Wildlife Refuge.*

The buffalo corral at the New York Zoo in October 1907 where the animals were crated for shipment to Oklahoma. *Courtesy New York Zoological Society.*

Boy feeds a buffalo calf belonging to Pawnee Bill's herd, c. 1915. Note the cow on right with raised tail, a sign that a buffalo is getting angry. *Courtesy Division of Manuscripts, University of Oklahoma Library.*

Yellowstone National Park buffalo, 1907. Roosevelt, king of the herd, is second bull in foreground; Carrie Nation is the cow just beyond and to the right. *Courtesy Library of Congress.*

Yellowstone National Park buffalo, c. 1903. Anderson Collection photo. *Courtesy William F. Whithorn.*

Modern-day buffalo holding pens at Houck Triple U Ranch near Pierre, South Dakota. *Courtesy Dana C. Jennings, Madison, South Dakota.*

Baynes became fascinated with the buffalo. He learned as much as he could about the animal and its history, and he soon realized that the buffalo's very existence was being threatened. When he left the preserve in June 1904, he wrote a series of articles about the Corbin herd for the *Boston Evening Transcript*.[12] By August 1904 Baynes was sure the time had come to make certain the buffalo was saved for future generations. He immediately began talking to his friends in hopes of gaining support for the move. One of them was Professor Franklin W. Hooper, director of the Brooklyn Institute of Arts and Science, who suggested that Baynes send letters to leading Americans pointing out the crisis and urging them to become concerned about the vanishing buffalo. Baynes liked the idea and sent the letters. One went to President Theodore Roosevelt, who knew the West and the buffalo. The President immediately took up the crusade and advised Baynes that he would mention the plight of the buffalo in his annual message to Congress. This Roosevelt did on December 5, 1905. In his remarks, President Roosevelt said:

> The most characteristic animal of the western plains was the great shaggy-maned wild ox, the bison, commonly known as buffalo. Small fragments of herds exist in a domesticated state, here and there, a few of them in the Yellowstone Park. Such a herd as that on the Flathead Reservation should not be allowed to go out of existence. Either on some reservation or on some forest reserve like the Wichita Reserve or some refuge, provision should be made for the preservation of such a herd.[13]

Baynes continued to write articles in magazines and newspapers, pointing out how and why the buffalo was in danger, and many newspaper editors took up the crusade on their editorial pages. In January 1905 Baynes took to the road on a lecture tour. In Boston, on January 18, someone suggested to him that a group of some sort, perhaps a society, should be formed to work for the permanent preservation of the buffalo. Baynes thought the idea a good one and discussed it with prominent persons, including William T. Hornaday, then director of the New York Zoological Park.

Hornaday told Baynes that such a society was needed and that he would be glad to be its president. Soon afterward, President Roosevelt accepted an invitation to become the honorary president of such a group, should it be organized. It was then that Baynes made his move. He sent notices of an organizational meeting to about 200 persons all over the United States, all of whom, at one time or another, had expressed an interest in preserving the buffalo. But when the meeting was held in New York City, December 8, 1905, only fourteen persons attended. That was enough, however, and at the meeting in the lion house at the New York Zoological Park, the American Bison Society was founded.[14]

William T. Hornaday was elected president and Theodore Roosevelt honorary president. A. A. Anderson and Dr. Charles S. Minot were elected vice-presidents; Ernest Harold Baynes, secretary; and Edmund Seymour, treasurer. In addition, nine men, including writer Ernest Thompson Seton, were elected to an advisory board. Hornaday then selected an executive committee to draft a constitution, which was adopted at the society's next meeting a month later.[15]

The constitution of the American Bison Society provided for a large membership, which, it was hoped, would enable the society to raise adequate funds for its work. Anyone could become an associate member by paying $1 annually. Active memberships were $5 a year and life memberships $100. Persons contributing $1,000 automatically became patrons of the society. The society's membership climbed to several hundred. Although it was open to anyone in the United States and Canada, most of its members came from the East, particularly from Massachusetts, Philadelphia, and New York City, where the society established its headquarters. The membership was large but the society's accomplishments during its first decade were the work of a small group of men, most of them scientists or wealthy businessmen. Of this group of fourteen men, all but four lived in New York City. Of the four, only one, C. J. "Buffalo" Jones, lived in the West.

When the American Bison Society was formed in early 1905, the U.S. Government watched over fewer than a hundred buffalo. A handful were in the National Zoological Park in Washington, D.C., operated by the Smithsonain Institution. They had first obtained a few head of buffalo in the late 1880s. Initially the animals were kept in a large pen at the Smithsonian building on the mall near the Capitol. The buffalo attracted much attention and helped Smithsonian officers obtain funds from Congress to establish the National Zoological Park in 1888. After the park was established in the northern area of the District of Columbia, the buffalo were moved to the new location. The only other buffalo under federal protection in 1905 were in Yellowstone Park, where about twenty-five wild mountain buffalo roamed over the rugged country far from the summer tourist areas and another twenty-five plains buffalo were in the fenced area near the park's headquarters at Mammoth.

Before the American Bison Society was formed, Hornaday had expressed concern for the captive plains buffalo at Yellowstone. He felt they would not perpetuate themselves for very long under close confinement. A few months before the society became a reality, Hornaday offered, on behalf of the New York Zoological Park, to give the federal government a small number of buffalo if the government would start a new herd on the National Wichita Forest Reserve in southwestern Oklahoma. Congress accepted the offer and approved the use of the land in Oklahoma for a wildlife refuge. Six months after the American Bison Society came into existence, Congress provided funds to enclose 8,000 acres of the reserve with a high wire fence and to build the necessary chutes, corrals, and sheds for the buffalo.

By October 1907, the Oklahoma range was ready for the buffalo. In New York City, Hornaday had fifteen of the healthiest buffalo at the zoo crated and shipped free of charge, thanks to the railroads involved, to southwestern Oklahoma. There, on October 17, 1907, the animals were turned loose in their new home.[16]

The American Bison Society then began to work toward establishing another federal buffalo range, not counting the one at Yellowstone. Hornaday proposed that this new range should be located on the Flathead Indian Reservation in western Montana. At that time, 1906, Michel Pablo's captive buffalo herd was still there. President Roosevelt and Interior Secretary Hitchcock were in favor of buying Pablo's herd, but, as we have seen, Congress refused to provide the funds. Some lawmakers in Washington felt that Pablo was asking too much for his buffalo. The attitudes in Washington changed quickly, however, after the Canadian government bought Pablo's herd. Congress felt embarrassed. Numerous Congressmen who had opposed the purchase of Pablo's herd had red faces. Several publicly regretted the loss of what was then considered to be the largest private buffalo herd in North America.

Within two years, Congress appropriated enough money to purchase land for a buffalo refuge in western Montana. But since the Pablo herd had been sold, it was up to the American Bison Society to provide the buffalo for the new range. The society decided to raise $10,000, and Hornaday sent letters to officials in more than 150 cities asking for financial support. The society received much editorial support from magazines and newspapers, but a handful of papers opposed the society. For instance, the *Indianapolis Star* said:

> The American Bison Society, with headquarters in Boston [actually it was in New York City], has secured from Congress a tract of land in Montana where a herd of buffaloes will be established and maintained by the Government, sacred from the predatory hunter. Why Boston people should take an interest in the buffalo and why any intelligent persons should care for the preservation of these moth eaten, ungainly beasts, when their room might much better be taken up by modern blooded cattle, beautiful to look at, are conundrums no one answered.[17]

Fortunately, not everyone held the same view as the *Star,* and the drive to raise funds was successful. $10,560 was gathered, most of it from the East, not the West, where millions of buffalo had been slaughtered for their hides. With this money the American Bison Society bought thirty-four buffalo for $275 a head from Mrs. Alicia Conrad of Kalispell, Montana. These buffalo had originally been part of the Allard-Pablo herd and were close to the new federal buffalo range.

By October 1909 the new range had been fenced, and the Conrad buffalo had been brought across Flathead Lake by barge to a railroad, where they were loaded and transported south to their new home. There were thirty-seven buffalo in all on the new range: thirty-four had been bought from Mrs. Conrad, and she gave two others as gifts; the thirty-seventh was a gift of rancher Charles Goodnight in Texas. With thirty-seven buffalo on the National Bison Range, as the new refuge was called, the federal government in October 1909 had a total of 158 buffalo: seven in the National Zoological Park in

Washington, D.C., about ninety-five at Yellowstone National Park, nineteen on the Wichita Game Reserve in Oklahoma, and the new herd in Montana. A year later, the total number had increased not only naturally, but Austin Corbin had given three buffalo from his New Hampshire game preserve to the National Bison Range in Montana.

The creation of the Wichita Mountains Reserve in Oklahoma and the Montana buffalo range marked the beginning of the U.S. Government's serious efforts to preserve the buffalo and other big game such as moose, elk, bear, antelope, and mountain sheep. In the next ten years wildlife refuges were established in other parts of the nation.

The American Bison Society had a hand in establishing a buffalo herd on the Fort Niobrara Reservation near Valentine, Nebraska. One Society member, T. S. Palmer, raised more than $2,000 to fence a pasture near Valentine for six buffalo donated by the Society to start the Niobrara herd in 1913. A few months later, the Society provided fourteen buffalo to start the Wind Cave National Game Preserve buffalo herd in the Black Hills of South Dakota.

The last federal buffalo herd that the American Bison Society helped to establish was near Asheville, North Carolina, in the Appalachian Mountains. After the federal government obtained a large area of land in 1915 and called it the Pisgah National Forest and Game Preserve, Edmund Seymour, an American Bison Society member, began to work toward establishing a buffalo herd there. Seymour, a Wall Street broker, got the government to agree to the herd if the Society would obtain the buffalo. This it did. Austin Corbin, Jr. donated six animals from his New Hampshire herd, and in January 1919 the buffalo were delivered to the North Carolina preserve. But the herd never prospered. The first year some of the buffalo died from eating a poisonous weed. By the late 1930s the last of the buffalo had died. No further attempts were made to restock the preserve.

Between 1920 and the mid-1930s, the American Bison Society remained active helping to establish buffalo herds in various states under state and county control. And it continued to inform the public about the buffalo, especially through its annual reports. From its beginning until 1930, the Society published annual reports and sent them to members, libraries, and anyone interested in the buffalo. But in the early 1930s the Society became less active. Although it continued to function during the depression, it concentrated its efforts on other North American big game. By 1935 its membership had dropped considerably. Then, dues were dispensed with altogether. During the early 1940s, the Society held no meetings, but some of the members kept the movement alive through correspondence. In 1949 Edmund Seymour died at ninety. He had been president of the American Bison Society for the preceding thirty-three years. Martin S. Garretson carried on for the Society until 1953, in which year the organization was listed for the last time in national directories as a group dedicated to wildlife protection.

The American Bison Society quietly died. But it had accomplished what its members had set out to do—preserve the American buffalo.

President Lyndon B. Johnson at his Texas ranch, November 8, 1966. *Courtesy UPI.*

XVII

In Captivity

One of the saddest sights today
Is watching a buffalo cow chew baled hay,
In a pen, in a zoo, like a sardine can;
It's a lump in the throat of an outdoor man.
J. R. Williams 20th century

After a two-year adventure which had taken him to the Pacific Ocean, Dick Wootton, mountain man, returned to Bent's Fort on the Arkansas River in 1840. The usual happy mood of the trading post, just east of the Rockies, had been replaced by gloom. Fur prices in the East had dropped. The beaver hat had been replaced by the silk hat, and fur trading had all but come to a halt in the Rocky Mountains.

Wootton reluctantly quit trapping and settled down at Bent's Fort. After loafing around for a while, he contracted to supply buffalo meat to the trading post. He liked hunting the shaggies. To Wootton, it was more fun than work. It was one of the finest sports.

One day about a month after he started his new job, Wootton killed nearly twenty buffalo. As he stopped to rest while his skinners finished the messy job of butchering, he noticed a set of twin buffalo calves nearby. Their mother had been killed by one of his shots. Now, Wootton had seen motherless calves before on the plains. Many times he had shot cows, leaving their calves to fend for themselves, which usually meant they would end up as a meal for the wolves. But this time he began to feel sorry for the calves, perhaps because they were twins. Instead of leaving them for the wolves, he took them back to Bent's Fort, where he persuaded a good-natured milch cow to adopt the little orphans. The little calves made themselves right at home and grew strong and healthy. In time, Wootton decided to

241

add more calves to his little buffalo herd. He told his friends at Bent's Fort, "I'm goin' into the business of buffalo farmin.' "

In what is now Pueblo, Colorado, Wootton built a corral for his buffalo and a cabin for himself, and he bought forty milch cows to provide nourishment for his young buffalo. Wootton became quite successful raising buffalo, and in time he trained some of them. He recalled many years later, "I had two or three pairs of them broke to work like oxen and they made first class teams." Wootton might have continued with his "buffalo farmin' " had it not been for the lure of the Rocky Mountains and his love for adventure. These meant more to Wootton than did his buffalo. Thus, in 1843, he decided to sell his buffalo, all forty-four, and return to the life he loved most.

After several weeks of preparations, Wootton, in what is perhaps the longest overland buffalo drive in history, herded his buffalo from Bent's Fort across what is today Kansas to Independence, Missouri, on the Missouri River. With his milch cows he was able to control the buffalo. At Independence, he sold his animals and made a nice profit for his three years of work.

Some of his buffalo were taken east and resold to showmen and others. Some ended up in the Central Park Zoological Gardens in New York City, where they were well cared for.[1] But others appear not to have been so fortunate. Some ended up in the hands of Dade and Wilkins, two Missouri showmen. Although the facts are scarce, it appears that Dade and Wilkins bought twelve buffalo. Then, with nine Osage Indians, an interpreter, and two Mexicans, Dade and Wilkins took the animals on a steamboat down the Missouri River to St. Louis, arriving in March 1844. What happened next is unknown, but, according to newspaper accounts, the showmen went as far east as Baltimore.[2] In the middle of June, the show headed west again. A few weeks later, a Baltimore newspaper, in a sketchy report, said: "Six of the exploited Osages, including a middle-aged chief, returned to St. Louis in July, emaciated and nearly naked, after being abandoned and left destitute at Louisville, Ky."[3]

Although there is no mention of the buffalo in this account, if the treatment of the Indians is any indication of how Dade and Wilkins cared for their buffalo, the animals' lives were probably short.

Long before Dick Wootton became a "buffalo farmer," other men had tried to domesticate buffalo. Fifty years after Coronado's expedition across the southern plains, Don Juan de Onate, governor of New Mexico, became fascinated by stories he had heard about buffalo. He ordered Vicente de Zaldivar to capture some buffalo to see whether they could be domesticated. At first, Zaldivar's expedition had little success. The men found only one buffalo, an old and decrepit bull. But as they traveled farther and farther onto the Staked Plains, they found many buffalo. When the party tried to capture some, Juan Gutierrez Bocanegra, secretary to the party, recorded, "They are remarkably savage and ferocious, so much so that they killed three of our horses and badly wounded forty." Zaldivar's expedition was unsuccessful and finally settled on killing a few buffalo for tallow before returning home.[4]

During the 1600s, other Spaniards were more successful with buffalo. The monks of St. Francis in Zacatecas, in what is today central Mexico, had two buffalo that pulled a two-wheeled cart. According to Jean Louis Berlandier, these buffalo were "suited to all work which could be expected of them. . . . There is no doubt that the buffalo becomes accustomed to man and that fact alone should get the attention of agronomists, for the more man multiplies his number, the more also will he have to increase his resources."[5]

About the time Berlandier was making his observations, another Spaniard, Antonio Barriero, was trying to stimulate interest in agriculture in what is today New Mexico. Barriero suggested that the buffalo be used in place of the ox to till the soil. He said the buffalo was twice as strong as the ox. But whether Barriero ever succeeded in training or even capturing any buffalo is unknown.[6]

Along the east coast of North America, the first efforts to raise buffalo did not begin until 1701, when Huguenot settlers at Manikintown, on the James River near modern Richmond, Virginia, began to raise some buffalo calves.[7] Later, "people of distinction" in the Carolinas and other provinces south of Pennsylvania raised buffalo calves as well. However, most were not trained to do the work of oxen but were used in crossbreeding experiments with domestic cattle.[8]

Probably the earliest detailed account of raising buffalo in captivity in the United States is that of Robert Wickliffe of Lexington, Kentucky. Starting with some buffalo calves about 1815, Wickliffe raised buffalo for nearly thirty years. And like the Franciscans nearly 200 years earlier, Wickliffe trained some of the buffalo to do the work of oxen. "I have broken them to the yoke, and found them capable of making excellent oxen; and for drawing wagons, carts or other heavily laden vehicles on long journeys they would, I think, be greatly preferable to the common ox," wrote Wickliffe. However, the Kentuckian failed to completely domesticate his buffalo. They remained too wild, capable of doing only a few tasks for which they had been trained. Wickliffe had to keep them in secure enclosures, and it was not uncommon for them to charge their keeper with all the fury of a trapped mountain lion. Wickliffe ended his experiments about 1845.[9]

As white men moved onto the plains and into buffalo country in the early 19th century, a few of them captured and raised buffalo calves. But the efforts were not always serious. They were often more for novelty than anything else. There is also the intriguing one-line report by Don Diego Gonzales Herrera about a non-white, one Castro, a Lipan Indian chief in Texas, who in 1840 had a pet buffalo cow "trained to follow his saddle animal."[10] Even George Catlin raised a few buffalo calves which he had obtained from Indians on the northern plains. He shipped several down the Missouri River by steamboat to friends at St. Louis, but only one of the calves survived the trip because "milk could not be procured."[11]

When the area of Kansas and Nebraska was opened to settlement during the mid-1850s, a few settlers captured buffalo calves and raised them. One settler in the Republican River valley of Kansas Territory trained a pair. Little is known about the man or his buffalo. Even the man's name has been lost in time. But Frank A. Root, an early Kansas writer, saw the rancher and his trained buffalo after they had traveled nearly 200 miles from the Republican River east to Atchison, Kansas Territory. Root wrote:

> It was in the early 1860's, during the Civil War. These were domesticated and yoked together, having been driven in by a ranchman from the Republican valley, hauling produce to the Atchison market. They attracted considerable attention from the business men. They seemed to travel all right, were extremely gentle, and, under the yoke, appeared to work quite as nicely as the

patient ox. Nothing particularly strange was thought about the matter at the time, when there were immense herds of buffalo roaming wild on the plains of western Kansas; but I never, after that year, saw any of the shaggy animals yoked and doing the work of oxen.[12]

The story of early captive buffalo would not be complete without Joseph G. McCoy's account of capturing buffalo on the Kansas plains. McCoy is credited with opening the Texas cattle trade to the railroad at Abilene, Kansas, in 1868, when he decided to send a carload of buffalo east to advertise the Abilene cattle market. What happened was described by McCoy in 1874:

> The frame or slats of an ordinary stock car were greatly strengthened by bolting strong, thick plank parallel with the floor, and about three feet above it, to the sides of the car. Putting in a camp outfit, and supplies abundant in one car, and a half dozen horses, well trained to the lasso, in another car, a party of half a dozen, departed for the buffalo regions, out into which the Kansas Pacific Railway was then being operated. Arriving at Fossil Creek siding [now Russell, Kansas], the cars were put upon the side track, and camp pitched. The horses were unloaded by means of an inclined plane or platform, temporarily improvised for that purpose. In the party were three or four Texan cow boys, also three California Spaniards, all experts with the lasso.
>
> After partaking of a hearty dinner, the party saddled up the ponies, and started out in quest of the buffalo. . . . [T]he time was brief before a huge old bull was spied, and immediately preparations to chase and lasso him, were made. Circling around, he was started in the direction of the railroad, and when within a few hundred yards thereof, a sudden dash was made upon him by two Spaniards, and in the twinkling of an eye their lariats were around his neck. So soon as the old monarch found himself entangled, and his speed checked, he became furiously enraged, and alternately charged first at one and then the other of his pursuers. It was noticeable how intensely angry he became; he would drop his head and stiffen his neck, set his tail erect over his back, and with eyes green with pent-up wrath, await the

near approach of his tormentors. So soon as one came near, he would plunge at him, and pursue at his utmost speed, so long as there was the least hope of overtaking him. Then stop and whirl about, and attack his nearest pursuer. After getting him quite close to the railroad track by stratagem, the third lasso was adroitly thrown around his hind legs, and in a jiffy the great behemoth was lying stretched, helpless upon the ground. It was vain for him to struggle, the well trained horses watched his every motion and kept the lariats as tight as fiddle-strings, shifting their positions dexterously, to check or counterbalance his every motion.

When he ceased to struggle, his legs were securely tied together with short splashes of rope or thongs previously prepared for the purpose, then the lassos were taken off, and after adjusting the inclined plane, a block and tackle were brought into requisition, one end of which was attached to his head, the other to the top of the opposite car door, and before the hot panting bison was aware of what was being done, he was aboard the car; his head securely bound to a post of the car frame, and his feet relieved. He would not bound up and show fight, but lay and sulk for hours. In two days ten full grown bull buffaloes were lassoed, but the weather being very hot, four of them died from the heat and the anger excited by capture. Three became sullen, and laid down before they could be got near the cars, so but three were got aboard in good condition.[13]

The sides of the buffalo train car were covered with large canvas broadsides advertising Abilene as a cattle shipping point. In St. Louis the train car received much attention. In Chicago, where the buffalo were turned loose in a pen at the stockyards, McCoy's cowboys showed astonished Chicagoans how they had roped the huge animals on the plains. Then, they were presented to a Professor Gamgee, an English veterinary surgeon, who later sent the buffalo's stuffed hides to London. For McCoy, the buffalo meant increased trade to Abilene.

About 1949 my granduncle, Charlie Engel, who had been a gunsmith at Manhattan, Kansas in the late 1800s, told me a story he had heard from Bill Rehfeld, who was city marshal at Manhattan in the late 1870s. Rehfeld described how cowboys and other persons frequently led buffalo calves into Manhattan in the early 1870s. Rehfeld said that a showman, Ruben Blood, bought three calves on one occasion for his traveling tent show which happened to be in town. Blood paid $3.50 for each calf. He had his hands full trying to keep the animals in tow. What eventually happened to the little buffalo is unknown.

Another oldtimer, Louis Charles Laurent, recalled a similar sight in Topeka, Kansas during the 1860s. During those years the buffalo range in Kansas began about forty miles west of Topeka. Said Laurent:

> Many times have I seen men pass through Topeka going after buffalo calves. They would have a wagon and sometimes as many as four fresh cows tied thereto. When they arrived where the buffaloes were they would single out a buffalo cow and shoot her, then take the young buffalo calf to their cows. I have seen many calves following the cows tied to the wagons.[14]

Often it was not necessary to shoot a buffalo cow to get her calf. Oldtimers have said that if a calf happened to be separated from its mother, which frequently happened, they would get the calf to follow them home by simply blowing their breath into the calf's nostrils. Others have said that if they placed one finger in the calf's mouth and let the animal suck it for a moment, the calf would follow them. Other men simply roped a calf, provided its mother was not around, and took the calf home.

Samuel Walking Coyote, Charles Goodnight, and "Buffalo" Jones were successful in catching and raising buffalo calves. And there were others. Sometime during the late 1860s, Joshua Shaffer and his son Lewis of Linn County, Kansas, went west to about what is now Wichita, Kansas, where they killed several buffalo, skinned them, and loaded their wagon with much meat and several hides. They also captured several buffalo calves which they took back to their farm, where they raised them and trained a pair to work in the yoke as oxen. The animals caused much excitement in the neighborhood and at nearby county fairs. Finally, a traveling showman bought the buffalo and shipped them to Coney Island, New York, where they were displayed for several years.[15]

Other men were not so successful. In May 1875 F. W. Lockard of

Norton, Kansas, and five other men, went hunting for buffalo along the Solomon River near present-day Hoxie, Kansas; they captured several calves, enough to fill their large freight wagon. "We had little trouble in breaking them to the milk pail, but that long ride in the wagon seemed to be the cause of their undoing. Most of them died that summer," recalled Lockard.[16]

During the late 1880s, "Buffalo" Jones successfully trained a pair of buffalo bulls to work in harness, as Dick Wootton and others had done years before. One of Jones' pair, named Lucky Knight, had been captured as a calf by Jones in May 1886. The other, a six-year-old when Jones bought him, was a killer. He was a big and powerful animal and had killed his previous owner, A. H. Cole of Oxford, Nebraska, when Cole tried to keep him from getting onto a road to fight a domestic bull that had just sounded the challenge. Cole succeeded in getting the bull to stop, but then turned his back on the animal to leave. At that point the bull, apparently thinking he had bluffed Cole, charged and drove one of his sharp horns into his owner's back. Cole died about ten days later.

After Jones bought the buffalo bull from Cole's estate, he successfully broke both Lucky Knight and the Cole bull to the harness. "They are gentle, and work well together, but their keeper is ever warned never to give the one that has murder charged to his account. an opportunity to repeat the crime," said Jones several years later.[17]

"Buffalo" Jones began training the buffalo by using strong hemp rope attached to heavy forged iron bits in the buffalo's mouths. These lines ran back to a windlass on a two-wheeled cart. Jones could control either line by simply turning one of two cranks. With his feet, he controlled brakes on the cart. "The teamster can draw harder with one hand and foot than four men could without the windlass," said Jones. After he felt that the buffalo were fully trained, Jones sometimes drove them using only his hands and reins. But if he was planning to pass near water on such trips, he always used the windlass. "When they desire to drink . . . the windlass must control them," warned Jones.[18]

After his experiences one day in the 1880s, a German farmer in the Tongue River valley of Montana probably wished he had had a windlass. He had trained two buffalo cows to pull a wagon and was using them to haul a load of potatoes to town. It was a long drive,

"Cowboys, roping a buffalo on the plains," Dakota Territory, 1880s, J. C. H. Grabill photo. *Courtesy Library of Congress.*

Captured wild buffalo being loaded onto a train. From Joseph G. McCoy's *Historic Sketches of the Cattle Trade. Courtesy Kansas State Historical Society.*

C. J. "Buffalo" Jones with his team of 7-year-olds, Lucky Knight (left) and the A. H. Cole killer bull. During the winter months, Jones used the team to take feed to other animals on his Kansas ranch; in the spring he sometimes used them to plow the fields. *Courtesy Kansas State Historical Society.*

Yoked buffalo team belonging to Charles Goodnight and driven by "Buffalo" Jones prepare to pull sled in an exhibition of buffalo strength, Texas Panhandle, late 1880s. *Courtesy Panhandle-Plains Historical Society.*

Verne Elliot on bucking buffalo at Cheyenne Frontier
Days, 1910. *Courtesy Library of Congress.*

At Montana rodeo, early 20th century. *Courtesy William F. Whithorn.*

Russ Greenwood, Sundre, Alberta, with two buffalo, Punch and Judy, after he trained them for the Calgary Stampede, 1957. *Courtesy Russ Greenwood.*

Dick Rock, Henry's Lake, Idaho, riding a buffalo bull he raised from a calf. The bull was quite tame, Rock thought, but one day it turned on Dick and gored him to death. The buffalo had to be shot before Rock's body could be recovered.

and the buffalo became very thirsty. When they spotted the Tongue River some distance away, or smelled the water, they took off running. The farmer did everything he could to stop them, but his shouts and efforts to halt the team failed. When the buffalo reached the edge of the bank above the river, over they went headlong into the water. With them went the farmer, his wagon, and the potatoes. As the story goes, that farmer never again used buffalo to haul his produce to market.[19]

Charles Goodnight never broke a buffalo to harness, but in a letter to Joe Miller, president of the Miller Brothers' 101 Ranch in Oklahoma, he said he believed it possible to tame a buffalo to drive. Goodnight pointed out that the buffalo would have to have the right temperament. He felt, however, that the losses would be too great to make the experiment profitable.[20]

In spite of Goodnight's warning, Miller one day decided to see whether he could break two young buffalo bulls to pull a buggy. On a cold Sunday afternoon in January 1905, Miller sent two cowhands, both expert ropers, out on their ponies to rope two buffalo bulls in the 101 Ranch herd. Both animals were wild, and when the cowboys charged in, the bulls thundered off. Over the low sand hills along the Salt Fork of the Arkansas, the buffalo raced. Turning and dodging, they tried to escape the cowboys' swinging lariats. In a little while, one of the bulls was roped around his neck. But the pony, used to working lighter weight cattle, got a shock. Pony and rider were pulled along by the buffalo. Then suddenly, the bull, irritated by the drag, turned and with tail erect gave a bellow, charging the pony and rider. Fortunately for the cowboy, his pony had seen this move before with domestic bulls and gracefully dodged the buffalo. Meantime, the other cowboy rode into position, tossed his lariat around the buffalo's horns, and pulled from the other side. The buffalo stopped in his tracks. The force from the ponies and riders pulling from opposite sides was too much for the bull.

When the riders got the buffalo back to the corral area near the cookhouse, the cowboys who had been watching the excitement from a distance began shouting and cheering. One yelled for Dutch, the ranch cook, to ride the buffalo. The other cowboys joined in, demanding that old Dutch try his hand as a buffalo rider. Well, Dutch, not wanting the cowhands to think he was scared, climbed down from the fence and headed for the buffalo. As he neared, the bull began kicking. Dutch turned tail and retreated to a safe distance. At that moment White Buffalo, a Sioux Indian and the best horseman on the 101 in those days, rushed past Dutch, jumped atop the buffalo, waved his hat in the air, and jumped to the ground before the buffalo knew what was happening. This restored Dutch's confidence. A few moments later he was on top of the buffalo, both hands deep in the animal's mane. A second or two later, the cowboys dropped the ropes that held the animal. As if struck by lightning, the bull took off across the sand hills. Dutch held on for dear life and was not shaken from his seat as most of the cowboys had expected. Instead, he rode the buffalo like a trooper for one or two minutes. By then he had had his fill of buffalo riding and jumped off. The cowboys returned Dutch and the buffalo to the ranchhouse area.

As Joe Miller watched his cook take the buffalo ride, an old cowboy walked up and reminded his boss what "Buffalo" Jones, up in Kansas, had once said, that when a buffalo became too greatly enraged, it seemed able to die of its own volition. "It will stiffen its limbs and in a moment or two fall dead," said the cowboy in his slow drawl. Miller listened but passed off the remark and ordered the other buffalo bull roped. Both buffalo bulls were tied securely to a corral fence while three or four cowboys began the job of harnessing the pair. The buffalo did not like what was happening. They pawed the earth and fought to get loose, but the ropes held.

Bridles were slipped over their heads and harness was thrown from a safe distance onto their backs. The tongue of a heavy freight wagon was slowly shoved between the bulls and fixed into a neck-yoke. The traces were fastened to the singletrees with a long hooked iron rod. Then a lariat was fastened around the horns of each bull and held by a mounted cowboy, one on each side of the buffalo, to prevent a general smashup should the buffalo stampede when turned loose. Seven cowboys climbed aboard the large freight wagon and the buffalo were released. Like the angry animals they were, they broke for open country. The cowboys in the wagon fired their pistols into the air and the wagon shot across the open country like a dog with tin cans tied to his tail. It was indeed a strange sight. A couple of times the buffalo tried to turn upon each other, apparently thinking the other was responsible for the state of affairs, but each time one turned he was held back by the mounted cowboys speeding alongside on horseback and holding tight to the ropes.

The buffalo grew tired, and gradually their speed slowed. As it did, their rage grew more furious. Then they stopped. When the cowboys tried to get them to move, they refused. The larger bull dropped to his knees, his forehead pushing down into the dirt. Prodding did not move him. He only rolled over on his side and glared with his two red eyes at his tormentors. The other bull turned sideways and tried to break loose from his harness, but he was held back. As a cowboy tried to unfasten the traces, the bull lunged forward and cornered the cowboy between the wagon tongue and the wagon box. With an ugly toss of his head, the bull grazed the cowboy's ribs with one of his horns. The cowboy fell backward across the tongue and rolled over the other buffalo to the ground. An inch nearer and the buffalo might have killed the man.

At this point, Miller rode up on horseback and ordered that the mad buffalo be corraled. The animal was, after some difficulty, unhitched, tailed to its feet, and led toward the corral. But as the bull neared the fences, he broke away and charged full strength against a large corral post. There was a loud thump as he hit. The shock was terrific. Blood burst from the animal's nose, and before the cowboys could grab the draglines, the buffalo hit the post two more times with equal force. The cowboys then subdued the animal. Miller ordered that the bull be placed in a box stall, where his wounds could be looked after. But as the cowboys got the huge animal into the stall, the bull fastened his horns under the feed box and stiffened his legs in a desperate attempt to escape. He remained in that position for about ten seconds and then shook himself and fell over. He was dead.

Joe Miller was stunned. The bull had seemingly "died of its own volition." Perhaps, said Miller, "Buffalo" Jones was right. Miller walked away from the stall and ordered hands off the other buffalo for the rest of the day. But that night Miller was heard to say, "I'll have a buffalo team if I have to buy every dad-blamed buffalo in the country."

The following morning, the other buffalo bull had quieted down. He was roped in the corral, led outside, and hitched to a buggy. About then a group of Ponca Indians who lived nearby arrived at the ranch. They had heard about the buffalo wagon ride the day before. They came to see it for themselves.

The Indians watched as Miller climbed aboard the buggy and, handling the animal carefully, drove the buffalo a short distance with no trouble. When he returned to the corral, Miller invited one of the Indians to join him in a ride. The Indian smiled and climbed aboard and off they went. The cowboys and Indians watching from the corral fence slapped their sides and laughed at the sight of a white man and an Indian riding side by side, being pulled by a buffalo. It was perhaps symbolic of the changes that had come to the plains.

After the ride, Miller climbed down from the buggy and ordered that the buffalo be released. When it was, the animal took off as fast as he could for the sand hills, where the ranch's buffalo herd was peacefully grazing.[21]

Another Oklahoman to raise buffalo was Major Gordon W. Lillie, better known as "Pawnee Bill." He was a widely known showman who at one time was a partner with William F. "Buffalo Bill" Cody in a wild west show. Lillie was also a successful businessman. He retired early this century to his two-thousand-acre ranch southwest of Pawnee, Oklahoma, where he raised, among other things, pure-blooded buffalo. He started with a herd he purchased from a Missourian named Casey. Then, whenever possible, he bought cows to increase his herd. When an imperfect calf was found, it was removed and sold for meat in the East at fancy prices. In the early 1930s Lillie claimed he owned the second largest pure-blooded buffalo herd in the United States. He continued to raise buffalo until he died in early 1942 as he neared his eighty-second birthday.

Bob Yokum of Pierre, South Dakota was another man who raised buffalo. He also trained them. But unlike Joe Miller in Oklahoma, Yokum did not try to break full-grown buffalo. He started with young buffalo calves. About 1905, Yokum bought a pair of four-month-old calves and, with patience and understanding, worked with them nearly every day. By the time they were two years old, he had them pulling a small two-wheeled cart around his ranch. One afternoon, he drove them into Pierre and caused quite a bit of excitement. The buffalo moved at a slick pace, turned sharp corners, and circled about as the townsfolk watched. Yokum had lots of fun that afternoon with his buffalo. They responded well to his training and convinced him that the best way to train a buffalo was to start when the animal was young.[22]

This is also the firm belief of Russ Greenwood of Sundre, Alberta, Canada, a modern buffalo trainer who has broken other animals, including horses and Highland steers. In 1956 Greenwood, then sixty-three years old, was asked to break a pair of yearling buffalo for the Calgary Stampede. Greenwood said he would try. When fall arrived, the two young buffalo were delivered to him at Sundre. He began to get to know the animals and worked with them almost daily. By spring he had them pulling a two-wheeled cart with ease. In the summer of 1957, Greenwood's trained buffalo were a big hit at the Calgary Stampede.

In 1969, seventy-seven years old and still breaking saddle horses, Greenwood told me how he trained his buffalo to drive. First, he said, you must get the animals used to halters. But "a regular halter is no good." He makes his own. Everything Greenwood used in training the buffalo was homemade and "all from my own design."

> My halter leads from over the top of the animal's nose. I put rings in the buffalo's noses to start with, but later take them out. I have broke lots of oxen in my day and I go about it a lot the same way with the buffalo. There's no one in the world who uses a yoke like mine. It's all my own planning. It's a cross between a Nova Scotia horn yoke and the homesteader's ox yoke.

After getting the buffalo used to wearing his homemade halters, Greenwood led them on horseback into Sundre, where they became familiar with the noise of cars, trucks, and people.

> Every animal—dogs and cats and horses—is usually scared of them buffalo, but people—they don't have brains enough to be scared of them. I don't have much trouble with the Indians. They know enough to keep back, but lots of white people will want to go up and pet the buffalo. That's dangerous. I don't think there's such a thing as a tamed buffalo. I don't figure I have my buffalo tamed, but I think I have made a very good job of training them to drive.[23]

Another man who trained buffalo was Buddy Heaton, who lived east of Hugoton, in southwestern Kansas. He trained buffalo for parades and rodeos from Canada to Texas. He started with a six-month-old calf named Grunter. Heaton made friends with the young buffalo, and the calf followed him around like a dog. When Grunter was about a year old, Heaton decided to break the buffalo to the saddle. With a friend's help, he snubbed the buffalo close to the side of a saddle horse and cinched a saddle on Grunter. Heaton and his friend then climbed atop the buffalo and the horse.

"We both got throwed off," recalled Heaton. "Boy, did the buffalo buck."

Eventually, Grunter got used to the idea of being ridden. About 1959 Heaton rode him up the steps of the Colorado statehouse in Denver and in doing so outraced a quarterhorse. In 1961 Heaton and his buffalo, by then called Old Grunter, gained national attention as Heaton rode him down Pennsylvania Avenue in the inauguration parade of President John F. Kennedy. In the winter of 1961, Old Grunter appeared as Clyde, a tame buffalo, in the "Wagon Train" series on television.

"He's as gentle as a saddle horse," asserted Heaton. "I have perfect control over him—except when I go to saddle him."

Once when Heaton was away from home, Old Grunter got out of his corral and wandered onto a highway. There he just stood. Cars and trucks were backed up for three quarters of a mile in short order. Someone called the Kansas Highway Patrol, but when a trooper arrived, Old Grunter would not let the man get out of his patrol car. The trooper finally went to Heaton's farm to get help to herd Old Grunter back to his corral.[24]

The first buffalo bull calf captured by Charles Goodnight on his Texas Panhandle ranch in 1878 grew to become an extraordinary animal. Old Sikes, as the bull became known, was as independent as Goodnight. Barbed wire meant nothing to Old Sikes. If he decided to go somewhere and barbed wire happened to be in his path, he went through it. Nothing could keep him in an enclosure. At feeding time it was not uncommon for Old Sikes to go to Goodnight's horse corral, stick his head under the gate, lift it off its hinges, and go in to eat his fill of corn. When that happened, Goodnight's horses were usually scared and scattered. But Goodnight put up with such problems. He liked the old buffalo.

One day in 1886, Old Sikes took off, and Goodnight sent two ranch

Don Hight, buffalo rancher from Murdo, S.D., leads 18-year-old Buffalo Bill down a Denver street. This steer reportedly is the only housebroken buffalo in captivity. *Courtesy Dana C. Jennings, Madison, S. D.*

Buddy Heaton riding Old Grunter in the presidential inauguration parade, January 20, 1961. *Courtesy Wide World Photos.*

Buffalo jumping into a pond. An old publicity postcard identifies the photo as having been taken at the Courtland Du Rand Big Elk Ranch near Martinsdale, Montana, where "Elk and Buffalo dive 40 feet into water; where guests ride swimming Buffalo, Elk and Horses." *Courtesy Conservation Library Center, Denver Public Library.*

Handling buffalo is always a potential problem. This animal is attempting to jump a 6′ fence. *Courtesy South Dakota Highway Department.*

hands to bring him back. They traveled several miles before finding him. At first, they tried to chase him back toward the ranch, but Old Sikes would not budge. They then picked up handfuls of buffalo bones off the prairie and kept tossing them at him until he decided to start for home. Back at the Goodnight ranch, Old Sikes was put in a corral with a six-foot fence. All thought the corral would hold Old Sikes, but when they checked the following morning, the buffalo was gone. During the night he apparently jumped the fence and headed for the Salt Fork.

Again, Goodnight sent some cowboys after him, but they lost the trail and gave up. It was not until the next summer that the old buffalo turned up during the roundup. Some cowboys returned the animal to the Goodnight ranch, but by then the rancher's patience was worn pretty thin. Old Sikes was dehorned and placed in a pasture that had been fenced for Goodnight's other buffalo. There the big buffalo seemed to be content.

When fall came, however, and Goodnight sent out Henry Taylor, one of his men, to pen Old Sikes for the winter, the buffalo turned on Taylor and chased him out of the pasture. Goodnight then decided he would do the job himself. But Old Sikes charged Goodnight. That was the last straw. Not long afterwards, Old Sikes disappeared from the buffalo pasture. On Christmas Eve, 1886, a 964-pound carcass hung next to the cookhouse. At Christmas dinner, everyone said the meat tasted better than beef.[25]

There seems to be little doubt in the minds of most experienced buffalo men that the shaggies have individual personalities much like dogs, cats, horses, and human beings. As many animal trainers are quick to admit, an animal's environment and treatment by man during the animal's early formative years play a big role in shaping the animal's personality, just as in human beings.

A captive buffalo owned by the sutler at Fort Hays, Kansas, in 1871, is a good example of this. The sutler came across the animal when it was a calf. Its mother had apparently been killed by hunters. He took it back to the fort and raised the little buffalo on a bottle. By the time the buffalo was two years old, he was spoiled rotten. Living just outside the sutler's store, where many a soldier quenched his thirst after a hot and dusty day on the prairie, the buffalo frequently got beer from the troops, sometimes by the bucketful. He developed a great fondness for it and sometimes would perform rather strange antics under its influence.

On one of these occasions, the buffalo decided to clear the "officers' room" in the sutler's store. He charged in and mounted a billiard table. The soldiers had a most difficult time trying to dislodge the buffalo, but after about an hour they succeeded.

Another time, after indulging in too much brew, the buffalo climbed the stairs leading to the second story. With great difficulty some of the soldiers and the sutler finally persuaded the buffalo to climb down the outside stairs.

But the animal's excesses—encouraged by man—plus the lack of proper care and a most unnatural diet finally brought the buffalo's downfall. He began to grow thin, and he died at the age of three.[26]

During the 1870s, other frontier communities on the edge of the buffalo range had captive buffalo. Those in Dodge City, Kansas provided material for many stories that are still told.

George Reynolds, the eldest son of Philander G. Reynolds, who owned and operated several stage lines out of Dodge, frequently drove a stage for his father down into Indian Territory. Whenever he observed any buffalo, he would stop the stage, unhitch one of the horses, and go riding out over the plains trying to rope a buffalo. Sometimes he would be gone for days at a time, leaving passengers stranded in the stagecoach.

By the early 1870s George Reynolds had roped two buffalo calves and raised them on bottles from infancy. They had the run of Dodge City, and, like any two-year-old child, they often got into trouble. They roamed through residents' backyards, poked their noses into kitchen doors and windows, and begged for bread or anything else they could get. Dodge City almost needed a buffalo catcher.

One day a roadshow came to Dodge City. It had a big flashy band with two dozen players. Their instruments, shined for the occasion, glistened in the sun. The band leader was a tall man. He wore a big bearskin cap, had a large baton, and was dressed brightly as were all members of the band. As they marched down Bridge Street, their music reverberated off the wooden buildings and flowed on across the prairie. The two buffalo, hearing the strange sounds, wandered up into a yard to see what was going on. They stopped in their tracks and gazed in wonder at the flashing instruments. When the band leader moved close to where the buffalo were standing, he waved his baton toward the animals' faces. That did it. Heads lowered, the buffalo shot toward the band, sending the surprised musicians scattering in all directions. The bass drummer dropped the bass drum, and the buffalo stopped to attack it with their horns. Then one of them espied the bearskin cap dropped by the fleeing band leader and hooked it on his horns as if it were a vicious grizzly.

When the dust finally settled, the band members were everywhere. Some had taken to roost on a picket fence. Others had jumped into a ditch full of water. The buffalo, meanwhile, quietly walked away, seemingly satisfied that the band was, after all, quite harmless.

That night, after cleaning themselves up, the band gathered again to perform in the street directly in front of the building where the show was to be held. As they began to play with broken and dented instruments, someone led the two buffalo to the rear of the band. With a swat, he let them loose. The animals charged through the assembled musicians. Although the seats of many musicians' trousers were ripped and instruments were damaged even more, no one was seriously injured. But that night, as one oldtimer related, two dozen nervous men swore off the musical profession. They claimed that in Dodge City it was far too dangerous an occupation.[27]

Officials in charge of the U.S. government wildlife refuges do not, as a general practice, try to tame or train buffalo on their refuges. Most consider it out of character with the refuges, whose aim is to raise animals in as near a wild state as possible. Yet, occasionally refuge personnel have found it necessary to do more than observe and protect their charges.

Victor "Babe" May, foreman of the National Bison Range in western Montana, has twice raised orphaned buffalo calves. If he had not, they probably would not have survived. In 1961 a young heifer calf was found on the range alone. She was taken to the refuge headquarters and placed on a bottle. Soon she became something like a mascot. Several of the refuge men called her Jezebel, but May nicknamed her Knothead. After several months the young buffalo was taken to the annual Boat, Trailer and Sport Show at Portland, Oregon, as part of a government wildlife display. May went along to watch after her.

At first, she found all of the visitors interesting and she would come up to the small corral fence and let people pet her. But after a day or so she became bored with everything. And from then on, she ignored the visitors even when they would call her by name—"Jezebel." But you know, I could walk over to the fence and whisper the word "Knothead," and you know, she would perk up and come over to me.

May's other experience with a bottle-fed buffalo came in April 1964, when range hands found an abandoned heifer calf about two days old. Like Knothead, the calf was taken to refuge headquarters and was bottle-fed four times a day. After a while, feeding was slackened off to only twice a day, once in the morning and once in the evening. The little heifer, raised in the fenced yard of the assistant refuge manager, became known as Klunk, and grew to be a healthy buffalo. At six months, refuge officials returned her to the exhibition pasture, where they could keep an eye on her. She grew, matured, mated, and gave birth to a calf. Later, there was a second calf, and in June 1969 when I visited the National Bison Range, Victor May and the others at the refuge were anxiously awaiting the birth of Klunk's third calf. Not only did the calf arrive, but Klunk has produced two more calves for a total of five.[28]

Raising buffalo in captivity is not all fun and games. Buffalo can and usually do cause problems. Most of the buffalo men with whom I have talked complain about the animals' wandering off their normal range. The buffalo is erratic—perhaps the word should be independent; he may, without any warning, decide the grass is greener on the other side of a fence and simply go through it. At other times, buffalo simply take advantage of broken or downed fences and wander off.

Many years ago at the Wichita Mountains Wildlife Refuge in Oklahoma a buffalo bull broke through a fence and wandered up to the refuge headquarters building. When he was spotted, two buffalo men got their horses and tried to drive him back to his pasture. But, as Earl Drummond recalled:

> The bull went into a grove of trees and would not come out. So one of the men went in on foot to run him out. But the bull charged him. The man would run around one tree then make a run for another and run around it. The bull chased him around

three trees, then ran out of the grove and headed for the pasture.[29]

In the late 1950s Julian Howard, longtime refuge manager of the Wichita Mountains Wildlife Refuge, got a call reporting a small bunch of buffalo on the loose about ten miles south. Assuming they were some of the refuge's animals, Howard took some men and headed south. As he later told me:

> When we got down there, southwest of Indiahoma, we found a yearling calf, a couple of cows, and a four-year-old bull. We tried for two days to herd them back up to the refuge. But the buffalo always ran on us. They tore up some fences and a corral in the process. Then one of the boys from the refuge got a close look at the calf and noticed it wasn't branded. All of the refuge buffalo are. Well, we started checking around and found out those buffalo weren't even ours. They belonged to some farmer near Frederick, Oklahoma, about thirty miles to the south. Those buffalo had broken out several days before and just wandered north. When we called the farmer he came up and tried to take them back to his place, but he couldn't. He finally had to shoot them.[30]

A Kansas farmer who raised buffalo as a hobby had about as much trouble with his animals, but he did not kill them. Owen C. McEwen kept three buffalo—a bull and two cows—on his farm near Augusta, Kansas. One day the bull jumped the fence and took off crosscountry. A few hours later, someone called the sheriff and reported seeing a buffalo. The sheriff went out, and, sure enough, it was McEwen's buffalo bull. For more than five hours the sheriff, McEwen, and some other men tried to corner the bull, but the animal wanted no part of it. The bull gouged a horse and caused general havoc until someone roped the 1,800-pound animal and quickly anchored him between two pickup trucks. After he had received two tranquilizer shots, he was loaded into a large horse trailer and returned to McEwen's farm. McEwen said later that his biggest problem in raising buffalo was to provide "adequate fencing" to hold the animals.[31]

Erwin H. Weder of Highland, Illinois had a similar problem with one of his buffalo. In 1950 Weder began raising buffalo commercially. Seven of them were transported to his farm near Decksprairie, Illinois, where he was building his herd. "We unloaded them into a barn, but one of the bulls went right through the side of the barn, over the fence and away," recalled Weder. Fortunately, he had four saddle horses standing by. With three other men, Weder gave chase. "We chased that buffalo for sixteen miles through and over barbed wire fences until we were finally able to rope him and bring him back in a trailer," he said.

Another time, as Weder and some hands were loading three bulls into a trailer to take them to his farm, the animals refused to load. Weder began shoving boards through the slots of the trailer's wall to prod the animals. Slowly the buffalo moved forward. Everything went smoothly until suddenly one old bull let go with a kick the force of dynamite. The board flew off the side of the trailer hitting Weder with such force that it knocked him sixteen feet, where he hit the ground and lay unconscious. Weder recovered. "The board hit me just across the ear and almost severed the lobe," he recalled.[32]

Is it possible that a small number of wild buffalo, descendents of the wild plains buffalo of a century ago and untouched by man, still exist in some remote area of the United States? Most unlikely. Yet, a stray buffalo or two, escaped from some small, private herd, such as the above stories relate, occasionally gives rise to the tantalizing question.

In October 1966 David Driskill struck and killed a buffalo bull with his auto on the northern outskirts of Turkey, in the lower Texas Panhandle. The bull was one of three buffalo seen running wild in Hall County that year. One of the animals was captured on the Mill Iron Ranch and taken to the Amarillo Storyland Zoo. No one came forward to claim the buffalo. Various theories about their origin were expressed by authorities, but their owners were never found.[33]

Six years later, in May 1972, L. B. Robertson of the Margaret community in Foard County, Texas, about eighty miles southeast of Hall County, reported seeing a wild buffalo. Sheriff Emmett R. Howard later located the animal in the middle of the night "running full blast on California Street headed west in front of the City Hall" at Crowell, Texas. Riding in the back of a pickup truck driven by Sheriff Howard, Jack Walker soon roped the buffalo. The animal was shot full of tranquilizers and loaded into a stock trailer. When no owner came forward, the buffalo was given to Ace Reid, the western cartoonist, who took the animal to his "Cowpoke City" near Electra, Texas.[34]

Whether these buffalo were the remnants of a "lost herd" is doubtful, yet their owners—if there were any—were not located. Such occurrences keep alive the lost herd legends. But private herd fugitives remains the logical answer.

Adrian W. Reynolds, a member of the Wyoming legislature from Green River, recalled in 1969:

> A few years back, a herd of buffalo, escaped from a private herd on the Sweetwater river north of Wamsutter started developing in the northeast corner of the Red Desert, and were reported by the game and fish boys to be, variously from eighteen to seventy-five head. No recent reports, and the rumor is that the uranium miners have enjoyed buffalo hump and tongue.[35]

An island in the Pacific with lush tropical growth, palm trees, wild orchids, volcanoes, and a never-changing climate seems to be an unlikely place to find buffalo. Such a setting is a far cry from the grass-covered buffalo plains that stretch southward in a wide swath from Canada to Texas. Yet in the state of Hawaii, captive buffalo have prospered. The Honolulu Zoo, like many other zoos around the world, maintains a small display herd of captive buffalo. It was started in 1951.[36]

In 1969 the Kahuku Ranch near Naalehu, on the island of Hawaii, had a dozen buffalo, each about a year old. The animals had come from North Dakota, where they were purchased in 1968, loaded aboard a large truck, and transported to San Francisco. There, a huge jet transport was waiting to fly the small buffalo herd to its new home in Hawaii. Plans called for the truck to unload the animals directly into three large crates, four buffalo to a crate. But when the truck arrived, workers found that it was too large to enter the building where the crates were resting on scales to determine the animals' weight. Workers tried a high lift, but soon found it would not work.

Freddy Rice, manager of the Kahuku Ranch, said workers then put together a "makeshift alley of crates forming a crude walkway" leading from the truck parked outside the building to the crates on the scales. Another precaution was taken. Rice had heard that buffalo did not like the dark. Rice and his men waited until after dark to begin unloading the animals. When darkness set in, all lights were turned off in the building except for a single floodlight at the end of the alley. The light attracted the buffalo. As Rice said:

> They went straight into the lighted boxes. They lay down in the boxes most of the way and were never upset. In Honolulu, they were transferred to new crates by a high lift and flown to Hilo, and then transported to the ranch without incident.

Several thousand miles from their old plains home, the buffalo on the Kahuku Ranch began to roam over 200 acres set aside for them. And they have fared very well. Rice noted:

> I found that the buffaloes' sense of survival is very strong, and they overcome fear. They will go into a pen or come to a man if it's the only way they can get feed or water. Cattle by comparison would starve rather than go into a pen with a man in it unless they were tame and used to the surroundings.

Rice said that the only problem with the buffalo was that they walked over the ranch's stone fences with ease.[37]

The oldest buffalo herd known to exist on any Pacific island is on Catalina, just off the coast of California near Los Angeles. The story of how that herd came to be is as fascinating as the island itself.

In December 1924 Paramount Pictures was preparing to film "The Vanishing American," one of the few early films dealing with the American Indian as a Western hero. One of the crew, Tom White, had the job of getting some buffalo on the island for part of the filming. White shipped fourteen buffalo to Catalina, where they were uncrated and slowly herded to the hillsides on the western end of the island. The buffalo made themselves at home. The following February, Famous Players-Lasky used the buffalo in the filming, directed by George B. Seitz. The picture starred Richard Dix, Noah Beery, and Lois Wilson, among others. When the filming ended, the movie company had no further use for the buffalo and did not want to go to the expense of rounding them up and shipping them back to the mainland. So the buffalo were left on Catalina to fare for themselves.

Since 1925, the original fourteen buffalo have prospered on the rich grassland that covers much of the island, and they have increased in number. In 1969 the herd numbered 300. Today, they are owned by P. K. Wrigley of chewing gum fame, and the animals range over 30,000 acres of Wrigley's Catalina Rock and Ranch Company property on the island. Several years ago, an improved management plan was begun in hopes of developing the Catalina buffalo herd into a commercial meat source.[38]

In 1970 Hollywood returned to Catalina Island. Producer Stanley Kramer used the island and the buffalo herd for portions of the Columbia Pictures movie "Bless the Beasts & Children," based on Glendon Swarthout's novel of the same title. It stars six teen-age boys, castaway sons of wealthy parents, who meet at a summer camp in Arizona. They witness the annual "kill" of buffalo in the state, whereby winners in a lottery may, for a fee, go to a state corral and shoot one animal. The "hunter" gets the head, the hide, and a quarter of the meat. The rest may be purchased by the public.

Kramer said the movie was an attempt "to humiliate a segment of our society." He continued:

> I was astounded and sickened by the barbarous way the buffalo were killed. Often the shooters are amateurs and wound the animals in several places. They stagger, bleeding, around the corral until a sharpshooter finally delivers the fatal shot. All of the buffalo are driven to within one hundred yards of the shooters. If the buffalo try to move away they are driven back into close range by riders. Real hunters wouldn't have deer driven to them by beaters. Cattle and hogs are killed humanely. But not these buffalo.[39]

When Kramer asked permission to film the picture in Arizona, the governor was enthusiastic, but members of the Arizona Game and Fish Commission took a dim view of plans to film the actual killing of their buffalo. Kramer appeared before the Commission in person. Members assured him that their annual shoot was "necessary in order to 'weed out' the buffalo, for 'feed purposes,' " but after a vote, the commission refused to grant Kramer permission. Thus, he went to Catalina Island, where the Arizona kill was re-enacted with the Catalina herd.[40]

"Bless the Beasts & Children" was a controversial film and provoked much pro and con response. It is interesting to note this news item that appeared in the aftermath of reaction to the movie:

> The New Mexico Department of Game and Fish announced here this week that it had abandoned plans to establish a huntable herd of buffalo in the state and that it would sell most of the animals during the next two months.
>
> A token herd of about 35 of 140 buffalo now at the Fort Wingate reservation will be kept for "viewing purposes" at the 5,000-acre fenced-in United States Army installation, according to Walter Snyder, the department's chief of game management.
>
> Mr. Snyder said that "anti-hunter sentiment and public resentment" had led to the decision to discontinue the pilot project begun six years ago.
>
> The department has been authorized by the State Game Commission to donate six buffalo to the Jicarilla Apache Indian tribe, which had requested the animals for "ceremonial purposes," and to give two buffaloes to the Albuquerque Zoo, he said.
>
> The remainder, as well as the surplus that are produced by the small herd at Wingate, will be sold, presumably to other zoos and owners of private reservations, Mr. Snyder said.
>
> "When the project was started at Wingate with 15 imported animals," Mr. Snyder said, "the buffalo in a wild state had been extinct in New Mexico for almost 100 years, largely because of their ruthless annihilation by meat and hide hunters who roamed the mountains and plains in the nineteenth century.
>
> "It had been expected that eventually the expanding Fort Wingate herd would be shifted to the vast stretches of the MacGregor Military Range in southern New Mexico. But when it became apparent that it would be most difficult to conduct sporting buffalo hunts, we approached organized sportsmen's groups, who were almost unanimous in their opinion, based on their knowledge of buffalo hunts staged in other states, that the plans for building up a huntable herd in New Mexico should be discontinued."[41]

Did Kramer's movie have any significant effect in developing the "anti-hunter sentiment and public resentment" that led to the cancellation of New Mexico's buffalo project, or was it sheer coincidence that the decision came in the wake of this powerful film? Perhaps one can only speculate on the answer. When the New Mexico Department of Game and Fish was asked specifically if the movie contributed to the public sentiment lying behind their decision, they replied: "We definitely feel that attitudes against hunting are influenced by many TV and movie programs. We cannot, however, weigh the effect that any single program would have in developing this sentiment."[42]

The buffalo owned by the state of Arizona were purchased from James "Uncle Jimmy" Owens in 1927 for $10,000. Owen's animals—nearly a hundred—were remnants of a herd established by C. J. "Buffalo" Jones in the late 19th century at House Rock Valley, north of the Colorado River in what is now known as the Grand Canyon National Game Preserve. New blood was introduced into the herd in 1941 when a dozen young buffalo were shipped from the Wichita Mountains Wildlife Refuge in Oklahoma. Eighteen more were added to the Arizona herd in 1945 and ten more in 1956.

Still another Arizona herd was established in 1949 on the Huachuca Wildlife Refuge in Cochise County. The site was historic Fort Huachuca, then abandoned by the military. Animals for this herd came from the National Bison Range in Montana, Yellowstone National Park, and the House Rock Valley herd.

In November 1954, after it was announced that the Signal Corps would reactivate Fort Huachuca, chances of maintaining the buffalo herd on the reservation appeared slim. Thus, the Arizona Game and Fish Commission in January 1955 held a hunt to remove most of the animals. A Commission bulletin summarized:

> Some 217 head were removed in the hunt, with 39 head trucked to Raymond Ranch, and 20 head given to the state of Sonora, Mexico. The herd had lasted but five years at Huachuca. It is the intent of the Arizona Game and Fish Commission to keep a herd averaging 150 head at Raymond Ranch and a herd averaging about 200 head at House Rock. . . . It has become the policy of the Arizona Game and Fish Commission to crop the excess over carrying capacity of the range every two years. To accomplish this, a hunt, restricted to legal Arizona residents only, is held on alternate years at each of the ranches.[43]

Eliminating the buffalo from the Fort Huachuca range left two state-owned buffalo herds in Arizona, one on the 15,717-acre Raymond Ranch between Flagstaff and Winslow, and the other on the 67,500-acre South Canyon Ranch better known as House Rock Ranch north of the Grand Canyon on the east side of the Kaibab National Forest.

Initially, the Arizona Game and Fish Commission charged a fee of $40 for each hunter wanting to shoot a buffalo. The fee entitled the hunter to the head, hide, and one quarter of the meat. The rest of the meat was sold to the public to help defray the cost of maintaining the herds. By the early 1970s the fee was increased to $45. But on March 3, 1973, the Game and Fish Commission issued a new policy statement, "Concerning the Harvest of Buffalo by Firearms":

> The Commission recognizes its two managed buffalo herds in the State of Arizona are a unique and valuable resource of the State. In order to perpetuate this resource for the benefit of the citizens of this State, it is necessary to manage the herds and the range which supports them. Proper management entails harvest of surplus buffalo.
>
> The Commission believes that shooting is a proper method to accomplish a regulated and controlled harvest.
>
> Therefore, the Commission hereby declares that it is its policy and direction to the Department that when firearms are utilized to harvest buffalo, the Department shall not permit the shooting of buffalo under any condition in which the buffalo, prior to and for the purpose of such shooting, are placed within an enclosure of such size so as to prohibit their movement from the practical range of a firearm.

The revised policy was adopted, in part, because of pressure from various organizations, including the Arizona Society for the Prevention of Cruelty to Animals.

In August 1973 Robert A. Jantzen, director of the Game and Fish Commission, wrote that much of the publicity about the Arizona buffalo harvests "has been grossly inaccurate as to the methods employed in the annual removal of the animals." He continued:

> Most of this started with the book and motion picture *Bless the Beasts and Children.* Although the story is pure fiction, many people seem to feel the buffalo shoot described in it is factual.

This is definitely not the case. The hunt scenes and the behavior portrayed in the book and film are as appalling to us as they are to any other civilized people. . . .

Being healthy animals, our buffalo reproduce, making some form of removal necessary. The range on which they exist is limited, and it would be physically impossible to simply allow their numbers to build up without any form of control. Some people have suggested to us that buffalo should *never* be killed, but the alternatives are impossible. The concept of giving them to various zoos, parks, etc., is a worthwhile idea, but unfortunately, the market for live buffalo is very limited. It would be impossible for us to find "new homes" for 80 to 120 or more animals each year, although this is the number which we feel should be removed from our herds for the best possible management of the animals and the range on which they exist.

Buffalo are not endangered in America. There are a number of other herds being maintained on the state, federal, and private level, so the popular idea that buffalo are on the verge of extinction is simply not true.

Another idea which has been suggested is that we simply feed the buffalo artificially. Because they reproduce steadily, however, this would soon take a haystack of gigantic proportions to meet their requirements for food. Even if we did follow such a procedure and could manage the staggering financial burden it would entail, we would eventually reach the point where some culling would be necessary.

This brings us to the point, then, where we have to accept the idea that some buffalo must be removed from the herd. There are a couple of ways this could be done. One would be to run the herd as a straight cattle ranch with a slaughterhouse operation when it came time to thin the population. This type of operation would involve a considerable amount of additional expense, and it would deprive sportsmen of the opportunity to participate in the removal. We are also unconvinced that a typical slaughterhouse operation would be more humane than a swift bullet in the brain. Another suggestion has been to let our own people do the shooting, but this would eliminate the sportsman from the scene entirely. In a sense, Arizona's sportsmen have a special interest in the buffalo herds' management, as they have had to support them all these years.

The sportsmen actually help the Department by participating in the hunt. For one thing, the fee they pay helps defray the cost of maintaining the herd. . . . The Arizona taxpayers — or the country's taxpayers for that matter—do not spend any money to support the buffalo herd. It is financed entirely by sportsmen.

After you consider these alternatives, I think you will reach the point where you agree there is no way to get around the idea that a buffalo harvest must be effected. There might be a more humane method of dispatching the animals, but so far no practical ideas have been heard. Buffalo are large, powerful animals—much more dangerous than beef cattle—so we avoid "handling" them any more than is absolutely necessary. The safety factor is of major importance—both for our personnel and for the participants and spectators, and this is reflected in our actual hunt procedures.

One point the critics of the hunt have missed completely is that the animals are shot at relatively close range to insure a quick kill. Some critics have complained of cruelty, and then in the next breath they complain that the animals are at close range in a fenced area. This admittedly doesn't offer the potential of a sporting hunt, but we have preferred to accomplish the job this way rather than run the risk of wounding a buffalo in an actual hunt situation. In effect, we have chosen a middle ground between a point blank shot and regular hunting. We feel the close range maintains humanitarian standards while still offering some small measure of sporting opportunity.

To make sure that the type of needless suffering portrayed in *Bless the Beasts and Children* does not occur, each hunter is accompanied by one of our personnel. If the hunter fails to kill the animal quickly and cleanly, our man immediately drops it. The average life span of a buffalo from the time the trigger is squeezed until it is lying dead on the ground is probably only a few seconds. There is, in our opinion, no more humane *and* practical way to handle the harvest.

Although many people become indignant over the manner in which we harvest the buffalo and personally do not feel they would care to participate, it would be well to bear in mind that had it not been for the sportsmen of Arizona, this state would never have had buffalo in the first place. They are not native here. Also, if the sportsmen had not been happy to finance the operation of our ranches down through the years, it would have been impossible for us to maintain these two herds and to provide Arizonans with an opportunity to see this living heritage of the American west.[44]

October 1973 was the starting date for the new buffalo hunting procedures at Raymond and South Canyon (House Rock) ranches. A hunter wanting to kill a yearling was asked to pay a $160 fee. It entitled him to the whole animal, and he was expected to do his own skinning and provide his own means of transporting the carcass home. A sportsman wanting to shoot a mature buffalo was asked to pay a $100 fee. It entitled him to one-fourth of the animal's meat. Officials of the Game and Fish Commission, who would skin and butcher the mature animals, planned to sell the surplus meat to the public. Bill Sizer, chief of information and education for the Commission in Phoenix, said that the surplus buffalo in 1973 included thirty yearlings and nineteen mature buffalo at Raymond Ranch, and twenty-nine yearlings and forty-eight mature buffalo at South Canyon (House Rock) Ranch. Only these animals would be killed, he said. Sizer pointed out that hunters could drive their cars, trucks, jeeps, or other vehicles into the hunting areas, but that such vehicles must be kept on the roads. "Most hunters probably will walk in to make their kills," he said, adding that the new hunting procedures in Arizona give the buffalo a better chance to escape. "It's really a buffalo hunt now," concluded Sizer.[45]

What is perhaps the largest buffalo herd—private or government—east of the Mississippi belongs to Bill Neff of Harrisonburg, Virginia. Neff's herd was started in 1968 when he purchased a three-year-old buffalo from another Virginia man for $300. Later he purchased other buffalo. By 1973 Neff's herd numbered more than fifty. Grazing in a large pasture between Interstate 81 and the western foothills of the Blue Ridge Mountains, Neff's buffalo startle tourists and truckers. So many began stopping to see the buffalo that the Virginia State Police were forced to erect a "No Stopping" sign next to the pasture to avert accidents. Neff put up his own sign to warn the curious. It read: ABSOLUTELY NO TRESPASSING — NO ONE ALLOWED ON OTHER SIDE OF THIS FENCE—BUFFALO ARE DANGEROUS. Neff, after five years' experience raising buffalo, observed that there is little difference in raising buffalo and beef cattle—with one exception, "When they get riled up they run like express trains. I've seen them run right through strong fences and they're amazing jumpers." Neff recalled that when one of his animals strayed from the pasture in 1973, it jumped over a pickup truck parked in its path. "Buffalo," Neff added, "are really very nice animals. My sole reason for getting into this is so that someday people will say the buffalo is back and that someone like me had the foresight and imagination to accomplish it."[46]

XVIII

Cattalo

The hump of the cattalo was the best damn meat I ever et.
John Simon 19th century

Credit for coining the word "catalo" about 1888 to identify a cross between domestic cattle and buffalo goes to C. J. "Buffalo" Jones.[1] Since then other spellings of the word have developed. Catallo, cattalow, and cattalo are also used to identify a cross between a buffalo and other domestic animals. As the title of this chapter indicates, I have chosen the last spelling. Webster says it is proper. It is also the spelling used by the late Charles Goodnight and many other writers.

During the late 19th century, Jones tried to give the impression that he was the first person to get the idea and to succeed in crossing a buffalo with a domestic cow: "I turned my attention to carrying out my original idea of producing a race of animals, by engrafting the buffalo upon our domestic cattle."[2] But the idea was hardly new in Jones' time. Though there is no reason to doubt that he created the word catalo, the idea of crossing buffalo with domestic cattle was, in 1888, already at least 200 years old. It may have been in the mind of Don Juan de Onate, governor of New Mexico, when, late in the 16th century, he sent Vicente de Zaldivar on the unsuccessful expedition to capture wild buffalo for domestication. Other Spaniards may also have had the thought, but whether any actually succeeded is unknown.

We have noted (in preceding chapter) that "people of distinction" living from the Carolinas northward to Pennsylvania raised buffalo calves during the mid-18th century and crossed many of them with domestic cattle.[3]

Domesticated buffalo also existed in some of the counties of Virginia in the 1780s. Albert Gallatin, secretary of the treasury under

271

Thomas Jefferson, wrote in 1837 that many buffalo were bred to domestic cattle and that the mixed breed was fertile. Gallatin told one story about a buffalo bull owned by a farmer "living on the Mononga-hela, adjoining Mason and Dixon's line":

> The bull was permitted to roam at large and was no more dangerous to man than any bull of the common species. But to them he was formidable and would not suffer any to approach within two or three miles of his own range. Most of the cows I knew were descended from him. For want of a fresh supply of the wild animals they have now merged into the common kind. They were no favorites, as they yielded less milk. The superior size and strength of the buffalo might have improved the breed of oxen for draft, but this was not attended to, horses being almost exclusively employed in that quarter for agricultural pursuits.[4]

Unfortunately, little else is known about the crossbreed experiments described by Gallatin. Even the names of the persons who conducted the experiments and where they lived have been lost in time.

The earliest crossbreeding account with some detail comes from Robert Wickliffe of Lexington, Kentucky. If it had not been for John J. Audubon, however, Wickliffe's experiences might never have been recorded. Hearing that Wickliffe was raising buffalo, Audubon wrote him a letter in 1843. Wickliffe replied in a letter dated November 6, 1854, which said in part:

> The herd of buffalo I now possess have descended from one or two cows that I purchased from a man who brought them from the country called the Upper Missouri; I have had them for about thirty years, but from giving them away and the occasional killing of them by mischievous persons, as well as other causes, my whole stock does not exceed ten or twelve. I have sometimes confined them in separate parks from other cattle, but generally they herd and feed with my stock of farm-cattle.
>
> On getting possession of the tame buffaloes, I endeavored to cross them as much as I could with my common cows, to which experiment I found the tame bull unwilling to accede, and he was always shy of the buffalo cow, but the buffalo bull was willing to breed with the common cow.

From the domestic cow I have crossed half-breeds, one of which was a heifer; this I put with a domestic bull, and it produced a bull-calf. This I castrated and it made a very fine steer, and when killed produced very fine beef. I bred from the same heifer several calves, and then, that the experiment might be perfect, I put one of them to the buffalo bull, and she brought me a bull calf, which I raised to be a very fine large animal, perhaps the only one in the world of his blood, namely, a three-quarter, half-quarter, and a half-quarter of the common blood.

After making these experiments, I have left them to propagate their breed themselves, so that I have had a few half-breeds, and they always proved the same, even by a buffalo bull. The full-blood is not as large as the improved stock, but as large as the ordinary stock of the country. The crossed or half-blood are larger than either the half-blood or common cow. The hump, brisket, ribs, and tongue of the full and half-blooded are preferable to those of the common beef, but the round and other parts are much inferior. The udder or bag of the buffalo is smaller than that of the common cow, but I have allowed the calves of both to run with their dams upon the same pasture, and those of the buffalo were always the fattest; and old hunters have told me that when a young buffalo calf is taken, it requires the milk of two cows to raise it. Of this I have no doubt, having received the same information from hunters of the greatest veracity. The bag or udder of the half-breed is larger than that of full-blooded animals, and they would, I have no doubt, make good milkers.

Wickliffe observed that the half-breed heifer "will be productive from either race. I have tested beyond the possibility of a doubt." However, Wickliffe noted the sterility of half-buffalo bulls. He said his experiments did not satisfy him that they would produce again.

About 1845, two years after writing to Audubon, Wickliffe gave up his buffalo and crossbreeding experiments. Though it had been interesting, he really had very little of permanent value to show for his efforts.[5]

Nowhere have I found any records indicating buffalo crossbreeding experiments in North America between 1845, when Wickliffe quit, and the early 1870s. About 1870, however, J. W. Cunningham of Howard County, Nebraska began to raise buffalo. His little-known account is tucked away in the November 10, 1876 issue of *Turf, Field, and Farm,* a newspaper published in New York City:

> The buffaloes on my ranch consisted of two young cows and one bull. I fed them carefully with the cows, but kept them confined at night. In the spring it was discovered that two of my cows were with calf by the buffalo bull. The calves proved to be both heifers. When three years old they became mothers, although showing some of the buffalo characteristics, proved to be very good milkers, quite gentle, giving an average of fourteen quarts of milk per day for at least five months, and such rich milk I never saw. This strain of buffalo stock extends through a considerable section of Howard County. I have a half-bred bull of this stock which proves to be both useful and attractive. There are others, I learn, in other sections of Nebraska who own half and quarter breeds that prove to be very hardy.

Whether C. J. "Buffalo" Jones or Charles Goodnight read of Cunningham's experiments is unknown. Jones began his experiments about 1888 to "produce a race of cattle equal in hardiness to the buffalo, with robes much finer, and possessing all the advantages of the best bred cattle," as he said. For several years Jones conducted his experiments and in 1899 declared:

> Catalo are produced by crossing the male buffalo with the domestic cow. Yet the best and surest method is the reverse of this. Only the first cross is difficult to secure; after that, they are unlike the mule, for they are as fertile as either the cattle or buffalo. They breed readily with either strain of the parent race — the females especially.

But Jones, like Robert Wickliffe three-quarters of a century earlier, found it difficult "to secure a male catalo." Jones reported: "I have never been able to raise but one half-breed bull, and he was accidentally killed before becoming serviceable."[6]

Texan Charles Goodnight, who also began experiments in the late 1880s, was more successful than Jones, although both men really did not keep dependable records. Goodnight's buffalo bull, Old Sikes,

C. J. "Buffalo" Jones'
cattalo cows, c. 1890.

Part of "Buffalo" Jones' buffalo and cattalo herd on his Garden City, Kansas
ranch, c. 1890. *Courtesy Kansas State Historical Society.*

E. A. Price's cattalo cow,
Sedgwick, Colorado, 1968.

"Buffalo" Jones' cattalo cow,
Ernest Thompson Seton drawing.

Frederic Remington drawing.

Ed Harrell driving a team of trained cattalo belonging to Charles Goodnight.
Courtesy Panhandle-Plains Historical Society.

whose story we have told, was, as a calf, put on milch cows. When he grew to maturity, he served many cows in Goodnight's domestic cattle herd. When Goodnight's cattle were rounded up in 1883, his men found a half-breed calf near the carcass of a Texas cow. Goodnight was fascinated by the little calf which he carefully watched as it grew to maturity.

When O. H. Nelson gave Goodnight two Polled Angus heifers in 1885, Goodnight knew what he would do. The Angus were then considered the hardiest breed of domestic cattle, and the buffalo the hardiest wild plains animal. Goodnight decided to experiment to see whether he could produce a beef breed that could stand the severe winters on the plains as did the buffalo. He invested much money and time in his experiments.

Many years later, Goodnight concluded that it was best to start with the male of the breed whose qualities you wish to establish in the new breed. He said he had placed a buffalo bull with domestic heifers and found that a cow bred like this had about one chance in four of delivering her calf. If that calf was a heifer, the cow stood a fifty-fifty chance of delivering it and living. A bull calf, he noted, never came on the first cross. The cow always slunk her calf or both died in attempted delivery. Why? He never knew.[7]

Although William Hornaday and others had said it was because of the hump, Goodnight knew better. He knew that both cattalo and buffalo calves were smaller at birth than normal domestic calves. The hump story was only a legend. The hump appears later as the buffalo grows.

In 1910 Goodnight recalled that by breeding buffalo with the Polled Angus, "we have them from $\frac{1}{16}$ Buffalo on up to the half-breed, or Cattaloes." He said he had been able to "produce in the breed the extra rib the Buffalo has which is 14 on each side." Goodnight concluded:

> They make a larger and more hardier cattle and will cut a greater per cent of net meat than any other cattle and they require less food and are a longer-lived cattle. We have a Buffalo Cow 28 years old with a young calf this year, and as yet no one knows just how long a Buffalo lives.[8]

Another who experimented in buffalo hybrids was Mossom M.

Boyd of Bobcaygeon, Ontario, Canada. He began in 1894 with what Alvin Howard Sanders, writer and authority on cattle, called in 1925 "the nearest approach to a scientifically adequate experiment." Sanders wrote: "In 102 successful impregnations of cows by Buffaloes, there were 63 abortions and 39 births. Of the 39 births, 6 were males, only two of which survived over 24 hours, and the one that became adult proved to be sterile."[9]

Boyd's experiments ended about 1915 when he sold sixteen female and four male hybrids to the Canadian government, which wanted to develop a range beef animal for western and northwestern Canada, where low temperatures and severe blizzards during the winter and heavy fly infestations during the summer produced conditions detrimental to improved domestic breeds under range conditions.

Boyd's animals were taken to the Dominion Experimental Station at Scott, Saskatchewan, and in 1916 the Canadian government began its experiments. None of Boyd's hybrid animals reproduced, however, and the experiment had to begin over again with original crosses between buffalo and domestic cattle. A few buffalo calves were raised on domestic cows and later used for crossing with domestic females of the Shorthorn, Hereford, and Holstein breeds. The mating of buffalo males with domestic females resulted in a very high mortality of both cows and calves at the time of birth and was discontinued. But the reverse mating was continued, using domestic males with buffalo females. The problem here, however, was getting domestic bulls to associate with female buffalo.

At one point a number of yak were included in the breeding. It was thought that through them fertile hybrid males might be obtained. However, yak blood complicated the experiment, and the yak was eventually eliminated.

After nearly fifty years of continuous experiments, the Canadian government gave up in the early 1960s.

H. F. Peters, a geneticist with the Canada Department of Agriculture in Ottawa, said in 1968:

> We found that the cattle x bison hybrid cow was very hardy and long-lived. The ¼-bison cattalo from hybrid cows also possessed a high degree of cold-tolerance and winter survival ability. In most winters, however, cold depressed growth rate for

only short periods, and as inherent growth rate of bison is low the cattalo did not have any advantage over range Hereford cattle in that respect. We found that even the Brahman, when crossed with British beef breeds, produced more productive and growthy hybrids.

In the early 1960's we initiated co-operative work with George G. Ross, Jr., of Manyberries, Alberta, to develop a new breed of range beef cattle from crossing of selected growthy Holstein and Brown Swiss bulls with range Hereford-Shorthorn cows and a bit of Red Angus thrown in. This project is proceeding well on Ross ranches at the present time. . . . We felt that this type of approach had more potential for future beef ranching on the prairies than cattalo."[10]

Although the future of buffalo hybrids looked bleak in the late 1960s, a California rancher, D. C. "Bud" Basolo, continued to experiment. Basolo, who began such experiments on his B-Bar-B Buffalo Ranch in Wyoming in the early 1960s, announced in 1973 that he had solved the problem of male sterility in hybrids. Basolo told me that he had produced "an absolutely fertile hybrid bull from a cross with a male buffalo and a female domestic cow. I used Charolais and Holstein cows. I won't tell people how to do it. That's a secret, since I can't patent my method, but the key is the genetic makeup of the animals used," said Basolo, who owns eight ranches in California. His main ranch is near Tracy. Basolo calls his animal the "Basolo Hybrid Beefalo." By August 1973 he had produced more than 5,000 of the hardy cow-buffalo hybrids. Pointing out their advantages, Basolo noted:

[They] are cheaper to feed, more resistant to disease, and stronger than standard breeds of domestic cattle. And they fatten faster than regular steers. It takes less than twelve months to reach the market weight of 1,000 pounds. And they are higher in protein. Their fat content is under 7 percent compared to 30 to 40 percent for trace cattle. Their meat tastes much like conventional beef but is slightly richer. Eventually, meat from these hybrids will be priced 25 to 40 percent cheaper than regular domestic beef.

Although Basolo refused to tell the secret of his successful hybrid, he was giving other breeders an opportunity to raise his "Basolo Hybrid Beefalo." In 1973 Basolo sold more than 100,000 vials of his male animals' sperm (one cubic centimeter per vial, $7 per vial) to interested ranchers in the United States, Canada, Central and South America, and Africa. Apparently not satisfied with his success in crossing the American bison with domestic cattle, Basolo said he plans to invest three million dollars between 1973 and 1976 in experiments with the water or Cape buffalo, the bulky ox-like creature that has the reputation of being one of the most dangerous of the African big-game animals.[11]

Whether Basolo's experiments were successful is not known, but Basolo continued to raise beefalos which he described as three-eighths buffalo, three-eighths Charolais and one-quarter Hereford. His beefalo were, in turn, used to impregnate still more cows, and his firm, Texas Meats Brokerage, Inc., has since reportedly sold thousands of vials of beefalo semen to ranchers, many of whom have since established their own herds of beefalo.

Charles Goodnight cattalo. *Courtesy Panhandle-Plains Historical Society.*

XIX

The American Buffalo as a Symbol

Oh, give me a home where the buffalo roam,
Where the deer and the antelope play,
Where seldom is heard a discouraging word,
And the skies are not cloudy all day.
Brewster Higley 1873

In 1913 the U.S. Treasury Department coined the buffalo nickel. James Earle Fraser, who designed the coin, was asked why he selected the buffalo. Fraser replied:

> My first objective was to produce a coin which was truly American, and that could not be confused with the currency of any other country. I made sure, therefore, to use none of the attributes that other nations had used in the past. And, in my search for symbols, I found no motif within the boundaries of the United States so distinctive as the American buffalo.[1]

Indeed, the American buffalo is distinctive. Today the use of the buffalo as a symbol is greater than ever before. For many persons the animal has become the picturesque symbol of vanished days when civilization was only a word in the West and not a way of life. For some the animal represents strength and endurance. For others the buffalo is symbolic of man's ignorance of conservation.

The buffalo, like the mustang, bear, cougar, bobcat, and other wild animals, has been institutionalized by colleges, universities, high schools, and junior high schools. Football, basketball, and baseball teams have borrowed the name. For most, the name apparently is

279

supposed to transmit to opponents a "you can't push us around" attitude.

The University of Colorado at Boulder is perhaps the best known institution of higher learning with the Buffalo athletic nickname. The university even has a live mascot. In 1973 it was a buffalo cow named Ralphie. She makes an appearance at most home games and many out-of-town football games. Civilization, however, has caused Ralphie problems. The installation of artificial turf at Folsum Stadium in Boulder and at other Big Eight schools presented problems to Ralphie and her handlers. In September 1971 Ralphie was first taken onto the new turf. She roamed around as frisky as ever, but handlers quickly moved in when she tried to eat the artificial turf. Stadium officials also discovered that Ralphie damaged their high-priced turf with her hooves. Her handlers set about to design special boots for Ralphie's feet to keep her hooves from tearing into the turf.

The owners of other buffalo, those drawn and painted by Charlie Russell, do not have to worry about such things. Russell, the Montana artist, painted and sketched countless buffalo, and used a buffalo skull beside his signature on nearly every piece of art he did.

Frederic Renner, an authority on Russell's art, said, "In addition to different outlines of a buffalo skull that Russell did in connection with his signature on more than three thousand paintings, watercolors and drawings, there are five individual drawings of a buffalo skull."

Renner owns one. He describes it as Russell's "Skull Number 4."

"I can only guess that it may have been executed about 1910," said Renner. "The original was given by Russell to his friend, Dick Jones of Great Falls, Montana, and I acquired it about 30 years ago from his widow."

Renner's Russell drawing has been used since 1956 by the Potomac Corral of The Westerners on their stationery and numerous publications.[2]

The Westerners, an international organization of people interested in the lore and history of the Old West, have many chapters, called corrals or posses, throughout the United States and in many foreign countries. At least seven in addition to the Washington, D.C. corral use the buffalo as their symbol.

For example, Kansas Westerners use a complete buffalo, which is also the official state animal. It was drawn by Gary Holman, who is on the staff of the Eisenhower Library at Abilene, Kansas.

The Chicago Corral, the first Westerners group (formed in 1944), uses a buffalo skull illustration drawn by Burleigh Withers, a native of Wyoming, who for many years was proprietor of Withers-McCallum-Stearns, a Chicago commercial art studio. The buffalo skull drawing used by the Los Angeles Westerners may have been drawn by the same artist, according to Don Russell.

The Chicago Corral also has a real buffalo skull which they use in their meetings. It is named Old Joe. Says Don Russell:

> Everyone stands in Napoleonic stance, right hand over the heart. As the Sheriff [President] counts, "One two, three!" the two newest members heist the "veil" from the bison skull on the wall, and all yell: "Hello *Joe* you old buff*alo!*" They grin, they sit, they grin some more—and then the meeting gets going. One dinner and a speech (with probably an argument or two) later, the ritual is repeated in reverse with "Adios *Joe,* you old buff*alo!*"

Recalling the history of Old Joe, the skull, Russell said:

> The Chicago Westerners salute to Old Joe dates back to 1956-57. One roundup night, co-founder Leland D. Case scribbled a note that was passed up to Sheriff John Jameson, suggesting that the white skull on the wall be named after a remembered sing-song childhood rhyme: "Joe, Joe, broke his toe, riding on a buffalo!" Sheriff Jameson, top-level idea man for a national advertising agency, was quick on the uptake. So the Old Joe opening gambit was adopted that very night. Some time later, no one seems to remember when, we added the closing "adios" bit.
>
> Old Joe himself—the actual bone, I now mean—got into our orbit as early as January, 1947. Edmund B. Rogers was telling Chicago Westerners about Yellowstone Park, of which he was superintendent. Observing the printed emblem on our *Brand Book,* he asked if we'd like a real bison skull. We did—and soon it came. Where Rogers got it I never learned.[3]

In June 1972 members of the Chicago Corral shook in their boots when they learned that a fire had destroyed much of Jason's Restaurant, the site of their monthly meetings. Old Joe was stored at the restaurant and most members feared the worst. But when the ashes cooled, Old Joe was found. He was battered and discolored, and the plaque on which he was mounted was in even worse shape. Ray Scott, a corral member, undertook the difficult restoration job, finishing in time for the corral's next meeting.

The U.S. Department of the Interior, under whose watchful eyes the American buffalo is today protected on numerous wildlife refuges in the western United States, uses the buffalo as a symbol. When the department was created in 1849, millions of wild buffalo still roamed the West, but it was not until November 1929 that it adopted the buffalo on its departmental seal "as being more significant of the early settlement of the West." The buffalo remained a feature on the department's seal until 1968 when it was replaced by one with a stylized pair of hands. But on October 12, 1969, Secretary of the Interior Walter J. Hickel ordered the department to return to its traditional buffalo emblem as its official seal, because "Department employees and public individuals who spoke out on the issue overwhelmingly urged a return to the buffalo emblem."[4]

In still other ways the government has commemorated the buffalo. Aside from the buffalo nickel,[5] it has issued three stamps and a $10 note containing buffalo illustrations.

On June 10, 1898, an engraving of an Indian on horseback hunting a buffalo appeared on a 4-cent stamp issued in the Trans-Mississippi Omaha Exposition series. The engraving was from H. R. Schoolcraft's *Indian Tribes of the United States,* published in 1860.[6]

Next came the $10 note. Issued by the government in 1901, it had a drawing of a buffalo on the front of the bill. This was the so-called "Buffalo-bill," issued in connection with the Pan-American Exposition in Buffalo, New York, May 1 through November 2, 1901. Although William Hornaday claimed that the stub-horn buffalo bull he killed in Montana and mounted for the Smithsonian was the model for the "Buffalo-bill," Charles R. Knight, the artist, wrote in 1939:

> This drawing was done at the Zoological Park in Washington and *not* from the Museum group. The story of that drawing, however, gives a little light on the idea that it might have been done from the mounted group. A certain Mr. Baldwin, an engraver for the Bureau, was trying to make a drawing from the

I use this illustration on my stationery.
O. H. Bacher drawing,
from *Scribner's,* September 1892.

National Zoo buffalo used by Charles R. Knight as model for his $10 bill drawing,
from Smithsonian *Annual Report* for 1914.

Some logos used by various Westerners groups.
Upper left: Westerners International; borrowed from Chicago Corral.
Upper right: Kansas Westerners; artist, Gary Holman, Abilene.
Lower left: Potomac Corral, Washington, D.C.; artist, Charles Russell.
Lower right: Chicago Corral.

The Alternative Press

Advertising booklet cut in the shape of a buffalo head: 9¼″ high, 7⅞″ wide;
32 pages; published in 1898 by Courier Co., Buffalo, N.Y. *Courtesy Don Russell,
Elmhurst, Illinois.*

C. M. RUSSELL

GREAT FALLS, MONTANA

World's largest buffalo,
Frontier Village, Jamestown, N.D.
Erected 1958 as a monument
to the vast herds of buffalo
that once roamed the prairies.
Made of 8″ steel H beams,
rods, wire mesh, concrete, stucco;
26′ high, 46′ long, 60 tons wgt.
Sculptor, Elmer Peterson.
*Courtesy Jamestown
Chamber of Commerce.*

Bumper sticker.

Bookplate of
Stanley Vestal (pen name).

Bronze commemorative medal (1½″ dia.)
issued in 1972, from a 36-medal National
Park Centennial series, 3 of which used a
buffalo motif. *Courtesy Roche Jaune, Inc.,
Kalispell, Montana, official medalist for
the National Park Service.*

Silver medallion (1 9/16″ dia.) issued in
1971, from a 36-medal America's Natural
Legacy series, to highlight threatened wildlife. Sculptor, Joseph DiLorenzo; Witt-
nauer Mint. *Courtesy Longines Sympho-
nette Society, New Rochelle, N.Y.*

Manitoba

The Shaw & Powell Camping Co., one of the earliest (1898-1917) and most important of the tourist concessionaires in Yellowstone, employed the buffalo in its logo—used on the company's brochures, letterheads, luggage tags, drivers' metal hat insignias, etc. *Courtesy Rose Shaw, Butte, and Montana Historical Society.*

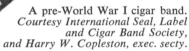

A BISON BOOK

A pre-World War I cigar band. *Courtesy International Seal, Label and Cigar Band Society, and Harry W. Copleston, exec. secty.*

Logo, Western Publications, Inc. George Phippen drawing, done for the first issue of *True West.*

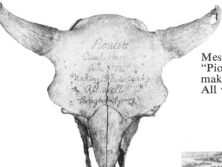

Message inscribed on this buffalo skull:
"Pioneers camped here June 3d 1847
making 15 miles today
All well Brigham Young"

Emblem of 10th Regiment,
U.S. Cavalry,
the "buffalo soldiers."

Advertising flyer, c. 1870.
Henry Worrall drawing.
*Courtesy Kansas State
Historical Society.*

Bronze buffalo head
on the Union Pacific Railroad bridge,
Council Bluffs, Iowa

State seals and flags.

Museum group when I offered to do this for him. Only too gladly he hustled me over to the Bureau and I was commissioned to do the drawing, which I promptly did—from the Zoo specimen.[7]

Knight's illustration of a buffalo was later used on a 30-cent postage stamp issued in 1922.[8]

The most recent U.S. philatelic commemoration of the buffalo was on July 20, 1970, where at Custer, South Dakota, near the large buffalo herd in Custer State Park, a 6-cent buffalo stamp was issued. Postmaster General Winton M. Blount said earlier:

> Indeed this would be a desolate world in the absence of wildlife, for these bring man in tune with Nature. I hope that issuance of this stamp will remind all Americans of the continuing need for protecting from extinction our birds and beasts and fish.

The 6-cent buffalo stamp, issued in the Wildlife Conservation series, was designed by Robert Lougheed of Newton, Connecticut, who made more than forty sketches of buffalo before settling on the single buffalo drawing used on the stamp. The initial printing of the stamp was 130 million; by November 1971 more than 142 million had been issued.[9]

Often when people think of the buffalo today they immediately associate the animal with the Plains Indian of a century or more ago. Few, however, associate the buffalo of today with the contemporary American Indian, and many are surprised to learn that the buffalo still plays an important role in the lives of many American Indians.

A buffalo skull is still used today by the Northern Cheyennes to represent *is si wun,* the sacred buffalo hat, a living symbol of life and power for the Cheyennes, especially the Cheyenne women. Today the Northern Cheyennes often must turn to the white man to obtain a skull for ceremonial use. However, the skull must not have been touched by white men's metal such as a bullet, and preferably not by any white person, but this purism often has to be overlooked.

Other Indian tribes such as the Crow and Shoshone use a buffalo head in their ceremonies today. But because of the scarcity of buffalo skulls they sometimes are forced to use a mounted buffalo head.

From Buffalo, New York to Buffalo, Kansas, and on west, buffalo place names also commemorate the animal. There are towns and cities, rivers and creeks, mountain peaks and hills, and even islands, lakes, springs, and counties named after the buffalo. Frank Gilbert Roe, in *The North American Buffalo,* listed 211 place names related to the buffalo. These he found in journals, narratives, atlases, history books, and other records. Many of these place names, however, do not exist today. Some have been changed, others lost in time.

The U.S. Board on Geographic Names is thereby happier. Back in the 19th century they protested the excessive duplication of *Buffalo* place names (along with, it should be added, *Bald, Beaver, Cottonwood, Elk*—"altogether too numerous.").

Yet, in nearly every state and province in the United States and Canada, at least one buffalo place name can still be found.

Indiana and Kansas have the buffalo in their official state seals. Wyoming has a white buffalo on a field of blue in its state flag. In Canada the official seal of the Royal Canadian Mounted Police contains a buffalo, as does the coat of arms of the province of Manitoba. An illustration of the buffalo's head is used on the beer and ale labels of the Calgary Brewing and Malting Company in Alberta.

Nor has our flora and fauna been neglected. We have a buffalo berry, a buffalo bird, buffalo clover, a buffalo fish, buffalo grass, etc.

Commercial use of the buffalo motif has been as legion as buffalo place names. In our own day there is a *long* list of examples ranging from buffalo bars (there is likely one in your home town) to the Denver bank that gave away stuffed toy buffalo as premium promotions for new savings accounts. Perhaps the ultimate in commercial exploitation was achieved by Richard F. Ferguson, who tells his own story:

> Not long ago I set out to prove there's a market for anything—if it's packaged right—even those odd-shaped little mounds that dotted our great American prairies in years gone by, odd-shaped little mounds that reflected temporary relief for the buffalo and fuel for the fires of pioneers trekking Westward. Call them what you will—buffalo chips, excrement or manure—the point is I proved people will buy anything.
>
> When the challenge first crossed my mind, I contacted a friend whose home is located reasonably close to a midwestern game

preserve stocked with . . . buffaloes. My request was simple. "Please pick up and send me about six good chips."

Several days later a large, un-insured parcel arrived. . . . [The chips] were carefully brushed, varnished and dried, then bedded in delicate white tissue paper.

An artist friend and I worked up package copy and design, based on the famous World War II fighting cry, "Gung Ho!" Only our merchandising plan called for the words to be changed to "Dung Ho!"

The first of our conversation pieces sold on sight to a jeweler who later presented it to the captain of his firm's "last place in the league" bowling team. . . .

My supply depends only upon my understanding friend and the continued eating habits of the thundering herd![10]

Finally, it should be noted that the word *buffalo* has even been verbed into the American language. However, if asked about the origin of the meaning, I must reply that I am buffaloed by your question.

XX

The Future of the American Buffalo

*When the last individual of a race of living things breathes
no more, another heaven and another earth must pass before
such a one can be again.*

William Beebe 1906

If we compare the buffalo population of the world a century ago to
the leaves of a large cottonwood tree, today's buffalo population—I
am referring to the American bison (*Bison bison* and *Bison athabascae*)—would be only a single leaf on the tree. Yet, the world's buffalo
population, over 99 percent of which is in North America, has increased considerably since *London Field,* a British magazine, reported on November 10, 1888 that only 1,300 live buffalo remained
in the world. How the *London Field* conducted its survey and arrived
at its figure is unknown, but it was apparently the first buffalo survey
ever taken. By calling it a world-wide survey, the magazine's editors
presumably felt it would have more impact than if they had called it
what for all practical purposes it was: a census of buffalo in North
America. Regardless of such possibly misleading labels (world figures vs. North American figures), numerous subsequent surveys (most,
so far as is known, conducted by mail) have been taken by individuals
and groups and give us a good picture of the decline and growth of
the buffalo during the past eighty-five years. Since only about 400
North American buffalo are known to exist outside of North America
today—and fewer in earlier years[1] — the table that follows reflects
what has been the North American buffalo population since 1888.

I can find no record between 1933 and 1951 of any survey or
census having been taken by either a private or a governmental
group. In 1951, however, Henry H. Collins, Jr., of Bronxville, New

North American Buffalo Population[2]

Year	Survey taken by	Number of buffalo
1888	*London Field* magazine (Miller Christy)	1,300
1889	Smithsonian Institution (William T. Hornaday)	1,091
1895	Ernest Thompson Seton	800
1900	*Boston Evening Transcript* (Mark Sullivan)	1,024
1902	U. S. Department of Interior (S. P. Langley)	1,394
1903	Smithsonian Institution (Frank Baker)	1,753
1905	Smithsonian Institution (Frank Baker)	1,697
1908	American Bison Society (William T. Hornaday)	1,917
1910	American Bison Society (William P. Wharton)	2,108
1911	American Bison Society (William P. Wharton)	2,760
1912	American Bison Society (William P. Wharton)	2,907
1913	American Bison Society (William P. Wharton)	3,453
1914	American Bison Society (William P. Wharton)	3,788
1916	American Bison Society (Edmund Seymour)	5,592
1918	American Bison Society (Martin S. Garretson)	6,466
1919	American Bison Society (Martin S. Garretson)	7,418
1920	American Bison Society (Martin S. Garretson)	8,473
1921	American Bison Society (Martin S. Garretson)	9,311
1922	American Bison Society (Martin S. Garretson)	12,053
1923	American Bison Society (Martin S. Garretson)	12,457
1924	American Bison Society (Martin S. Garretson)	14,369
1926	American Bison Society (Martin S. Garretson)	16,417
1929	American Bison Society (Martin S. Garretson)	18,494
1933	American Bison Society (Martin S. Garretson)	21,701
1951	Henry H. Collins, Jr.	23,154
1972	David A. Dary	30,100
1978	David A. Dary (estimated)	65,000
1982	David A. Dary (estimated)	83,000
1989	David A. Dary (estimated)	98,000

York, undertook a new survey, reporting in April 1952 that 23,154 buffalo were living in North America. Of these, 13,902 were in Canada, 9,252 in the United States.[3]

When I began research for this book in 1964, Collins' survey was thirteen years old. I felt certain the buffalo population of North America had continued to increase during that time. But when I asked officials of the Department of the Interior in Washington. D.C., what the latest figures were, they said they had not taken any such survey and could only guess at the buffalo population.

Thus, I decided to take my own survey. I began by obtaining the names and addresses of every known private and government buffalo herd in Canada and the United States, including Hawaii and Alaska. I used Collins' 1951 census and an incomplete list of buffalo herds in the United States compiled by the Animal Disease Eradication Division of the U.S. Department of Agriculture, then the only U.S. government agency to have such a list.[4] Other names and addresses were obtained from known buffalo men, and numerous letters revealed still other buffalo owners. Also, several state governments were able to provide names and addresses of buffalo herds, as were officials at several federal wildlife refuges who had sold surplus buffalo to private citizens.

In Canada, officials of the Canadian Wildlife Service provided not only addresses of all private buffalo herds but also a complete list of government herds and the current population figures for each.[5] By the time my mailing list was compiled there were 490 known private and government buffalo herds in North America ranging in size from one or two animals to nearly 2,000.

My survey was taken by mail. A one-page questionnaire was prepared (see Appendix A). It asked how many buffalo were in each herd and such questions as what was the oldest buffalo in the herd, had buffalo twins ever occurred, what was considered the biggest problem in raising buffalo, what was the weight of their largest bull, when was the herd started. The questionnaires were sent out with stamped and addressed return envelopes over a four-month period. When they began coming back, I learned that many buffalo herds reported in Collins' 1951 survey and in the government report no longer existed. Most such herds had been privately owned. The owners had presumably tired of raising a few head of buffalo or had found it too expensive to continue and had sold their animals.

Of 490 questionnaires mailed out, 210 (slightly less than 43 percent) were returned by January 1, 1970. When the results were tabulated, 10,131 living buffalo had been reported in the United States and 15,114 in Canada, or a grand total of 25,245 buffalo in North America, including Hawaii.

However, because only 43 percent of the questionnaires were returned and because many new buffalo herds may have been started since previous surveys were taken, my survey total was no doubt lower than the real. And this proved to be true. During the next two years I located more than 4,000 additional buffalo owned by private citizens in the United States. Some of the herds were new. Others had existed for many years but had never been included in any known survey or report. By January 1, 1972, I had accounted for nearly 15,000 living buffalo in the United States or about 50 percent more than had been confirmed in my 1969 survey. In Canada, the figure of 15,114 is felt to be reasonably accurate for 1972 since nearly all buffalo in Canada are owned by the government and the herd sizes are controlled; private buffalo owners in Canada have less than 400 buffalo.

As 1972 began, the buffalo population of North America was slightly more than 30,000, or about 30 times the number of buffalo that existed in 1889 and almost 50 percent more than in 1933. By January 1, 1973, it was estimated that the buffalo population of North America was approximately 33,000. Since then the number of buffalo has increased significantly because of the increasing number of private citizens in the United States and Canada raising buffalo for profit and pleasure. Today (1989) there are nearly 100,000 buffalo throughout North America. Nearly 80,000 head are in the United States; about 15,000 in parks or wildlife refuges and about 65,000 in private hands. Near 20,000 other buffalo including Wood Bison are in government and private hands in Canada.

An interesting finding from my survey was how many private citizens, especially in the United States, were raising buffalo. Many said they were doing so because of the novelty, or to help preserve the species, or, as in the case of one elderly man, "simply because I find them fascinating to watch."

Of more significance was the fact that many persons were raising buffalo for commercial purposes. The largest private herd turned up in the 1969 survey was owned by the Durham Meat Company of San Jose, California. They owned more than 1,700 buffalo in 1969; in August 1973 they reported 2,250 head. The animals are maintained on a 50,000-acre ranch near Gillette, Wyoming. Durham raises buffalo as a meat source providing supermarkets, restaurants, private clubs, and other concerns with buffalo steaks and other buffalo meat cuts. They reported having customers in all fifty states.[6]

Armando "Bud" Flocchini, vice president and general manager of the Durham Meat Co., said his company marketed about 150,000 pounds of buffalo meat in 1968. This jumped to 350,000 pounds by 1972. "We expect to average an increase in sales of about 50,000 pounds each year during the next few years," said Flocchini. He pointed out that in 1968 a 500-pound buffalo carcass sold for around $500, compared with $300 for a similar-sized beef carcass in 1968. In 1973 a 500-pound carcass sold for about $625, compared with $425 for a similar-sized beef carcass.[7]

But the Durham Meat Company was not alone in the field in 1968, nor is it today. Several other buffalo owners had begun to produce buffalo meat for commercial use when the market began to develop in the late 1950s. Into the 1960s the market increased as many grocery stores, particularly specialty houses in the larger cities, began stocking buffalo steaks. By 1966, food buyers noted that the demand was greater than ever, and buffalo steaks were selling for $2.95 a pound in New York City.[8]

In 1968, as the buffalo meat market was growing in the United States, Tommy's Joynt, a popular San Francisco restaurant, used 600 pounds of buffalo meat a week in its special buffalo stew. The price: $1.88. "Tourists are attracted to it because it has so much romance," said owner Tommy Harris. "They know the pioneers lived on it, and it gives them a feeling of going back in history. Besides, it makes good eating."[9]

In New Orleans in the late 1960s Tom Pittari, Jr., co-owner of T. Pittari's restaurant, added a pan-sauteed buffalo steak to the restaurant's menu. The price: $8.25. Pittari said buffalo meat tastes something like beef, but is richer, coarser, and more filling.[10]

By 1973 four buffalo ranchers—Flocchini and his Durham Meat Co.; Ron Gregory, a Frontier Airlines pilot from Longmont, Colorado; Roy Houck of Pierre, South Dakota; and Don Hight of Murdo, South Dakota—were producing the bulk of commercial buffalo meat. When the meat shortage developed during the summer of 1973, these ranchers could not meet the demand for buffalo meat. William McDonald, meat buyer for Albertson's, a large western grocery chain centered in Denver, looked at buffalo meat as an attractive substitute for beef steaks and cuts, but the supply of buffalo meat was limited. McDonald had difficulty locating a hundred buffalo he had hoped to butcher and sell in Albertson's twenty-five stores from Casper, Wyoming to Santa Fe, New Mexico. "The primary reason we are turning to buffalo is to find some meat for our customers," said McDonald.[11] Another August 1973 news story reported:

Two West Coast chains are offering buffalo meat for the adventurous. "It's quite good and very much like beef," explained a spokesman for one store.

Buffalo meat used to be higher than beef because there aren't very many buffalo herds. But cattlemen have been withholding livestock from market because of the continued ceiling on beef prices and now it's steers that are scarce.

A check of one store showed buffalo burger and chuck roast for $1.09 a pound; the same cuts of beef brought $1.19 a pound.[12]

The market for buffalo meat did not grow as rapidly in Canada as it did in the United States. It began to develop in Canada in 1966, but the buffalo meat supply was limited because most buffalo in Canada were owned by the government. By 1969, however, seven private citizens or companies in Canada had purchased buffalo from the government either for breeding or as a commercial meat source.

Perhaps the most active Canadian company dealing in buffalo meat was Canada Packers Limited at Toronto. G. E. Speers told me there was a "fair market for a reasonable quantity provided it is priced in line with other meats, but I do not think the production will ever be large."[13]

As the U.S. market for buffalo meat increased during the '60s, many owners of commercial buffalo herds felt they would do better if they worked together. On March 11, 1967, twenty-nine buffalo breeders gathered at the Brown Palace Hotel in Denver, Colorado, to form the National Buffalo Association (NBA). The Association's purpose became fourfold:

1. To promote buffalo and buffalo products, namely meat and hides.
2. To seek fair and equitable regulations in the control of disease and in the movement of buffalo and buffalo products, both intrastate and interstate.
3. To promote and develop recreation where it is connected with buffalo.

4. To advance any and all ventures wherein we can better our
position in the promotion of the buffalo.[14]

By 1969 the National Buffalo Association had about seventy members and several associate members, persons who did not raise buffalo but were interested in them. By 1973 there were 163 active members.

Roy Houck, president of the National Buffalo Association in 1973 and a man who owns more than 2,000 buffalo on his Triple U Ranch near Pierre, South Dakota, told me:

> The buffalo is again coming back into his own, not only as an attraction, but from an economical standpoint in the production of meat and various other products made from the hide, hooves and horns.
>
> It has been found that when the young animals are grain-fed, their meat is equal to if not superior to grain-fed beef; and after the initial investment, the cost of production is much less. While I realize that everyone cannot handle and raise buffalo, I predict many more people will get into this business in the near future.[15]

Houck considered the buffalo practically immortal compared to domestic cattle on his South Dakota ranch:

> They harbor few parasites, don't catch all the diseases domestic cattle have, and they get along nicely on the open range in the winter when most cattle have a hard time surviving. Buffalo apply their own insecticide by rolling in the dust and they lick alkali soil if someone has forgotten to put salt blocks out. In a storm they'll take care of their young the way cattle won't. I've seen buffalo calves born in storms at 10 to 20 degrees below zero get up, and as soon as they got some milk into 'em, follow right along with their mothers.[16]

Houck spoke of another advantage buffalo have over cattle: "They don't take as much feed to put on weight, as they forage some weeds that cattle pass up. That's why three buffalo can live on range that would support only two cows."[17]

But Houck is not typical of the private buffalo owners today. Most private buffalo men have fewer than fifty animals.

D. L. Wallace is typical of the men who entered the buffalo business during the 1960s. He started raising a few head on his farm near Winchester, Kansas, and became the prime supplier of buffalo meat for the Savoy Grill in Kansas City, Missouri. When the Savoy opened in 1903, buffalo was on the menu, but it was taken off around 1928. Forty-three years later, in 1971, it was placed back on the menu. Donald Lee, the owner, was going back through the restaurant's records when he noticed that buffalo steaks had once been served at the Savoy. Lee got in touch with Wallace, and the tradition of serving buffalo steaks at the Savoy was revived.

When Charles Goodnight was in his nineties, he told J. Frank Dobie that the best tonic he had ever found was a mixture of whiskey and the extract of buffalo meat. Goodnight considered buffalo meat more conducive to longevity than Texas beef, a belief shared by many in the Old West. The extremes of such belief were caricatured in the early 19th century by Rufus B. Sage:

> The *voyageur* is never more satisfied than when he has a good supply of buffalo-beef at his command. It is then his greasy visage bespeaks content, and his jocund voice and merry laugh evince the deep-felt pleasure and gratification that reign within.
>
> Talk not to him of the delicacies of civilized life,—of pies, puddings, soups, fricasees, roast-beef, pound-cake, and dessert, —he cares for none of these things, and will laugh at your *verdancy!*
>
> He knows his own preference, and will tell you your boasted excellencies are not to be compared with it. If you object to the sameness of his simple fare, he will recount the several varieties of its parts, and descant upon each of their peculiar merits. He will illustrate the numerous and dissimilar modes of so preparing them, that they cannot fail to excite by their presence and appease by their taste the appetite of the most fastidious. And then, in point of *health,* there is nothing equal to buffalo-meat. It, alone, will cure dyspepsy, prevent consumption, amend a broken constitution, put flesh upon the bones of a *skeleton,* and *restore a dead man again to life!*—if you will give credence to one half of the manifold virtues he carefully names in your hearing.[18]

Modern science in fact has proved that buffalo meat does have certain advantages over domestic beef and that there is some basis to the oldtimers' belief.

At their national convention in Oklahoma City in March 1971, members of the National Buffalo Association were told that analysis of buffalo meat by scientists at South Dakota State University had indicated that buffalo meat was lower in cholesterol and higher in polyunsaturated fats than domestic beef. Buffalo fat was a favorite of early westerners even before science validated it. One traveler observed in the 1840s:

> The agreeable odor exhaled from the drippings of the frying flesh, contained in the pan, invited the taste,—a temptation claiming me for its subject. Catching up the vessel, a testing sip made way for the whole of its contents, at a single draught,— full six gills! Strange as it may seem, I did not experience the least unpleasant feeling as the result of my extraordinary potation.
>
> The stomach never rebels against buffalo-fat. Persons, subsisting entirely upon the flesh of these animals, prefer an assortment of at least one third solid dépouille [fleece-fat].[19]

Fat must be distributed throughout domestic beef if the meat is to be tender. Buffalo meat, however, is just as tender without the marbling and without the saturated fatty acids. And it does not have to be cooked as long as domestic beef to make it tender. As a meat source, then, buffalo provides several advantages over domestic beef. And allergy specialists say that allergy to buffalo meat is almost unheard of.

"Now maybe we're learning why the Indians and mountain men of the last century, who lived largely on buffalo meat, were so hardy and long-lived," said Roy Houck.[20] John Palliser gives an informative comment from that last century:

> Old Mr. Kipp, at Christmas [1847], thinking to give all the employés and voyageurs of the Furr Company at Fort Union a great treat, had for some time previously been fattening up a very nice small-boned heifer cow, which was killed in due time, in prime condition. All who had been reckoning on the treat this would afford them, sat down in high expectation of the ensuing feast; but after eating a little while in silence, gradually dropped off one by one to the bison meat which was also on the table, and were finally unanimous in condemning the beef, which they

said was good enough, but nothing remarkable, and the fat sickening. A plate-full of it was also given, as ordinary buffalo beef, to an Indian woman in another room at the Fort, on the same occasion; she pronounced it good food, but said she, "it is both coarse and insipid," and the fat, if she were to eat much of it, would make her sick.

I mention these circumstances, having been one of the very few who have seen the comparative merits of the two meats tested by Europeans, Americans and Indians at the same time, and heard the unanimous verdict in favour of the wild bison.[21]

Research begun by the National Buffalo Association has shown that all buffalo have one blood type. Domestic cattle, however, have more than fifty different blood types. Some researchers have put forth the theory that buffalo blood, high in gamma globulin, might some day be used in humans when a single large transfusion is required as in the case of heart surgery or a serious accident.

But there are many questions still to be answered about the buffalo, especially concerning the disease brucellosis, commonly called undulant fever. Brucellosis was first diagnosed in buffalo in 1917, when blood samples were taken from buffalo in Yellowstone National Park, but it was not until 1935 that brucella tests were occasionally given buffalo. Such tests were limited until the 1960s, when they were conducted on larger samplings of federal and private herds in the United States.

As a larger market developed for buffalo meat, hides, and breeding stock during the 1960s, the renascence of the buffalo created much concern over brucellosis because, in part, the federally sponsored brucellosis eradication program for domestic cattle was approaching success. A running controversy developed among buffalo men, cattlemen, and government veterinarians. Some buffalo men said there was no evidence that buffalo harbored brucellosis. Cattlemen, especially those living next to a buffalo herd, contended that the buffalo were responsible for the brucellosis in their cattle.

Tests indicated that many buffalo herds, government and private, were infected with what appeared to be brucellosis. Some of these herds were observed. Then a surprising discovery was made. Evidence was found indicating that the buffalo had the ability to fight off the disease. At first, buffalo showed symptoms similar to those observed in domestic cattle. During the early period of the disease, many buffalo pregnancies ended in abortion, but once the animals became conditioned to brucellosis, calves born later were often as healthy as those in an uninfected herd. What caused buffalo men real problems were the obtaining of blood samples and vaccinating buffalo against brucellosis. Since buffalo are wild animals, handling them is difficult, especially in the large herds. The techniques used by cattlemen to obtain blood samples and give vaccinations on domestic cattle were not as successful on buffalo. (22)

During the late 1970s and early 80s, however, buffalo raisers working with veterinarians learned a great deal more about handling and treating buffalo until today the procedures are more sophisticated than in the early 1970s. The second edition of the American Bison Association's *Bison Breeder's Handbook* published in 1988 includes detailed information on treating buffalo injuries and diseases including anaplasmosis, anthrax, bloat, brucellosis, coccidiosis, clostridium chauvei (Blackleg), Clostridium Tetani (lock jaw), and eye problems such as pink eye. The attention now being given to buffalo medicine is making buffalo raising more of a science than at any time in history.

The increasing interest in the buffalo during the early 1970s resulted in the formation of the American Buffalo Association (ABA) during the summer of 1975. Similar in nature to the National Buffalo Association (NBA) formed eight years earlier, both organizations work for the preservation, production and marketing of buffalo and the animal's products. In 1987, both the ABA and NBA changed the word *buffalo* in their names to *bison* to better represent the animal and the industry while at the same time correcting any misconceptions that the organizations were promoting the African buffalo or water buffalo of Asia. Both organizations have members in the United States, Canada and foreign countries, hold regular meetings, and publish newsletters and magazines. Today (1989) the mailing addresses for these organizations are:

National Bison Association   American Bison Association
10 East Main Street   Stockyard Station
Fort Pierre, South Dakota   Box 16660
           57532   Denver, Colorado 80216

Today (1989) about 10,000 buffalo are slaughtered and processed annually to meet the demand for buffalo meat. The number of firms selling buffalo meat is perhaps at the highest point since any time during the last century when countless buffalo roaming the western plains were killed for their meat. One firm, The Buffalo Tasters Club, P.O. Box 236, Rose Hill, Kansas 67133, began early in 1986 to promote the preservation of the animal and better health through nutritional education, in particular, the non-allergic and low cholesterol benefits offered by buffalo meat. Membership costs $25 with each new member receiving the Club Package including a buffalo cookbook, a 10 percent discount on all retail buffalo meat purchases and a regular newsletter. The Club was featured by Bloomingdales in New York City in the spring of 1988.

Beginning by the late 1970s, many private buffalo breeders recognized a market for buffalo related products aside from meat. The market for buffalo robes, mounted heads, and skulls increased as many people sought such decorative items.

The increasing buffalo population, the growing market for buffalo meat and related products, and breeding stock, and the continued efforts of governments and private citizens to preserve the buffalo as part of our heritage are signs that suggest the buffalo's future is bright. But the ultimate decision as to the fate of the buffalo rests with the creature who nearly exterminated him and then paradoxically saved him—man.

"FUNNY — I REMEMBER WHEN THEY ROAMED OVER THE PRAIRIE BY THE HUNDREDS"

A

Appendix

North American Buffalo
Survey Questionnaire

Name of person completing questionnaire _____

Title (owner, manager, foreman, director) _____

Address _____ Zip_____

Date questionnaire filled out _____

Total number of buffalo in herd (actual or approximate) _____

Number of bulls_____ cows_____ calves (under one year old)_____
        (If exact numbers are unknown, please approximate.)

Total acreage supporting herd _____

Purpose of your raising buffalo _____

_____

_____

History of your buffalo herd: Year herd was started_____

Where were the animals obtained? _____

How many buffalo were first obtained? _____

What is weight (actual or approximate) of the largest bull in herd? _____lbs.

Have you ever had a larger bull in your herd?_____ If so, what weight?_____

What is the age of the oldest buffalo in your herd? _____yrs.

Have you ever attempted to tame a buffalo?_____ If so, how successful

were you? _____

Have any cows in your herd ever given birth to twins? _____

If so, when?_____ How did the twins do?_____

_____

_____

Have any of your buffalo ever been bred to domestic cattle?_____

If so, what were the results? _____

_____

Do you consider the buffalo's eyesight good?_____ bad?_____

Explain _____

What do you consider the biggest problem or difficulty in raising buffalo?

_____

_____

I am interested in any other comments concerning the buffalo, including stories and anecdotes from your experiences with the animal, which could be included in my book. Please use the space below, the back side of this questionnaire, and additional paper, if necessary, for answers.

Any assistance and help you can provide me will certainly be appreciated.

B

Appendix

So You Want to Raise Buffalo!

This is for the reader who by now may have the desire to raise buffalo. The information that follows is excerpted from *Buffalo Management,* Wildlife Leaflet 212, issued by the Bureau of Sport Fisheries and Wildlife, United States Department of the Interior, Washington, D.C.

ENCLOSURES

Buffalo, like most other big-game animals, thrive best where not closely confined. Large, open, grass pastures with a plentiful supply of clean, fresh water are best adapted to their needs. Although shade may not be an absolute survival requirement, herds may spend many midsummer hours in the shade where available. Also, the presence of trees or large rocks adds to their well-being by providing rubbing sites.

Fences for confining buffalo should be strongly constructed. Minimum recommended fencing consists of 47" (Style #1047-6-11 or #1047-6-9) woven wire set about 10" above ground level extending to the top of 7' heavy duty posts spaced not more than 1 rod apart. Under normal conditions, buffalo are not likely to test the fence very severely, but when excited they may charge blindly into it, and then even the strongest fences may fail.

Corrals and chutes should be made of planks or heavy poles so the animals can readily see them and tend to avoid them. Such fences also obscure the view of activities outside the corral and thus reduce disturbance within. Plank corrals are easy for a man to climb, a distinct advantage when he is seeking escape from excited animals.

Barbed wire should not be used in buffalo fences. The barbs are ineffective in deterring the animals from attempting to escape and are a source of injury and infection. The range should be kept free of all loose wire, nails, etc. to prevent injury or infection.

Regardless of the extent of handling and of apparent domestication,

300

buffalo are *dangerous wild animals* of uncertain temperament and should never be trusted. Supposedly tame animals have attacked owners caught off guard. *Buffalo are not suitable for pets or mascots.*

Ropes should not be used around the neck of the buffalo because the windpipe is easily crushed by such a form of restraint.

FORAGE REQUIREMENTS

The foraging habitats of buffalo are similar to those of cattle. On southern ranges, buffalo grass *(Buchloe dactyloides)* and grama *(Bouteloua gracilis* and *B. hirsuta)* provide excellent pasturage. These grasses appear to be preferred to the taller and coarser species. On northern ranges tall grass such as wheatgrass *(Agropyron* spp.), bluestem *(Andropogon* spp.), bluegrass *(Poa* spp.), and the smaller fescues *(Festuca* spp.) are generally preferred. Tame grasses also provide good pasturage for them. Since buffalo are more selective in utilization of grasses, and because of their habit of compact herding, portions of their range may become denuded, thereby necessitating a reduced rate of use. Buffalo normally weigh slightly more than domestic cattle of equivalent age and require proportionately larger amounts of forage to compensate for this difference. Where conditions permit, rotating pastures will result in better utilization of the forage with less danger of depleting the cover and less likelihood of contaminating the ground. Unlike cattle, buffalo make and use wallows. Heavily used wallow areas produce little feed.

In most parts of the United States, buffalo may be ranged throughout the winter without supplemental feeding or artificial shelters, provided adequate natural forage is available. When it is necessary to feed buffalo for special reasons, they readily consume hay and concentrated feed of the same kinds used for domestic cattle. Quantities required would be comparable as for cattle of the same weight.

MINERAL REQUIREMENTS

To insure proper development, buffalo need salt and other minerals in their diet. Many of these are supplied in their forage, but on pasture soil deficient in essential minerals the forage likewise will be lacking in them. These deficiencies can be relieved by supplemental mineral compounds that are prepared for cattle. The salt requirement varies with the composition of the forage consumed, but generally buffalo will use as much salt as cattle.

C

Appendix

Buffalo for the Gourmet

For those readers who have tasted a good buffalo steak or ribs or stew, or enjoyed a chunk of buffalo hump cooked in coffee, or bitten into a large slice of buffalo barbecue, the following recipes may cause you to drool. If you have never tasted buffalo, you are missing a treat. Your local grocer may be able to supply you with buffalo meat or tell you where you can buy it.

To determine the number of servings for most of these recipes, keep in mind that one pound of boneless meat equals four servings, one pound of meat with a small amount of bone equals three servings, and one pound of meat with a large amount of bone equals two servings.

The recipes included in this appendix are only a sampling of those that are available. For the serious buffalo cook I recommend the *Buffalo Cook Book,* published by the National Buffalo Association. Copies can be obtained by writing Crane Publishing Company, Inc., P.O. Box 822, Rapid City, S.D. 57701.

BUFFALO SWISS STEAK

As with beef, there are some cuts of buffalo that are less tender. These need long, slow cooking and are ideal for Swiss steaks. Try this recipe for simplicity.

2 steaks, 1½″ to 2″ thick	1 jar or can of your favorite
¼ cup seasoned flour	spaghetti sauce
Cooking fat	Salt and pepper
	Hot water (if necessary)

Pound as much seasoned flour as possible into both sides of the steaks. Brown on both sides in hot fat. Add the spaghetti sauce. This should cover the meat. If it does not, add enough hot water to cover. Simmer 1½ to 2 hours or until the meat is tender.

BUFFALO STROGANOFF

1½ lbs. buffalo top sirloin or
    top round
1 clove garlic, cut in half
¼ cup flour
1 tsp. salt
¼ tsp. pepper
⅓ cup butter or margarine

½ cup minced onion
½ cup water
1 can condensed cream-of-
    chicken soup, undiluted
1 lb. mushrooms, sliced
1 cup commercial sour cream
Snipped parsley, chives, or dill

Rub meat on both sides with garlic halves. Combine flour, salt, and pepper. Pound flour mixture into both sides of meat until meat is quite thin. (Use either a mallet or saucer edge for pounding.) Cut meat into 1½"x1" strips. In hot butter in Dutch oven or deep skillet, brown meat strips, turning them often; remove meat. In same butter in Dutch oven, sauté onions until golden. Add water and stir to dissolve brown bits in bottom of Dutch oven. Add soup and mushrooms; cook, uncovered, over low heat about 20 minutes, stirring occasionally, until mixture is thickened. Combine buffalo, mushroom mixture, and sour cream; heat through, but do not boil. Sprinkle with snipped parsley, chives, or dill.

This can be prepared in advance by following the above recipe and refrigerating the buffalo meat after browning and the mushroom mixture after it thickens. About 30 minutes before serving, combine buffalo, mushroom mixture, and sour cream; heat, but do not boil, about 25 minutes or until heated through.

Serve with noodles, mashed potatoes, or fluffy rice.

BUFFALO HUMP

Take 3 to 5 lbs. of buffalo hump meat. Cut slits completely through the meat. Insert slivers of garlic down into the slits. If no garlic, use onions. Pour 1 cup of vinegar over meat, making certain it runs into slits. Place in refrigerator and leave for 24 to 48 hours.

When ready to cook, place in large, heavy pot. Brown in oil until nearly burned on all sides. Then pour 2 cups of strong black coffee over the meat. Add 2 cups of water and cover. Simmer on top of stove for 4 to 6 hours. Season with salt and pepper 20 minutes before serving.

SWEETGRASS BUFFALO AND BEER PIE

(This recipe from *Canadian Cuisine,* published by the Canadian Government Travel Bureau, Ottawa.)

4 lbs. buffalo meat
Salt, pepper, sage, oil, and flour
3 medium onions
3 carrots
3 stalks celery
3 potatoes
3 tbsps. flour

2 pints beef stock
2 tbsps. tomato purée
1 pint beer
Herb bag containing a garlic
   clove, a bay leaf, parsley,
   3 cloves, a pinch of thyme

Cut meat into 1″ cubes. Season with salt, pepper, and sage. Roll in flour and brown in oil. Transfer meat to a heavy saucepan. Cut vegetables into ½″ cubes. Sauté in the same oil and add to meat. Add 3 tbsps. flour to oil and let it brown. Add stock, tomato purée, and beer and blend. Add herb bag and all the meat and vegetables. Simmer until tender. Remove the herb bag. Place stew in a casserole or individual pot pie oven dishes, cover with pastry, brush with milk, and bake until golden brown.

BUFFALO TIPS A LA BOURGOGNE

(This recipe was used in August 1964 when Mrs. Lyndon B. Johnson visited Green River, Wyoming in connection with the dedication of the Flaming Gorge Recreation Area. It was provided by Mrs. Gale McGee, wife of the U.S. Senator from Wyoming.)

5 lbs. buffalo meat
1 cup olive oil
2 large onions, diced
2 glasses Burgundy wine
½ quart tomato purée
3 bay leaves

1 tbsp. monosodium glutamate
3 cups chicken broth
Salt and pepper
½ lb. butter
2½ lbs. pasta

Barbecue buffalo meat over a slow open fire. Smoke until done. Cut meat into 1″ cubes. Sauté buffalo meat with diced onions in olive oil. When onions are golden brown, add 2 glasses of fine Burgundy wine. Reduce wine to ⅓, then add ½ quart of tomato purée, 3 bay leaves, 1 tbsp. monosodium glutamate, 3 cups chicken broth. Simmer for 1 hour, season with salt and pepper to taste. Serve with buttered pasta. Pasta should be boiled until tender.

BUFFALO ROAST

3-5 lb. buffalo roast
1 slice bacon, cut into small
  pieces
2 cloves garlic, sliced

Salt and pepper
1 bay leaf
2 cloves
1 cup orange juice

Cut slits in buffalo meat and insert small pieces of bacon and garlic. Salt and pepper well. Sear meat on all sides. Place meat in roaster and place bay leaf and cloves on top. Baste with orange juice. Roast in 325° oven, 25 minutes to the lb., basting frequently.

BUFFALO STEW

2 lbs. buffalo meat
Cooking oil
1 medium sized onion, chopped
1 can tomato sauce, 8 oz. size
6 carrots, peeled and sliced

Salt and pepper
3 medium potatoes, peeled and
  sliced
1 package frozen mixed
  vegetables

Cut meat into 1″ cubes. Brown in small amount of cooking oil. Add onion and cook until golden. Add tomato sauce, carrots, and salt and pepper to taste. Cover and lower heat to simmer. Cook for 1 hour. Add potatoes. Add a little water if needed. Cook for another ½ hour. Add frozen vegetables and simmer for another ½ hour.

If you want to dress this up just top with biscuits and bake in 450° oven till biscuits are browned.

SAUCE

This sauce can be used on roast, steaks, and ground buffalo:

1 pint olive oil
1 pint red dry wine
1 tbsp. oregano
1 tbsp. black pepper
1 tbsp. sweet basil

1 tbsp. garlic powder or
  3 cloves fresh garlic
Salt
3 tbsps. monosodium
  glutamate
2 cans tomato sauce

Blend all ingredients in blender. Baste on meat on all sides while cooking. Do not overcook, as buffalo meat is lean and juicy with very little fat. This amount of sauce is good for 5 lbs. of buffalo meat.

D

Appendix

Where to Find the Buffalo

The following is a list of wildlife refuges, parks, zoos, and private citizens in North America who watch over most of the world's buffalo population. The list is not complete nor 100 percent accurate; it cannot be. Buffalo die, change hands, or are slaughtered for meat, and no central agency keeps track of their number or location. Nonetheless, this is a reasonably comprehensive listing of government and private owners in the U.S.A. and Canada (47 states and 6 provinces and Northwest Territories) compiled during the late 1980s, and anyone wanting to go and see buffalo can use this as a guide. To doublecheck on the location of buffalo in a given area, contact the county agricultural agent, the director of the nearest zoo, government fish and game offices, and/or local residents. Buffalo are not always *in* the towns listed below but should be nearby.

U.S.A.

Alabama
Birmingham
  Zoo
Huntsville
  Ernest R. Schwarze
Selma
  Jamie Etheredge

Alaska
Anchorage
  Joe and Diane Hayes
Big Delta
  Alaska Dept. of Fish & Game
Chitina
  Alaska Dept. of Fish & Game

Copper River
  Alaska Dept. of Fish & Game
Farewell
  Alaska Dept. of Fish & Game
Healy Lake
  Alaska Dept. of Fish & Game
Kenai
  Bill and Brenda Zubeck
Kodiak
  Kodiak Cattle Co.
  Bill and Kathy Burton
Palmer
  Bob Milby

Arizona
Arivaca
  Loren and Beulah Baker

Arizona City
  Yavapai Ranching Co.
  M. C. Davis
House Rock
  Arizona Fish & Game Commission
  South Canyon Ranch
    (Grand Canyon Nat'l Game
    Preserve)
Laveen
  Sanford Aldon Estes
Mesa
  Triple Bar S Ranch, Inc.
Paulden
  Chino Valley Land & Cattle
  Merwyn C. Davis
Phoenix
  Zoo
St. Johns
  Municipal Park
Sierra Vista
  Apache Pointe Ranch
  Paul E. Kiefer
Snowflake
  James Martindell
Two Guns
  Arizona Fish & Game Commission
  Raymond Ranch
    (between Flagstaff and Winslow
    south of Interstate at exit sign
    "Buffalo Park")
White Mountain Lake
  Buffalo Ridge Land & Cattle Co.
  Chris McIntyre

Arkansas
Brinkley
  Bob Fuller
Fayetteville
  University of Arkansas
Little Rock
  Zoo
McRae
  William Shaver

Melbourne
  Flight Dollar Farm
  Jack M. Haley
Mountain Home
  Joe and Rose Kern
Nenana
  YK School District
  Bruce A. Kleven

California
Alturas
  Diamond C Ranch
  Doran W. Goulden
Angels Camp
  Dam Ranch
  Kenneth and Sall Lowry
Avalon (Catalina Island)
  Doug Propst
Blythe
  Jack and Susan Schuringa
Brawley
  Steve Seabolt
Canoga Park
  Timothy Matonak
Elk Grove
  Ray & Eugenia Steele
Encinitas
  Oak Springs Ranch
  Jerry O'Brien
Estacada
  Donald McCulloch
Fallbrook
  Ralph Peters
Fresno
  Zoo
Graeagle
  Vision Quest Ranch
  Robert and Jerry Galarza
Grass Valley
  James Bostick
Joshua Tree
  R. L. Hammond Buffalo Ranch
Lancaster
  David Whiteside

Lockeford
  Tony Meath
Lompoc
  Bar Y Ranch
  John and Marilyn Mosby
Los Angeles
  Zoo
Lost Hills
  Florisa Harris
Marysville
  Helder Mfg. Inc.
National City
  Romney Hayden
Newberry Springs
  High Desert Bison Ranch
Newbury Park
  David T. Lehto
Newcastle
  J. C. Hatfield
Newhall
  William S. Hart Park
Ojai
  Herman Rush
Ontario
  Jack Selle
Orange
  Bob Schwarze
Orinda
  Mysterious Valley Ranch
  Norman C. Pease
Ramona
  Kings Villa Bison Ranch
  Art and Linda Thomsen
  Smith's Buffalo Acres
  Ron and Linda Smith
  Star B Ranch
  Ken and Denice Childs
Red Bluff
  Buffalo-Beefalo Ranch
  Martha Louise Harper
Romoland
  John and Linda Frey

San Diego
  Zoological Garden
San Francisco
  Martin Dias
  Zoological Garden
San Jose
  Durham Ranches
  Armando Flocchini
Santa Maria
  Zoo
Saratoga
  Inquieto Ranch
  Helen Pastorino
Sepulveda
  Herbert and Jane Boeckmann
Sunnyvale
  Richard Lawson
Stockton
  Jack Williams Ranches
Yermo
  Brer Ranch
  Roy and Dian Hare
Yorba Linda
  James Bostick

Colorado
Arvada
  Phil and Matilda Black
Avondale
  Garry Tatsch
Berthoud
  Wilfred E. Dodds
Boulder
  University of Colorado
Brighton
  Orrie and Lois Klassen
Byers
  Gene and Shirley Linnebur
Center
  D&D Limousin Ranch
  Dale and Sherry Dilley
  Perrin Farms
  Woody and Buna Perrin

Rocky Mountain Bison, Inc.
Brian and Diane Ward

Colorado Springs
Mark Basman

Commerce City
Mountain Man Buffalo Ranch
Del and Cam Hensel

Craig
Lay Valley Bison Ranch
Daniel Martin

Denver
The Bisonsmiths
Ron Smith

Duncan Cattle Co.
Mike Duncan

Alvin Stjernholm

Zoo

Erie
Bob and Laurie Dineen

Evergreen
Evergreen Memorial Park

Fort Collins
Rawhide Power Plant

Buck M. Purcell

Fort Lupton
Rocking RH Ranch
Robert Heinz

Fruita
Colorado National Park Monument

Greenwood Springs
Larry R. Schmueser

Hartsel
Elk Mountain Cattle Co.
Monte Downare

Hudson
Ross Shandy

Indian Hills
Dr. James D. Gibson

Hidden Valley Ranch
Tory Meyer

Jefferson
Columbine Ranch

Kiowa
Carl and Janice Herbst

Longmont
B-J Acres

McCoy
C Bar C Ranch
Terry Ivie

Mead
Colorado Bison Co.
Merle and Susan Maass

Meeker
Devil's Hole Land and Livestock
Bob and Gayle Crawford

Nathrop
Don and Karin Adams

Olathe
DeVries Buffalo Ranch
George and Liz DeVries

Parker
Lazy R. K. Ranch
R. J. and Karen Cochran

Platteville
Boulter Buffalo
Bill Boulter

Sedgwick
Price Brothers

Steamboat Springs
High Wire Ranch
Dave and Sue Whittlesey

Sterling
Sherwin's Shiftin Sands Ranch
Carl and Effia Sherwin

Stratton
Gary and Cindy Brachtenbach

Wiggins
Bijou Creek Buffalo
Paul Kennedy and Pat Helstro

Yoder
Tatanka Buffalo
Robert Paul

Connecticut

Bridgeport
Beardsley Park Zoo

Collinsville
  Chet and Gerri Thomen
Greenwich
  Eugene J. Rabbitt
Southington
  Roy S. Florian
Stamford
  Stamford Museum and Nature
  Center
Watertown
  American Buffalo Co.
  Gregory Hostetler

Delaware

Dover
  El Ensueno Farm
  Frank DiMondi
Frederica
  G. Wallace Caulk, Jr.
Greenwood
  John Collins
Milton
  J. Wayne Melvin

Florida

Altamonte Spring
  Ronald Scarlata
Archer
  Robert E. Price
Bushnell
  JB Acres
  Beverly J. Darbyson
Clearwater Beach
  John S. Taylor, III
Christmas
  James and Landa LaFemina
Ft. Pierce
  Buffalo Creek Ranch
Jacksonville
  Charles H. Barco
  James E. Davis
Key Biscayne
  Crandon Park Zoo
Lake City
  Hugh Buie

Longwood
  David Cayton
  Frenchman Family Trust
  John and Pamela Yurick
Narcoossee
  Richard L. Murphy
Ocala
  Charles R. Forman
Opa Locka
  Chi-O-Ranchero, Inc.
  Arnold Oper
Panama City
  Jim Graham
Seffner
  Robert H. Humphreys
Tampa
  Busch Gardens
Tavernier
  Ned and Connie Maze
Trenton
  Endangered Species
Vero Beach
  Edward K. Halsey
Walnut Hill
  George E. White, Jr.
Zolfo Springs
  Peace River Growers
  Donnis A. Barber

Georgia

Augusta
  Buffalo Ranch
  Cecil F. Miller
Atlanta
  R. E. "Ted" Turner
Bowdon
  Dr. Wilbert R. Tutsch
Dry Branch
  Mark Durden
Fayetteville
  G/B Buffalo
  George and Becky Black
Lilburn
  Stone Mountain Game Ranch

Lyons
  S&K Bison
  Steven G. Coleman
Marietta
  John S. Johnson
Milledgeville
  Poin D. Galloway

Hawaii

Hanalei
  Bill and Marty Mowry
Honolulu
  Zoo

Idaho

Aberdeen
  Foster Family Farms
  Bruce and Kathy Foster
Boise
  Jerry Whitehead
Caldwell
  Harvey Saxton
Cascade
  Frank D. Callender
Coeur d'Alene
  Dr. Richard Morris
Filer
  Rockie R. Egner
Idaho Falls
  King B., Inc.
Pocatello
  Zoo
Rexburg
  John D. Ferguson
  Bill and Pat Moore
Troy
  Tatonka Whitepine Ranch
  Mike and Debi Kerley
Twin Falls
  Joe Miller

Illinois

Annawan
  The Rolling K
  Lyle R. Kuelper

Astoria
  Charles B. Briney
Batavia
  Fermi Nat'l Acc. Lab
Belleville
  Terry L. Hein
Champaign
  John Mitchler
Chicago
  Lincoln Park Zoo
Coal Valley
  Charles R. Akers
Dallas City
  Lyle and Judy Matejewski
Decatur
  Burton and Elizabeth Stollard
Des Plaines
  DoAll Company
Dixon
  Dean E. Considine
Du Quion
  Denny Harsy
Galesburg
  M&M Bison
  Mark and Mike Runbom
Geneseo
  John F. Edwards, Jr.
Highland
  Steven W. Enterprises
Lanark
  Lawrence L. Derrer
Marseilles
  Maynard Myer
Milan
  Edward and Barbara Garrett
Mundelein
  Lester's Nat'l Service
Metropolis
  Fort Massac State Park
Minooka
  Glenn Eggleston
Ottawa
  Buffalo Rock State Park

Pawnee
  Jennings Buffalo Ranch
  William and Cind Jennings
Park Ridge
  Long Star Ranch
  E. I. Wenske
Quincy
  Soldiers and Sailors Home
Reynolds
  Kenneth Ziemer
Riverside
  Rancho de Aurea
  Frank and Aurea Stemberk
Sugar Grove
  Lone Grove Buffalo Ranch
Wheaton
  Bison Ridge Ranch and Game
  Ranch
  James D. Atten
Villa Grove
  Timothy M. Berry

Indiana
Attica
  Patricia F. Lee
Avilla
  Jerry L. Cochran
Battle Ground
  Erich Klinghammer
Brazil
  Donald L. Price
Elizabeth
  Needmore Buffalo Enterprises
  Art and Jane Stewart
Elwood
  J&J Farms
  John E. Morse
Hoagland
  Larry M. Frecker
Madison
  Clifty Acres Zoo Farm
Monticello
  Carl Van Meter
Salem
  Ernest Deaton

South Bend
  Homer W. Fittering
Terre Haute
  Ernest L. Furnas
Warsaw
  Larry and Betty Boggs
  Thomas B. Maze
Williamsport
  R. H. Gillespie

Iowa
Atlantic
  Gary Anderson
Bloomfield
  Herbert Buffalo Farms
  Dwayne and Carol Herbert
Bode
  Ralph and Donna Olson
Burt
  Rodney A. Schmidt
Clarksville
  Neal and Arlene Wedeking
Davenport
  City Park
Denison
  Laubscher & Son
Dysart
  David Vogeler
Estherville
  Scott Benjamin
Everly
  James Hartmann
Fairfield
  John Huebner
Ft. Madison
  Marvin F. Strunk
Granger
  Jester Park
Independence
  Steve L. Miller
Iowa Falls
  Everette F. Gehrke
Janesville
  John and Esther Folkerds

Kensett
  Jack Norland
Larrabee
  Tom Mummert
Manilla
  Kevin L. Boyens
Monticello
  Stephen A. McDonell
Mount Pleasant
  Lloyd Wonderlich
Muscatine
  Robert Harbough
New Hartford
  Francis Scribner
Nora Springs
  Francis L. Sherman
Othio
  Jackson Brothers
Oxford
  Marvin P. Swenka
Promise City
  Jackson Brothers
Sigourney
  Wilmer L. Clingan
Sioux City
  Floyd Meyer
St. Ansgar
  Raymond and Arlyn Priem
Strawberry Point
  Charles W. Seedorff
Tama
  James and Joyce Dolezal
Thayer
  Beverly Houghton
Toddville
  Lyle E. McBurney
Waterloo
  Clarence Scribner
Webster City
  Bob Wearda
Woodbine
  Robert and Carolyn Probasco

Kansas
Auburn
  Wayne A. Copp
Barnes
  Brian and Kathy Wendland
Belleville
  Clyde and Betty Makalous
Beloit
  Butterfield's Buffalo Ranch
Brownell
  North Box X Ranch
  Harley and Donna North
Canton
  Maxwell Game Preserve
  Kansas Dept. of Wildlife and Parks
Codell
  Flying G Buffalo
  Don and Sharolyn Gramm
Clyde
  Lester C. Lawrence
Ellsworth
  Ralph Bruning
Ensign
  Thunder of the Plains Buffalo
  Doug Hinshaw
Fort Riley
  U.S. Army Buffalo Corral
Great Bend
  City Zoo
Haddam
  Allyson L. Novak
Hays
  City Park
Hill City
  Loren E. Chalfant
Hugoton
  Steve Kinser
Hutchinson
  Zoo
Kendall
  Soaring Hawk Ranch
  Joe and Kathy Rishel
Leavenworth
  U.S. Penitentiary

Leona
　Lee Simmons
Ludell
　Bob and Berdean Simminger
Moran
　E. J. Siefker
Peck
　Wesley Wilbur
Pittsburg
　LB Buffalo Ranch
　Lavon Barthelme
Pratt
　Mark Rice
　Crawford County State Lake
Rose Hill
　Best Butcher
　John Reed and Pam Morris
Salina
　Charles Walker
Scandia
　Scandia Buffalo Park
Scott City
　Rich and Susan Duff
Troy
　Larry D. Handley
Tyro
　Art & M Co.
　Mikel E. Melander
WaKeeney
　Del E. Wiedman
Wallace
　Wilber Williams
Wichita
　Bluestem Buffalo
　David and Barbara Palmer

Kentucky
Benton
　Dr. C. R. Freeman
Falmouth
　A. Dale and Linda Adams
Fancy Farm
　Bobby Joe Tyler

Golden Pond
　Tennessee Valley Authority,
　Land Between the Lakes
　(along U.S. highway 68)
Horse Cave
　Kentucky Buffalo Park
　Bill and Judy Austin
Mayfield
　George Knight
Richmond
　Earl Puckett

Louisiana
Cottonport
　Edgar B. Bain
Monroe
　Berstein Park Zoo

Maine
Falmouth
　Charles Payson

Maryland
Chesapeake City
　Richard C. duPont, Jr.
Churchville
　Paul D. Hines
Germantown
　Bill Robertson
Harwood
　C. Gerald Aben, Jr.
Lutherville
　Dr. Hans R. Wilhelmsen
Phoenix
　Paul Steffan
Riverdale
　Howard G. Sams
Upper Marlboro
　William L. Bondurant

Massachusetts
Essex
　Augustus Means

Jefferson
  Alta Vista Farm
  Howard and Nancy Mann
Worcester
  City Park

Michigan
Alto
  Richard Hawkins, II
Ann Arbor
  Domino's Farms
  Chris Nelson
Cheboygan
  Renegade Ranch
  Walter and Marilyn Romanik
Coldwater
  Pete Barone
Drafter
  Alex Frank Wieczorek
Dryden
  Charles Wright
Grand Rapids
  John Ball Zoological Gardens
Haslett
  Diana Many
  James Sikarski, DVM.
Jackson
  Dr. A. J. Kiessling
Lansing
  Fenner Arboretum
  Potter Park Zoo
Montrose
  Delbert C. Bamberg
Owosso
  David S. Couturier, DVM.
Oxford
  George Balch
Pinckney
  J. and A. Shirley
Toivola
  Ted and Janet Johnson
Traverse City
  Clinch Park Zoo
  Jerry Oleson

Ypsilanti
  Mark C. Dekarske

Minnesota
Bagley
  H. E.'s Buffalo Ranch
Bloomington
  Jon Malinski
Center City
  Hidden Acres
  Ed and Pam Eichten
Cotton
  Twin Rivers Buffalo Ranch
Deer River
  Laurence Paulsen
Gibbion
  Kenneth Briese
Hastings
  Sundance Ranch
  Terry Clark
Hibbing
  Joseph M. Shea
Lindstrom
  Bob and Virginia Luger
  Critter Haven
  Kitty Watson and B. Ruecker
Makinen
  Simon and Pat Ratzlaff
Mankato
  Dan Williams
Minneapolis
  Samuel H. Ordway
Olivia
  Greg O'Halloran
Pelican Rapids
  Rolling R Ranch
  Dale Rengstorf
Redwood Falls
  Alexander Ramsey City Park
Renville
  C. L. and Ann Van Zee
Sauk Centre
  John H. Flowers

Sleepy Eye
  Stanley Seifert
St. Michael
  Joseph and Mariene Zachmann
St. Paul
  Como Zoo
Warroad
  War Road Buffalo Ranch
  Michael Marvin
Watertown
  Sunset Crest Farm
  Eugene P. Sweet
Willmar
  J&L Buffalo Ranch
  John and Leila Arndt
Winona
  Ken and Arlene Westrud

Missouri

Bloomdale
  Donald Waser
Bonne Terre
  Paul and Lee McDowell
Clifton Hills
  Dan Shepherd
El Dorado Springs
  E. C. and Betty Witt
Kansas City
  Swope Park Zoo
Mansfield
  Robert H. Carr
Moberly
  Shepherd Farms, Inc.
  Jerrell Shepherd
Noel
  Buffalo Meat Co.
  Doug Cory
Peculiar
  Dr. Raymond Cook
Rayville
  Richard and Kelli Holm
Saint Louis
  Grant's Farm
  H. M. Sayers
  Zoo

Seymour
  J&J Buffalo Ranch
  Jim and Julie Fox
Springfield
  Garrett's Buffalo Ranch
  Midwest Fibre Co.
Washington
  Wilbert W. Tretmann
Washington
  Wilbert W. Tretmann
Willard
  Terry & Barbara Goff

Montana

Bainville
  Fay Crusch
Belfry
  Bill & Pat Moore
Belgrade
  Bud Sandquist
Biddle
  Russell & Rumph Ranch, Inc.
  Robert Rumph
Billings
  June Edgmond
Bozeman
  Buffalo Jump Ranch
  Dennis Rowe
Browning
  Blackfeet Fish, Wildlife & Parks
  Dept.
Butte
  Joyce and Page B. Pratt
Cohagen
  Sundance Ranch
  Mike and Karen Frost
Coldstrip
  Western Energy Co.
Dillon
  Ben and Faye Holland
Lame Deer
  Northern Cheyenne Indians
Livingston
  Carter Ranch
  Howard Carter

Luther
    Burnett Ranches
    Jim and Betty Burnett
Moiese
    National Bison Range
Sand Springs
    Steven W. Ranch
Twin Bridges
    Wink and Jerry Nyhart

Nebraska
Bassett
    Dennis and Kathy Spence
Battle Creek
    Rick and Shelia Renner
Beatrice
    Butch Malchow
Brule
    Rex and Jean Harris
Culbertson
    Jerry & Nancy Kautz
Emerson
    Michael and Connie Hassler
Fremont
    John G. Poehling
Gering
    Nebraska Game and Parks
    Commission
Hardville
    Stevens Farms
Lincoln
    George Eager
    Zoo
Mullen
    4-O Ranch
    Eric P. Ericksen
Newport
    Georgia Peterson
Omaha
    Henry Doorly Zoo
    Jackson Buffalo
    Bob and Nancy Jackson
    RJ Buffalo Farm
    Dick and Kathy Caradori

Polk
    Gaylerd Stevens
Potter
    Buffalo Bend
    Jon and Pauline Wright
Seward
    T. R. and Virginia Huges
Valentine
    Fort Niobrara Nat'l Wildlife Refuge

Nevada
Las Vegas
    Hershel Leverton
Reno
    L & L Ranch
    Dave & Laurie Lang
Schurz
    Pancho Quintero

New Hampshire
Milford
    Gervaus Castonguay

New Jersey
Mount Freedom
    Kohola Farms
    George Thomas Smith
Vineland
    Lois Martelli
Whiting
    Charles and Mary Menzer
Woodstown
    Stoney Harris

New Mexico
Albuquerque
    Rio Grande Zoo
Cimarron
    Philmont Scout Ranch
Clovis
    Zoo

New York
Binghamton
    City Park
Buffalo
    Zoological Gardens

Catskill
  Catskill Game Farm, Inc.
Clifton Springs
  Buttonwood Farms
  I. A. Morris
Ellicottville
  B&B Buffalo Ranch
  Glenn D. Bayger
Hillsdale
  Joe Flood
Holcomb
  Scott-T-Ranch
  Oliver B. Scott
Holtsville
  Explorer Post #10
  Kenneth A. Schwindt
Ionia
  Keith Bennett
Melrose
  Charac W. Case
Norfolk
  Buck Jones
Riverhead
  North Quarter Farm
  Ed Tuccio and Dee Muma
Skaneateles
  Tiffany Buffalo Ranch
  Brad Tiffany
South Schodack
  George Mesick
Stormville
  Homestead Farms
  Charles Tucker, Jr.
Troy
  Don Fane
Victor
  F. Keith Bennett

North Carolina
China Grove
  Country Pines Farm
  Edward and Candy Frye
Raleigh
  Rufus Wade Holden, Jr.

North Dakota
Backoo
  Terry Koropatnicki
Bismarck
  Dakota Zoo
  Harold Schafer
Bowdon
  Greg Schander
Carrington
  Hawks Nest Buffalo
  Bruce and Michelle Willyard
  Roger Miller
Dering
  Duane Woodall
Drayton
  H. B. Farms
Edmore
  Buffalo Phil's
  Phil and Tammy Ivesdal
Flasher
  Rossow Buffalo Ranch
  Clifford and Lloyd Rossow
Fort Totten
  Sullys Hill State Game Preserve
Garrison
  Herbert W. Mautz
Grand Forks
  Siouxland Buffalo
  Doug and James Earl
Grandville
  Ralph R. Carty
Hazen
  Herbert & Ruth Oster
Heaton
  Tim Schander
Hettinger
  Roy McFarland
Hillsboro
  Stuart Inc.
Maddock
  Brent and Debbie Evje
  Minnie Ha Ha Bison Ranch
  Greg Maddock

Mendora
  Theodore Roosevelt Nat'l Park
Michigan
  Ken and Kristi Shirek
Minot
  David Llewellyn
New Rockford
  RX Ken Mar Buffalo Ranch
  Ken Throlson, DVM.
Park River
  Kenneth Porter
Pingree
  Dakota Plains Buffalo
  Mark and Linda Ivesday
Rhame
  Double Diamond Ranch
  Calvin Miller
Valley City
  Thundering Herd Buffalo
  Ranch
  Mike and Cindy Schwehr
Wahpeton
  Bill Krump
Washburn
  Wicklanders Thundering
  Herd
  Ray and Dan Wicklander
Williston
  City Park

Ohio

Austinburg
  Kusar Farm
  John and Eileen Kusar
  Miller's Buffalo Ranch
Berlin Center
  Windy Knoll Acres
  William Ripple
Bowling Green
  David and Karen Apple
Bucyrus
  William Leitzy
Chillicothe
  Robert E. Litter

Circleville
  Roe K. Riegel
Coalton
  Baisden's Bison
  Ron Baisden
Columbus
  Bill and Janelle Kimmel
  Zoo
Dover
  Wild Bill Cody Gourmet Buffalo
Galion
  Charles Kline
Gibsonburg
  Timothy M. Younker
Hartville
  William A. Tocci
Logan
  Keith Fox
Malvern
  Gary Juwell
Medina
  Bar S Buffalo Farm
  Willard A. Stephenson
  Jack R. Barensfeld
Orange
  Barrister Farms
  Joseph A. Dubyak
Rochester
  Jack Hodgekin
Urbana
  Johsua Green Farm
  Curtis H. Springer
Valley City
  Lester Klooz
Willard
  Huron River Bison
  Fred and Lillian Sharp
Wooster
  K.D.K. Farm
  Kenneth D. Franks

Oklahoma

Adair
  Windsong Ranch

Bartlesville
Ken-Ada Ranches, Inc.
Ken and Diane Adams
Don and Kathy King
Woolaroc Museum and Ranch
Canute
Spurin "F" Ranch
Steve and Betty Finnell
Cushing
T Bar D Buffalo Ranch
Tom and Jeanne Davenport
O. S. Anderson
Fairview
Therrel Martens
Guthrie
Pollock Acres
Carl and Helen Brzozowske
Guymon
City Park
Hennessey
Lambert Kusik
Hinton
Edwin Opitz
Keota
Blue Chip Ranch
Carl and Alice Veley
Oklahoma City
Herman Meinders
Zoo
Logan
Smith Ranch
Bert and Carol Smith
Muskogee
Tom Caster
Pawnee
Pawnee Bill Museum
Stuart
P. S. and A. T. Smith
Sulphur
Platt Nat'l Park
Tulsa
Rambling Creek Ranch
Stan and Barbara Grogg

Oregon
Baker
John Bennett
Bend
Dan C. Boone
Burns
Glenn A. Brown
Cave Junction
Hancock Buffalo Haven
B. W. Hancock
Colton
H. T. "Ted" Eurick, Jr.
Enterprise
Robert Stangel
Heppner
Harold A. Wright
Hermiston
John and Marge Walchli
Umatilla Buffalo Ranch
Stuart F. Bonney
Lincoln City
Yellow Tail Buffalo Co.
Peter H. Gamroth
Mt. Angele
Robert Rash
Pendleton
Bob and Mary Lou O'Rourke
Riddle
Don R. Johnson
Roseburg
Bill and Geneva Pruitt
Vernonia
Bar "A" Ranch
Karl and Mary Altenhein

Pennsylvania
Albion
Richards Farm
Clarion
Diamond M Farm
James A. Mays
Carversville
Nicholas Zelenevich

Duncansville
  Richard T. Curry
East Earl
  Omar M. Graybill
Edinburg
  P. L. Miller
Hershey
  Hershey Park Zoo
Julian
  Martha Richards
Langhorne
  A. J. Kutney
Lilly
  Charles M. Cooney
Lock Haven
  Ponderosa Bison Ranch
Meshoppen
  Kutney's Korner Buffalo
  Ranch
  Andrew and Marge Kutney
Millville
  James F. DeVoe, DVM
New Castle
  Joan and Hugh Forbes
  Bob Mahoney
New Hope
  Robert J. Blanche
Northumberland
  Marlin E. Reedy
Moshannon
  Mountain Laurel Farm
  Larry and Karen Frisbee
Tidioute
  JMK Buffalo Ranch
  Larry and Geny Brown
Valley View
  A. R. Williamson
West Middlesex
  Joseph A. Mastrianno, Sr.

Rhode Island

East Greenwich
  Alastair C. Gowan

South Dakota

Aberdeen
  Oliver Ericksen
  Melgaard Park
Buffalo Ridge
  Buffalo Ridge Corp.
  Dean and Ruth Songstad
Britton
  Cyril Hastings
Buffalo
  Lyle Gunderson
Castlewood
  Robert Raasch
Custer
  Custer State Park
  Ron Walker
  LaVern and Joyce Busskohl
Fairfax
  RL Buffalo Ranch
  Ray and Loreen Ebsen
Fort Pierre
  Triple U Enterprises
  Houck Family
Fulton
  Mahoney Buffalo and Cattle
  Ranch
Gettysburg
  Buffalo Run Game Reserve
  Bronce LeBeau
Hot Springs
  Wind Cave Buffalo Ranch
Hurley
  Eugene Friman
Interior
  Badlands Nat'l Monument
Iroquois
  Eugene Hein
Keldron
  Bob and Kathleen Petik
Keystone
  Leo H. Toskin
Longlake
  Merle M. Meier

McLaughlin
Flying H Ranch
Monty and Adolph Hepper
Morristown
Lloyd Cleo Goeres
Piedmont
Black Hills Bison Co.
Presho
Trails End River Ranch
Larry and Dixie Byrd
Pukwana
Frank Sharping
Rapids City
Triple Seven Ranches
Duane and Rose Lammers
Martin F. Collins
Rockham
Heim Buffalo Ranch
Sioux Falls
City Park
Wessington Springs
John and Peggy Christensen

Tennessee
Bristol
John Kirk
Dandridge
Robert G. Finchum, Jr.
Hendersonville
Ace-Allen Cattle Enterprises
Duane Allen
Lawrenceburg
Diamond C Land and Cattle Co.
Jerry Cash
Waynesboro
LA's Buffalo Paradise
Linda Arvin

Texas
Amarillo
Storyland Zoo
Ben Wheeler
3-T Ranch
Bill and Jean Thompson

Bevins
John W. Corley
Bowie
R. D. Proctor
Breckenridge
Kuhn Ranch
Charles and Helen Kuhn
Brownwood
Herman Bennett Ranch
Canutillo
Cruz Tierra Farms
BJ and Margarete Crossland
Chapman Ranch
John D. Orr
Childress
Zoo
Clyde
John Hawk
Dallas
Zoo
Darrouzett
Bob and Sharon McIlhattan
Dublin
Bob Colwell
El Paso
Charles H. Dodson, Jr.
Eustace
Sam Jock
Evadale
Housetop Mountain Ranch
Mike Bruce
Forestburg
T. P. Reynolds
Fort Worth
David M. Ryan
Gainesville
Leighton & Janette Smith
Harper
E. Dean Hopf
Henderson
Keith Price
Houston
C. F. Ranch
Charles Farley

Boyd J-C Texas Longhorns
Crawford and Joyce Boyd
Justiceburg
Riley and Mary Miller
Lewisville
Leon Cross
Menard
Mrs. Frances Merrick
Midland
Wilson Ranch
William and Kathleen Wilson
Nacogdoches
Gerald and Weslynn Larsen
Naples
Walter Cox
Odessa
Cedar Mountain Buffalo Co.
C. L. and Peggy Lee
D&S Supply
Dale Leverett
Round Rock
Mark Basman
San Felipe
Buffalo Express
Richard J. Filip
Slaton
Greg Scott
Stonewall
LBJ State Park

Utah

Farmington
Antelope Island State Park
Helper
Utah Dept. of Fish & Game
Park City
JJNP Ranch
R. J. and Ginny Pinder
Panguitch
Jerrald and Laur Mosdell
Salt Lake City
Terry R. Giles
Roosevelt
ADS Ranch
Cary and Judy Smith

Vermont

Burlington
Anthony Perry
Granby
Circle B Ranch
Shelburne
Anthony Perry

Virginia

Abingdon
Mac R. Clifton
Annandale
David, Mike and Pat Shalap
Chantilly
Thistlewood Farm
Wray S. Dawson
Clifton Forge
Jefferson C. Morris
Dayton
Carlyle Meade Farm
Harrisonburg
William C. Rader
Marshall
Summerhill Farm
Ron Korn
McLean
Kinney Horn
Norfolk
Zoo
Stafford
Eugene Locklear
Suffolk
Chet Ehrenzeller
Virginia Beach
Chet Ehrenzeller

Washington

Ariel
Michael N. Cook, DVM
Chewelah
Beaver Valley Ranch
Deerharbor
Peter Whittier

Ellensburg
  JRL Ranch
  C. E. and Inez Lippencott, Jr.
Ferndale
  G. Landcastle
Mount Vernon
  Jack Chrysler
  John D. Chrysler
Pomeroy
  Gordon Killingworth
Port Angeles
  Robert and Annie Baublits
Selah
  Loren R. Wilson
Shelton
  Ray T. Coleman
Snohomish
  Lloyd W. Wibbelman
Spokane
  Marland Ray
Usk
  Kalispel Indian Tribe
Vader
  Gerald K. Brooks

Wisconsin
Black River Falls
  Thomas R. Bible
Birnamwood
  Vilas Cinaski
Burlington
  Buffalo Creek Ranch
  Jim and Barb Mangold
Cadott
  Dean and Marsha Moats
Cornell
  C-V Buffalo, Inc.
  Gary Vanden Heuvel
Darien
  Ernest and Virginia Kalb
Deer Park
  Waidelich Wilderness
  Jesse J. Waidelich

Forestville
  Stony Creek Buffalo Ranch
  Mike and Vicki Torp
Fountain City
  Mike Fogel
Green Bay
  J. J. Welhouse
Hartland
  Joe and Dorothy Becker
Kansasville
  Brighton Bison
  Jerry and Barb Daniels
La Crosse
  Richard E. Abraham
Lyndon Station
  Buffalo Valley Farms
  Harold M. Gall and Sons
Madison
  Vilas Zoo
Marathon
  Hank's Bison Farm
  Henry Bauman
Mauston
  Ray C. Feldman, Jr.
Medford
  Voss Valley Buffalo
  George and Dani Voss
Milwaukee
  Lazy K Farm
  Richard E. Kuhn
Mishicot
  Donald M. Thompson
Oconomowoc
  C-N Ranch
  John Connell
Onalaska
  San Lake Bison
Park Falls
  Howard Vander Veen
Phelps
  Smoky Lake Reserve
  P. C. Christiansen
Poynette
  Wisconsin Conserv. Dept.
  Poynette Game Farm

Racine
  Zoological Park
Randolph
  M. J. and Allen Cornford
Seymour
  Wisconsin Bison Co.
Sheboygan Falls
  Myrtle's Buffalo Ranch
  Marvin Feldmann
Tomahawk
  Buffalo Spirit Farms
Waupaca
  Elwyn West
Waupun
  David H. Bronkhorst
Wausau
  O.K. Buffalo
  Dave and Sandra Osterbrink
Withee
  Black River Ranch

Wyoming
Albin
  Tin Cup Ranch
  Gerald Olson
Big Horn
  Bar Eleven Livestock
  Brant and Helen Hilman
  Zane and Elaine Hilman
Casper
  William G. Hurley
  Mary C. Velous
Centennial
  Ron and Dotti Gregory
Cheyenne
  Iron Mountain Bison
  Ron, Dan and Janice Thiel
Gillette
  Little Buffalo Ranch
  Mrs. Toots Marquiss
  Armando Flochinni III
Kinnear
  Flying Goose Ranch
  Robert J. Clark

Lander
  Bob and Bette Thomsen
  David R. Raynolds
Lovell
  Tillett Ranch
  Hip Tillett
Moose
  Grand Teton Nat'l Park
Powell
  Bax X Ranch
  Martin and Hilda Kimmet
Riverton
  Wooden Nickle Ranch
Yellowstone Nat'l Park

CANADA

Alberta
Banff Nat'l Park
Bezanson
  Ross Adams
Calgary
  Calgary Stampede
Cardston
  Gordon Sherman
Cherhill
  A. McClunie
Clyde
  Nilsson Livestock Ltd.
  Bill Nilsson
Elk Island Nat'l Park
Elnora
  Canadian Commercial Buffalo
  Clay Curry
Gleichen
  Blackfoot Indian Band
High Level
  Wood Buffalo Ranch
  Carl Harris
Jasper Nat'l Park
Peace River
  Daco Buffalo & Cattle Ranch
  George and Gloria Underwood

Red Deer
  H. Kenneth Boake
Sundre
  Albert Walters
Taber
  Prairie Buffalo
  Len W. Ross
Vimy
  Bob Maas
Waterton Lakes Nat'l Park
Winterburn
  Nilsson Livestock Ltd.
  Bill and Ruby Nilsson
Wood Buffalo Nat'l Park

British Columbia
Christian Lake
  Peter Barg
East Pine
  M&R Farms
  Gary D. McCullough
Fort St. John
  Fern C. Mertens
West Vancouver
  Tara Investments Ltd.

Manitoba
Hamiota
  Bison Bison Meats
  Alan Wright
Oak Lake
  Jack L. Winters
Riding Mountain Nat'l Park
Teylon
  Dick and Irene Fish

Northwest Territories
Wood Buffalo Nat'l Park

Ontario
Aurora
  Ballymore Farm
  Harold & Yvonne Hadley

Blenheim
  Buffalo Head Ranch
  Romoe Restorick
Dryden
  Northern Buffalo Ranches
Earlton
  Bisons DuNord, Inc.
Frankville
  Kenneth R. Sheehan
Hanover
  B. K. Wiggins
Ingersoll
  Harvey Lawler
Kirkland Lake
  Howard Lovell
Moonbeam
  Randi August
Oro Station
  LBK Buffalo Ranch
  Bert Schumacher
Peterborough
  W. W. Belch
Sturgeon Falls
  R. G. Vaillancourt
Summerstown
  Archie J. MacDonald
Timmins
  Kidd Creek Mines Ltd.
  P. G. Coupland

Quebec
Sherbrooke
  Gervaism Bisson

Saskatchewan
Beaubier
  Prairie View Bison Hybrids
  Morris Johnson
Chagoness
  Barrier River Bison
  Richard Bintner
D'Arcy
  Badlake Animal Farm
  Richard Smith

Eston
  S. E. Holmes, DVM.
Prince Albert Nat'l Park
Sheho
  Almer & Eleanor Medvid

Swift Current
  Sawki Ranch
  Dennis Bradley
Wilcox
  Buffalo Wallow Ranch
  Roy McGann

Notes

Introduction

1 Dan D. Casement, *Random Recollections,* 102.
2 J. Frank Dobie, *Life and Literature of the Southwest,* 157-58.

Chapter I   The First Buffalo

In addition to the sources listed below, several authorities on prehistoric buffalo have been consulted; they include: Walter W. Dalquest, Midwestern University, Wichita Falls, Texas; Claude W. Hibbard, University of Michigan, Ann Arbor; David Webb, University of Florida, Gainesville; William A. Fuller, University of Alberta, Edmonton; and N. S. Novakowski, Canadian Wildlife Service, Ottawa.

1 Handel T. Martin, "On the Occurrence of *Bison latifrons* in Comanche County, Kansas," The University of Kansas *Science Bulletin,* XVII, 7 (September 1927), 397-400.
2 Letter to author from W. A. Fuller, July 11, 1968.
3 Thomas Townsend, *The History of the Conquest of Mexico by the Spaniards,* 326-27.
4 W. W. Davis, *The Spanish Conquest of New Mexico,* 67.
5 Ibid., 206-07.
6 Rev. Samuel Purchas, *Hakluytus Posthumus, or Purchas his Pilgrimes: Contayning a History of the World in Sea Voyages and Lande Travells by Englishmen and Others,* IV, 1765.
7 Ibid., V, 347.
8 George Bird Grinnell, "The Bison," in *Musk-Ox, Bison, Sheep and Goat* by Casper Whitney, George Bird Grinnell, and Owen Wister, 119-20.
9 Louis Hennepin, *A New Discovery of a Vast Country in America,* edited by Reuben G. Thwaites, I, 146.
10 William Byrd, *The Westover Manuscripts,* I, 172, II, 24-25, as cited in William T. Hornaday, *The Extermination of the American Bison,* 376.
11 T. Salmon, *The Present State of Virginia,* 14, as cited in J. A. Allen, *The American Bisons, Living and Extinct,* 483.
12 Hornaday, *Extermination of the Bison,* 378-79.
13 Reuben G. Thwaites, *Daniel Boone,* 90-91.
14 Allen, *American Bisons,* 506-07.
15 Henry W. Shoemaker, *A Pennsylvania Bison Hunt,* 30-35.
16 Ibid., 40.

Chapter II   Millions on the Prairies and Plains

1 [Nicholas Biddle (ed.),] *The Lewis and Clark Expedition,* I, 66.
2 Ibid., III, 789.
3 Elliott Coues (ed.), *The Expeditions of Zebulon Montgomery Pike,* II, 438.
4 Ibid., 548.
5 Thomas J. Farnham, *Travels in the Great Western Prairies . . .,* Vol. XXVIII of *Early Western Travels 1748-1846,* edited by Reuben G. Thwaites, 96.
6 George Catlin, *Letters and Notes on the Manners, Customs, and Condition of the North American Indians,* II, 13. As powerful and influential as were Catlin's written renderings, his paintings and drawings of the West and the Indians were even more so. We do well to recall Bernard DeVoto's conclusions about Catlin:

> In general he can hardly be praised too much. He gave the Plains Indian to the American eye for the first time.
>
> Moreover, he gave that eye much else for the first time. . . . He drew animals better than he customarily drew human figures and he drew buffalo best of all, as well as anyone was to draw them for another generation. He painted buffalo in a great variety of circumstances. Wounded, at bay, charging, galloping, dying on the prairie or in the snow, they established a durable convention. . . . These are the first paintings of the buffalo herds which travelers and explorers had described so vividly, and, as I have said, the individual Catlin buffalo becomes the buffalo of American iconography for a generation. . . .
>
> In 1832 . . . the American people were preparing to possess their West. Their imagination had acted on the printed word to create a multiform and fantastic West: somehow images must be formed for their avid eyes. On the tablet of their mind must be drawn figures that would truly represent the reality. Catlin had no predecessors of any influence or effect at all. He began the long job and to this day something of the images he made remains; he has never been entirely erased from the West seen by our eyes and by our imagination. [*Across the Wide Missouri,* 396, 395]

7 William D. Street, "The Victory of the Plow," *Kansas Historical Collections,* IX (1905-1906), 42-44.
8 Richard I. Dodge, *The Plains of the Great West,* 120.
9 A. A. Taylor, "The Medicine Lodge Peace Council," *Chronicles of Oklahoma,* II, 2 (June 1924), 101.
10 Robert M. Wright, "Personal Reminiscences of Frontier Life in Southwest Kansas," *Kansas Historical Collections,* VII (1901-1902), 50, 78.
11 L. C. Fouquet, "Buffalo Days," *Kansas Historical Collections,* XVI (1923-1924), 341-42.
12 Henry Inman, *Buffalo Jones' Forty Years of Adventure,* 225.
13 Ernest Thompson Seton, *Life-Histories of Northern Animals,* I, 292.

Chapter III   The Wild Plains Buffalo

1 John J. Audubon and Rev. John Bachman, *The Quadrupeds of North America,* II, 46.
2 Letter to author from E. Franklin Phillips, March 3, 1968.
3 Remsen [Henry Remsen Tilton], "After the Nez Perces," *Forest and Stream,* IX, 21 (December 27, 1877), 403-04, as cited in Mark H. Brown, *The Flight of the Nez Perce,* 400.
4 George F. Ruxton, *Adventures in Mexico and the Rocky Mountains,* 286.
5 Unpublished manuscript by the late Earl Drummond, Cache, Oklahoma. The manuscript was provided to the author by Earl Drummond's son, Edwin, a second generation buffalo man at the Wichita Mountains Wildlife Refuge, Cache, Oklahoma.
6 *McPherson Daily Republican,* June 3, 1932. Original source cited by newspaper is an old atlas written by H. B. Kelly; I am unable to locate the atlas.
7 *Kansas City Star,* May 28, 1911.
8 J. Evetts Haley, *Charles Goodnight,* 442.
9 Henry Inman, *Buffalo Jones' Forty Years of Adventure,* 50.
10 Cora Dean, "Early Fur Trading in the Red River Valley. From the Journals of Alexander Henry, Jr.," *Collections,* State Historical Society of North Dakota, III (1910) 364.
11 Joel Palmer, *Journal of Travels over the Rocky Mountains . . . 1845 and 1846,* Vol. XXX of *Early Western Travels 1748-1846,* edited by Reuben G. Thwaites, 53.
12 Edwin James, *Account of an Expedition from Pittsburgh to the Rocky Mountains . . .,* Vol. XVI of *Early Western Travels,* edited by Thwaites, 228; see also Vol. XV, 255-56.
13 George Catlin, *Letters and Notes on the Manners, Customs, and Condition of the North American Indians,* II, 57.
14 Inman, *Buffalo Jones,* 260-62.
15 Elliott Coues (ed.), *The Henry-Thompson Journals,* II, 618.
16 George Bird Grinnell, *The Fighting Cheyennes,* 209.
17 Robert M. Wright, "Personal Reminiscences of Frontier Life in Southwest Kansas," *Kansas Historical Collections,* VII (1901-1902), 79-80.
18 Tom McHugh, *The Time of the Buffalo,* 243.
19 *Beatrice Express,* January 20, 1872.
20 John Palliser, *Journals, Detailed Reports, and Observations Relative to Palliser's Exploration of British North America,* 122.
21 Richard I. Dodge, *The Plains of the Great West,* 129-30.
22 L. R. Masson, *Les Bourgeois de la Compagnie du Nord-Quest,* I, 294, as cited in Frank Gilbert Roe, *The North American Buffalo,* 168.
23 Coues (ed.), *Henry-Thompson Journals,* I, 174-77.
24 Henry Y. Hind, *Report on the Assiniboine and Saskatchewan Exploring Expedition of 1858,* 107.

25 D. S. Rees, "An Indian Fight on the Solomon," *Kansas Historical Collections,* VII (1901-1902), 471-72.
26 Coues (ed.), *Henry-Thompson Journals,* I, 253-54.
27 Catlin, *Letters and Notes,* II, 249-250.
28 *Leavenworth Daily Commercial,* March 19, 1869.
29 Wright, "Personal Reminiscences," op. cit., 79.
30 Inman, *Buffalo Jones,* 235-36.
31 Haley, *Charles Goodnight,* 443.

Chapter IV   Those of the Mountains and Woods

1 Richard I. Dodge, *The Plains of the Great West,* 146-47.
2 Washington Irving, *The Adventures of Captain Bonneville,* 343.
3 John C. Frémont, *Report of the Exploring Expedition to the Rocky Mountains, Oregon and California,* 414, 418.
4 George F. Ruxton, *Adventures in Mexico and the Rocky Mountains,* 285.
5 W. H. Brewer, *Animal Life in the Rocky Mountains of Colorado,* 58.
6 Dodge, *Plains of the Great West,* 144.
7 Ibid., 145.
8 Dr. R. W. Shufeldt, "The American Buffalo," *Forest and Stream,* June 14, 1888, as cited in William T. Hornaday, *The Extermination of the American Bison,* 411-12.
9 G. O. Shields, *Hunting in the Great West,* 88-90.
10 Hamlin Russell, "The Story of the Buffalo," *Harper's New Monthly Magazine,* LXXXVI (April 1893), 798.
11 Glen F. Cole, supervisory research biologist, Yellowstone National Park, in interview with author at Park Headquarters, Mammoth, June 1969.
12 Ibid.
13 Samuel Hearne, *A Journey from Prince of Wale's Fort in 1770, 1771, and 1772,* 255-57.
14 William Albert Fuller, *The Biology and Management of The Bison of Wood Buffalo National Park,* 5.
15 N. S. Novakowski and W. E. Stevens, "Survival of the Wood Bison *Bison bison athabascae Rhoads* in Canada," a paper (mimeo) presented June 20, 1965, before the Forty-Fifth Annual Meeting of the American Society of Mammalogists, Winnipeg, Manitoba.

Chapter V   For the Indian

1 Joe Ben Wheat, "The Olsen-Chubbuck Site: A Paleo-Indian Bison Kill," *American Antiquity,* XXXVII, 1, Pt. 2 (January 1972); for briefer version see Joe Ben Wheat, "A Paleo-Indian Bison Kill," *Scientific American,* 216 (January 1967), 44-52; for journalistic account see *Kansas City Star,* November 8, 1967.

2 Richard G. Forbis, "Early Man and Fossil Bison," *Science,* 123, 3191 (February 24, 1956), 327-28.

3 Until the mid-20th century, American Indian population data set forth by James Mooney and A. L. Kroeber were generally accepted (for bibliographical information and detailed discussion of Mooney's and Kroeber's figures, see Frank Gilbert Roe, *The North American Buffalo,* Appendix G, 742-803). In the 1960s and 1970s, the whole concept of Indian population and methodology has been altered by such scholars as Woodrow Borah, Shelburne Cook, Henry Dobyns, Harold Driver, and others. See, for example, Dobyns, "Estimating Aboriginal American Population, An Appraisal of Techniques with a New Hemispheric Estimate," *Current Anthropology: A World Journal of Sciences and Man,* 7 (October 1966), 395–449; and Wilbur R. Jacobs, "The Indian and the Frontier in American History—A Need for Revision," *The Western Historical Quarterly,* IV, 1 (January 1973), 43–56.

4 As cited in Bethel Coopwood, "Route of Cabeza de Vaca," *Quarterly of the Texas State Historical Association,* III, 4 (April 1900), 231; much on early Texas buffalo can be found in this article.

5 F. W. Hodge (ed.), *Handbook of American Indians North of Mexico,* II, 283.

6 William E. Connelley, *A Standard History of Kansas and Kansans,* I, 286. Those tribes whose cultures related most integrally with the buffalo were the Apache, Arapaho, Arikara, Assiniboin, Bannock, Blackfeet, Blood, Cheyenne, Comanche, Crow, Gros Ventre, Kansa, Kiowa, Mandan, Nez Perce, Omaha, Osage, Pawnee, Piegan, Shoshone, Sioux, Ute, and Winnebago.

7 Richard I. Dodge, *Our Wild Indians,* 286.

8 W. F. Butler, *Wild North Land,* 62-63.

9 *Salt Lake Tribune,* March 18, 1923.

10 John Ewers, *The Horse in Blackfoot Culture,* 150-51; see also Thomas Mails, *The Mystic Warriors of the Plains,* 190-91.

11 Warren K. Moorehead, "The Passing of Red Cloud," *Kansas Historical Collections,* X (1907-1908), 298.

12 Francis Parkman, *The Oregon Trail,* 81, 100.

13 William T. Hornaday, *The Extermination of the American Bison,* 451.

14 Sir Richard F. Burton, *The City of the Saints,* 48.

15 John B. Dunbar, "The White Man's Foot in Kansas," *Kansas Historical Collections,* X (1907-1908), 71.

16 Clark Wissler, *Indians of the United States,* 14.

17 George Catlin, *Letters and Notes on the Manners, Customs, and Condition of the North American Indians,* I, 124-25.

18 Frank Gilbert Roe, *The North American Buffalo,* 631-36.

19 John B. Dunbar, "Letters Concerning The Presbyterian Mission In The Pawnee Country, Near Bellvue, Neb., 1831-1849," *Kansas Historical Collections,* XIV (1915-1918), 603-04.

20 Richard I. Dodge, *The Plains of the Great West,* 355.
21 Henry Inman, *Buffalo Jones' Forty Years of Adventure,* 97-102.
22 Frank Gilbert Roe, *The North American Buffalo,* 636-42.
23 Frank Gilbert Roe, *The Indian and the Horse,* 347.
24 H. M. Chittenden, *The History of the American Fur Trade of the Far West,* II, 857.
25 Joseph Medicine Crow, "The Crow Indian Buffalo Jump Legends," Symposium on Buffalo Jumps, Montana Archaeological Society *Memoir,* No. 1 (May 1962), 35.
26 Richard G. Forbis, "A Stratified Buffalo Kill in Alberta," ibid., 3.
27 Crow, "The Crow Indian Buffalo Jump Legends," ibid., 35-39. This legend was told to Joseph Medicine Crow by his grandmother when she was ninety-seven years old. Just after 1900, according to Crow, his grandmother and her sister helped to dig up the bones and to gather up the buffalo heads at this particular buffalo jump. Crow said that was probably the last time the heads were ritualistically reassembled. As late as 1962 there were still some skulls left at the jump.
28 Correspondence between the author and Waldo Wedel, September 1971; see also *Kansas City Star,* July 1, 1971.
29 Catlin, *Letters and Notes* I, 127-28.
30 Claude E. Schaeffer, "The Bison Drive of the Blackfeet Indians," Symposium on Buffalo Jumps, op. cit., 30-31.
31 Buffalo Child Long Lance, *Chief Buffalo Child Long Lance,* 62-79.
32 George Bird Grinnell, *The Fighting Cheyennes,* 67-69; see also Peter J. Powell, *Sweet Medicine: The Continuing Role of the Sacred Arrows, the Sun Dance, and the Sacred Buffalo Hat in Northern Cheyenne History.*
33 James H. Birch, "The Battle of Coon Creek," *Kansas Historical Collections,* X (1907-1908), 409-13.
34 A. McG. Beede, "In the Days of the Buffalo," *Forest and Stream* (June 1921), 248.
35 Letter to author from Frank Gilbert Roe, July 15, 1968.
36 John D. Hunter, *Manners and Customs of Indian Tribes,* 287.
37 Letter to author from Frank Gilbert Roe, July 15, 1968.
38 Moorehead, "The Passing of Red Cloud," op. cit., 298.

Chapter VI   When the Killing Began

1 Levette Jay Davidson, "Colorado Folklore," *Colorado Magazine,* XVIII (January 1941), 7.
2 Alexander Ross, *The Red River Settlement,* 69-72; see also Gordon Charles Davidson, *The North West Company,* 119, 143.
3 Katharine Coman, *Economic Beginnings of the Far West,* II, 34.
4 L. R. Masson, *Les Bourgeois de la Compagnie du Nord-Quest,* I, 331, as cited in Frank Gilbert Roe, *The North American Buffalo,* 886.

5 George Catlin, *Letters and Notes on the Manners, Customs, and Condition of the North American Indians,* I, 256.

6 H. M. Chittenden, *The History of the American Fur Trade of the Far West,* I, 338.

7 John C. Frémont, *Expedition to the Rocky Mountains,* 145.

8 Rev. Joab Spencer, "The Kaw or Kansas Indians: Their Customs, Manners, and Folk-Lore," *Kansas Historical Collections,* X (1907-1908), 379-80.

9 Ross, *Red River Settlement,* 241-71.

10 Alice Ford (ed.), *Audubon's Animals: The Quadrupeds of North America,* 47.

11 George F. Ruxton, *Adventures in Mexico and the Rocky Mountains,* 255. Rufus B. Sage, another western traveler in the 1840s, described his introduction to buffalo meat:

> Towards night, several buffalo bulls having made their appearance, our hunter, mounting a horse, started for the chase, and in a brief interval, returned laden with a supply of meat. Camp had already been struck, and preparations for the new item of fare were under speedy headway.
>
> The beef proved miserably poor; but when cooked, indifferent as it was, I imagined it the best I had ever tasted. So keen was my relish, it seemed impossible to get enough. Each of us devoured an enormous quantity for supper, —and not content with that, several forsook their beds during the night to renew the feast, —as though they had been actually starving for a month.
>
> The greediness of the "greenhorns," was the prolific source of amusement to our *voyageurs,* who made the night air resound with laughter at the avidity with which the unsophisticated ones "walked into the affections of the old bull," as they expressed it. "Keep on your belts till we get among cows," said they, "then let out a notch or two, and take a full meal."
>
> It was equally amusing to me, and rather disgusting withal, to see the "old birds," as they called themselves, dispose of the only liver brought in camp. Instead of boiling, frying, or roasting it, they laid hold of it *raw,* and, sopping it mouthful by mouthful in *gall,* swallowed it with surprising *gusto.*
>
> This strange proceeding was at first altogether incomprehensible, but, ere the reader shall have followed me through all my adventures in the wilds of the great West, he will find me to have obtained a full knowledge of its several merits.
>
> The beef of the male buffalo at this season of the year, is poorer than at any other. From April till the first of June, it attains its prime, in point of excellence. In July and August, these animals prosecute their knight-errantic campaign, and, between running, fighting and gallantry, find little time to graze, finally emerging from the contested field, with

hides well gored, and scarcely flesh enough upon their bones to make a decent shadow.

It is nowise marvellous, then, that our lavish appropriation of bull-meat at this time, when it is unprecedentedly tough, strong-tasted, and poor, should excite the mirth of our better-informed beholders. [*Rocky Mountain Life,* 64]

12 David Lavender, *Bent's Fort,* 207.
13 P. St. George Cooke, *Scenes and Adventures in the Army,* 336.
14 J. A. Allen, *The American Bisons, Living and Extinct,* 563.
15 J. A. Allen, "History of the American Bison," *Ninth Annual Report of the United States Geological and Geographical Survey for 1875,* 563.
16 Letters of Peter Bryant, Jackson County, Kansas pioneer, in the manuscript section, Kansas State Historical Society, Topeka.
17 F. Geo. Heldt, "Sir George Gore's Expedition, 1854-1856," *Contributions,* Historical Society of Montana, I (1876), 128-31; see also William Clark Kennerly, *Persimmon Hill: A Narrative of the Far West,* 147-48.
18 *The Commercial Gazette,* December 1, 1860.
19 *Alma Signal,* August 27, 1892.
20 Edward Secrest, "An Invasion of Buffaloes," in *Log Cabin Days,* 55-56.
21 "Statement of Theodore Weichselbaum, of Ogden, Riley County, July 17, 1908," *Kansas Historical Collections,* XI (1909-1910), 569-70.
22 Musetta Gilman, "Pump on the Prairie," unpublished manuscript on the history of the Gilman road ranch (c. ten miles east of Ft. McPherson) in Nebraska, 110.
23 George Bird Grinnell, *The Fighting Cheyennes,* 124.
24 William T. Hornaday, *The Extermination of the American Bison,* 478.
25 E. N. Andrews, "A Buffalo Hunt By Rail," *Kansas Magazine* (May 1873), 453, 454.
26 J. Marvin Hunter (ed.), *The Trail Drivers of Texas,* I, 102-03.
27 Ibid., 391.
28 Theodore R. Davis, "The Buffalo Range," *Harper's New Monthly Magazine,* XXXVIII (January 1869), 163.

Chapter VII   The Slaughter

1 *Ellis County News,* November 5, 1925.
2 J. Wright Mooar interview with Earl Vandale, J. Evetts Haley, and Harvey Chesley, Snyder, Texas, March 2, 3, 4, 1939. Transcript in Vandale Collection, University of Texas Library, Austin; microfilm copy at Kansas State Historical Society, Topeka.
3 Ibid. See also W. C. Holden, "The Buffalo of the Plains Area," *West Texas Historical Association Year Book,* II (June 1926), 12; Ida Ellen Rath, *The Rath Trail,* 99.
4 *Ellis County News,* November 5, 1925.
5 Ibid. See also John R. Cook, *The Border and the Buffalo,* 134-35.

6 *Ellis County News,* November 5, 1925.

7 *Wichita Eagle,* November 7, 1872.

8 *Forest and Stream,* February 1873.

9 Frank H. Mayer and Charles B. Roth, *The Buffalo Harvest,* 87.

10 Richard I. Dodge, *The Plains of the Great West,* 140.

11 *Kansas Daily Commonwealth,* November 26, 1872.

12 Joseph W. Snell (ed.), "Diary of Henry Raymond," *Kansas Historical Quarterly,* XXXI, 4 (Winter 1965), 350-51.

13 Rex W. Strickland (ed.), "The Recollections of W. S. Glenn, Buffalo Hunter," *Panhandle-Plains Historical Review,* XXII (1949), 23.

14 William E. Connelley, "Life and Adventures of George W. Brown," *Kansas Historical Collections,* XVII (1926-1928), 120-21.

15 Mayer and Roth, *Buffalo Harvest,* 51, 51-52, 49, 49-50, 61-62, 35.

16 W. F. Thompson, "Peter Robidoux: A Real Kansas Pioneer," *Kansas Historical Collections,* XVII (1926-1928), 289.

17 Connelley, "Life and Adventures of George W. Brown," op. cit., 117.

18 *Ellis County News,* November 19, 1925.

19 George Douglas Brewerton, *Overland With Kit Carson,* 268.

20 *Hutchinson Herald,* February 26, 1928.

21 Don H. Biggers, "Buffalo Butchery in Texas Was National Calamity," *Texas Farm and Ranch,* November 14, 1925.

22 J. Wright Mooar, "The First Buffalo Hunting in the Panhandle," *West Texas Historical Association Year Book,* VI (1930), 109-10.

23 George Bird Grinnell, "Bent's Old Fort and Its Builders," *Kansas Historical Collections,* XV (1919-1922), 42-44.

24 Robert M. Wright, "Personal Reminiscences of Frontier Life in Southwest Kansas," *Kansas Historical Collections,* VII (1901-1902), 82.

25 In addition to Wright's reminiscences (ibid.), see the following on the Adobe Walls incident: Robert W. Wright, *Dodge City, the Cowboy Capital; Leavenworth Times,* November 17, 1877 (reprinted in *Dodge City Times,* November 24, 1877); Billy Dixon, *Life and Adventures of Billy Dixon of Adobe Walls* (one of the best accounts of the battle); Bill Knox, "The Miracle of the Ridgepole," *Southwest Heritage,* I, 2 (Spring 1967), 38-50; narrative manuscript by J. W. McKinley in the Panhandle-Plains Historical Museum Library, Canyon, Texas.

26 James M. Day, "A Preliminary Guide to the Study of Buffalo Trails in Texas," *West Texas Historical Association Year Book* (1960), 148-49.

27 Ernest Lee (ed.), "A Woman on the Buffalo Range: The Journal of Ella Dumont," *West Texas Historical Association Year Book* (October 1964), 146-47; spelling and dots in original. (After the death of Tom Bird in 1888, Ella Bird married A. Dumont.)

28 Strickland, "Recollections of W. S. Glenn," op. cit., 25-31. An interesting description of butchering comes from Rufus B. Sage in the 1840s when concern was less with the hides and more for the meat:

The carcase was first turned upon the belly, and braced to a position by its distended legs. The operator then commenced his labors by gathering the long hair of the *"boss,"* and severing a piece obliquely at the junction of the neck and shoulders, —then parting the hide from neck to rump, a few passes of his ready knife laid bare the sides, —next paring away the loose skin and preparing a hold, with one hand he pulled the shoulder towards him and with the other severed it from the body; —cutting aslant the uprights of the *spina dorsi* and "hump ribs," along the lateral to the curve, and parting the "fleece" from the tough flesh at that point he deposited it upon a clean grass-spot.

The same process being described upon the opposite side, the carcase was then slightly inclined, and, by aid of the leg-bone bisected at the knee-joint, the "hump-ribs" were parted from the vertebrae; after which, passing his knife aside the ninth rib and around the ends at the midriff, he laid hold of the dissevered side, and, with two or three well directed jerks, removed it to be laid upon his choicely assorted pile; a few other brief minutiae then completed the task.

Meanwhile, divers of the company had joined the butcher, and while some were greedily feeding upon liver and gall, others helped themselves to marrow-bones, *"boudins,"* and *intestinum medulae,* (choice selections with mountaineers,) and others, laden with rich spoils, hastened their return to commence the more agreeable task of cooking and eating.

The remaining animal was butchered in a trice, and select portions of each were then placed upon a pack-horse and conveyed to the waggons.

The assortment was, indeed, a splendid one. The "depouille" (fleece-fat) was full two inches thick upon the animal's back, and the other dainties were enough to charm the eyes and excite the voracity of an epicure. [*Rocky Mountain Life,* 66]

By the 1880s skinning buffalo had become bureaucratized and William T. Hornaday was writing a government manual for field workers supplying the Smithsonian with specimens, "How to Collect Mammal Skins for Purposes of Study and for Mounting":

It is a simple matter to prepare the skin of an ordinary quadruped, provided the operator is not afraid of getting a little blood on his hands, and is not naturally indisposed to physical exertion. A few minutes' work suffices for the skin of a small mammal, and a few hours for a large one, up to the size of a buffalo. With a sharp knife, detailed instructions, some cheap preservatives, and a little patient labor, the thing is done. . . .

The principal difference between the manner of skinning a small terrestrial quadruped and a large one, like a bear, deer, or buffalo, is that the skin of each leg is slit open from the bottom of the foot up the back of the leg nearly the first joint and from thence up the inside of the leg . . . until it meets the opening cut which has been made along the center of the body. In preparing a skin as large as that of a buffalo ... it is best to cut off the leg bones at the first joint above the foot, tie them up in a bundle with the skull, and *forward them with the skin,* properly labeled. [Smithsonian *Annual Report* for 1886, II, 659, 662-63]

29 William T. Hornaday, *The Extermination of the American Bison,* 502.
30 W. Henry Miller, *Pioneering North Texas,* 157, as cited in Day, "A Preliminary Guide to . . . Buffalo Trails," op. cit., 148-49.
31 J. Wright Mooar interview with J. Evetts Haley, January 4, 1928. Transcript in Vandale Collection, University of Texas Library, Austin; microfilm copy at Kansas State Historical Society, Topeka.
32 James B. Power, "Bits of History Connected with the Early Days of the Northern Pacific Railroad," *Collections,* State Historical Society of North Dakota, III (1910), 337.
33 *Yellowstone Journal,* September 1879.
34 C. Gordon Hewitt, *The Conservation of the Wild Life of Canada,* 120.
35 Frank Gilbert Roe, *The North American Buffalo,* 467.
36 Hornaday, *Extermination of the Bison,* 504.
37 *Sioux City Journal,* May 1881.
38 Hornaday, *Extermination of the Bison,* 510.
39 *Prose and Poetry of the Live Stock Industry,* 30.
40 Hornaday, *Extermination of the Bison,* 513.
41 Usher L. Burdick, *Tales from Buffalo Land: The Story of Fort Buford,* 166.
42 *Prose and Poetry of the Live Stock Industry,* 308.
43 Hornaday, *Extermination of the Bison,* 513.

Chapter VIII   Early Attempts to Save the Buffalo

1 *Natural History of Florida,* 174, as cited in J. A. Allen, *The American Bisons, Living and Extinct,* 560.
2 Edwin James, *Account of an Expedition from Pittsburgh to the Rocky Mountains . . .,* Vol. XV of *Early Western Travels 1748-1846,* edited by Reuben G. Thwaites, 256-57.
3 George Catlin, *Letters and Notes on the Manners, Customs, and Condition of the North American Indians,* I, 263.
4 Josiah Gregg, *Commerce of the Prairies,* II, 213.
5 Alice Ford (ed.), *Audubon's Animals: The Quadrupeds of North America,* 48.
6 Allen, *American Bisons,* 560.
7 Ibid., 560-61.
8 T. S. Palmer, "Chronology and Index of the More Important Events in American Game Protection, 1776-1911," U.S. Department of Agriculture *Biological Survey Bulletin,* No. 41, 17.

9 *The Congressional Globe,* 42 Cong., 2 Sess., Pt. 1, 80.

10 *Kansas Daily Commonwealth,* February 2, 1872; Hazen's letter of January 20, 1872 was reprinted in *Harper's Weekly,* February 24, 1872.

11 *Kansas State Record,* June 26, 1872.

12 *Newton Kansan,* November 28, 1872.

13 *The Congressional Globe,* 42 Cong., 2 Sess., Pt. 2, 1006.

14 Ibid., 1063.

15 Ibid., Appendix, 179-80.

16 The bill to protect buffalo was H.R. 921, the bill to tax buffalo hides H.R. 1689, both 43 Cong., 1 Sess.

17 *The Congressional Globe,* 43 Cong., 1 Sess., Pt. 3, 2105.

18 Ibid.

19 Frank H. Mayer and Charles B. Roth, *The Buffalo Harvest,* 29.

20 Field Office File, Letters Received, Central Superintendency of the Office of Indian Affairs, Lawrence, Kansas, now housed at Bureau of Indian Affairs Area Office, Muskogee, Oklahoma. Also, these two letters by Williams were printed by a Wichita newspaper in February 1874 and in July 1875.

21 John R. Cook, *The Border and the Buffalo,* 113.

22 *Helena Weekly Herald,* December 25, 1873, as cited in H. Duane Hampton, *How the U.S. Cavalry Saved Our National Parks,* 214n17.

23 Ibid., 39.

24 Ibid., 40.

25 As cited in Orrin H. Bonney and Lorraine Bonney, *Battle Drums and Geysers,* 127.

26 Hiram M. Chittenden, *The Yellowstone National Park,* 111; see also 107-12.

27 William T. Hornaday, *The Extermination of the American Bison,* 521-22. For a more recent and extensive treatment of the role of the military in the administration of Yellowstone, see Hampton, *How the U.S. Cavalry Saved Our National Parks.*

28 Hamlin Russell, "The Story of the Buffalo," *Harper's New Monthly Magazine* LXXXVI (April 1893), 798.

29 *Report of the Acting Superintendent of the Yellowstone National Park to the Secretary of the Interior, 1894,* 9-10.

30 Emerson Hough, "Forest and Stream's Yellowstone Park Game Exploration," *Forest and Stream,* XLIII (May 5 to August 25, 1894), a series of thirteen articles. See also Freeman Tilden, *Following the Frontier with F. Jay Haynes, Pioneer Photographer of the Old West,* Chapter XVII, "With Emerson Hough on the Poacher's Trail."

Chapter IX   After the Kill

1 Arthur C. Bill, unpublished manuscript, Kansas State Historical Society, Topeka.

2 Frank Root, *The Overland Stage to California,* 36.

3 *Wakeeney Weekly World,* July 5, 1879. The "bone black" or animal charcoal was used as a filtering material in purifying sugar syrup.

4 J. A. Allen, "History of the American Bison," *Ninth Annual Report of the U.S. Geological and Geographical Survey for 1875,* 572.

5 Richard I. Dodge. *The Plains of the Great West,* 140. This is a good place to help correct a typographical error that crept into Ralph K. Andrist's book, *The Long Death: The Last Days of the Plains Indians.* In discussing the bone trade, Andrist noted that "the Santa Fe alone hauled seven million tons of them [bones] in 1874." (182n) *Tons* should read *pounds.* Andrist says that he missed this typo in his reading proofs probably because he was engaged at the time in trying to calculate the total weight of buffalo bones harvested during the entire bone trade period, and those figures were indeed in tons. Incidentally, Andrist's conclusion was that a maximum of three million tons of bones were gathered and sold; another researcher, LeRoy Barnett (infra), has calculated a little over two million tons. Andrist recognizes that any figure is an unprovable estimate, but he has tried to incorporate into his calculations most of the necessary factors and contingencies: average weight of a buffalo skeleton, number of buffalo slaughtered, bones lost to fire and flood and other accidents, etc. (Letter from Ralph K. Andrist, September 19, 1973) On the matter of statistical estimates, a word about Colonel Dodge's figures should be added. They are valuable and likely fairly valid; they have certainly been accepted and used for years. However, the figures themselves force some skepticism upon us. Although Dodge, as he said, obtained the statistics, in part, from the railroads, he obviously juggled them (if even slightly). Note that the figures for the UP, KP, and others are *precisely double* those for the AT&SF; the same exact 2:1 ratio is true for Dodge's hides and meat table I used in Chapter VII. Coincidence? No. Another way to wonder about Dodge's data is to use it in calculating the number of buffalo represented by his 32,380,050 pounds of bones. Since the bones of one adult buffalo weigh from forty-five to fifty-five pounds (depending upon animal's age and size), I use fifty pounds as average; therefore, one ton (2,000 lbs.) of buffalo bones represents forty dead buffalo. Projecting this into Dodge's figures, the bones of just under 70,000 buffalo were transported in 1872 by the railroads he mentioned, slightly less than 165,000 in 1873, and nearly 445,000 in 1874. The total for the three years would, then, be about 680,000 buffalo, but this seems low inasmuch as Dodge reported (see my Chapter VII) that 1,378,359 hides were shipped east by these same railroads during the same three-year period—and those hides did not represent all of the buffalo killed. In other words, Dodge's own figures do not add up properly, and any time lag between slaughter and bone harvest does not go far toward explaining the discrepancy. One additional comment is needed on the statistical data above, specifically the weight of a buffalo's bones. In *The Old Santa Fe Trail,* written late in the

19th century, Henry Inman observed, "It required about one hundred carcasses to make one ton of bones." (203) If Inman's figures are correct, the bones of one buffalo would weigh about twenty pounds. But LeRoy Barnett, who has made a careful study of the 19th century buffalo bone trade, believes it was higher. Barnett cites as evidence an account of a Minot, North Dakota resident, Clement Lounsberry, who wrote in 1890 that the bones of a buffalo, "after bleaching on the prairies, weigh from 45 to 52 pounds." ("The Buffalo Bone Commerce on the Northern Plains," *North Dakota History,* XXXIX, 1 [Winter 1972], 36) Lounsberry had actually weighed bones in a large stack beside the railroad tracks in Minot. His figures are in line with a present-day estimate. Ray Ashton, a naturalist with the Kansas State Museum of Natural History at the University of Kansas, says that the bones of an average-size adult buffalo, one weighing between 1,600 and 1,900 pounds, weigh about fifty pounds. (Conversation with author, September 1973).

6 I. D. Graham, "The Kansas State Board of Agriculture: Some High Lights of History," *Kansas Historical Collections,* XVII (1926-1928), 795.

7 L. C. Fouquet, "Buffalo Days," *Kansas Historical Collections,* XVI (1923-1925), 347.

8 John R. Cook, *The Border and the Buffalo,* 135.

9 Major I. McCreight, *Buffalo Bone Days,* 79-80.

10 E. Douglas Branch, *The Hunting of the Buffalo,* 224.

11 McCreight, *Buffalo Bone Days,* 82-83; see also Barnett, "The Buffalo Bone Commerce," op. cit.

12 Carl P. Gauthier, "The Bone Trail," unpublished manuscript, 41.

13 William E. Connelley, *A Standard History of Kansas and Kansans,* III, 1316.

14 *Forest and Stream,* July 22, 1886.

15 Ibid.

16 Ibid.

Chapter X   Battling Bulls

1 George Catlin, *Letters and Notes on the Manners, Customs, and Condition of the North American Indians,* I, 249.

2 John C. Frémont. *Report of the Exploring Expedition to the Rocky Mountains,* 33.

3 Letter to author from Marvin R. Kaschke, April 7, 1969.

4 Earl Drummond manuscript.

5 *Brown County World,* n.d., 1891, article entitled "A Bull Fight" contained in a scrapbook in the files of the Kansas State Historical Society, Topeka.

6 Conversations at various times during 1968 between the author and R. V. "Tex" Shrewder, plus letter to author from Shrewder, March 3, 1969.

7 W. E. Webb, *Buffalo Land*, 252-53.
8 J. A. Allen, *The American Bisons, Living and Extinct*, 462.
9 Catlin, *Letters and Notes*, I, 257-58.
10 This particular story appeared in *Forest and Stream*, July 25, 1908. Earlier, however, George Bird Grinnell related a similar tale in "The Last of the Buffalo," *Scribner's Magazine*, XII, 3 (September 1892), 277. In Grinnell's version the buffalo bull made certain the bear was dead before leaving with the cow.
11 William A. Allen, *Adventures With Indians and Game or Twenty Years in the Rocky Mountains*, 218-25.
12 John Palliser, *Solitary Rambles and Adventures of a Hunter in the Prairies*, 281-82.
13 Ernest Thompson Seton, *Lives of Game Animals*, III, Pt. II, 688-89.
14 Conversations between the author and Elmer Parker at the Wichita Mountains Wildlife Refuge, Cache, Oklahoma, March 1969.
15 R. G. Carter, "Buffalo vs. Bulldog," *Outing*, XI (October 1887), 91-93.
16 George Philip, "South Dakota Buffaloes Versus Mexican Bulls," *South Dakota Historical Review*, II (January 1937), 51-72. Same article also in *South Dakota Historical Collections*, XX (1940), 409-30.

Chapter XI   How They Ran

1 Will E. Stoke, *Episodes of Early Days in Central and Western Kansas*, I (there was never a Vol. II), 24-28.
2 See J. Frank Dobie, *On The Open Range*, 17-22, for good description of a buffalo stampede in the Texas Panhandle. Still another fine description can be found in J. Evetts Haley, *Charles Goodnight*, 145. Numerous other stories of buffalo stampedes can be found in the following: Robert M. Wright, *Dodge City, the Cowboy Capital*, 78-79; Theodore Roosevelt, *The Wilderness Hunter*, II, 14-17; *Kansas Historical Collections*, VIII (1903-1904), 488-89; George F. Ruxton, *Adventures in Mexico and the Rocky Mountains*, 284-85.
3 Henry Inman, *Buffalo Jones' Forty Years of Adventure*, 232.
4 Ibid., 231-32.
5 The stories mentioned in note #2 above are good examples; see also Richard I. Dodge, *The Plains of the Great West*, 121; Wayne Gard, *The Great Buffalo Hunt*, 11-12; Marie Sandoz, *The Buffalo Hunters*, 231-32; Homer W. Wheeler, *Buffalo Days*, 162-63.
6 Josiah Gregg, *Commerce of the Prairies*, Vol. XX of *Early Western Travels 1748-1846*, edited by Reuben G. Thwaites, 270.
7 Elliott Coues (ed.), *The Henry-Thompson Journals*, II, 618.
8 Gordon A. Badger (ed.), "Recollections of George Andrew Gordon," *Kansas Historical Collections*, XVI (1923-1925), 497-504.

9 S. D. Myres (ed.), *Pioneer Surveyor, Frontier Lawyer: The Personal Narrative of O. W. Williams, 1877-1902*, 56-59.

10 William T. Hornaday, *The Extermination of the American Bison*, 421-22.

11 John R. Cook, *The Border and the Buffalo*, 152-53.

12 [Nicholas Biddle (ed.),] *The Lewis and Clark Expedition*, I, 237.

13 George Catlin, *Letters and Notes on the Manners, Customs, and Condition of the North American Indians*, II, 57.

14 J. A. Allen, *The American Bisons, Living and Extinct*, 472.

15 Richard I. Dodge, *The Plains of the Great West*, 121-22.

16 Ibid.

17 George Bird Grinnell, "The Last of the Buffalo," *Scribner's Magazine*, XII, 3 (September 1892), 280-84.

18 Dodge, *Plains of the Great West*, 126-27.

19 Randolph B. Marcy, *Thirty Years of Army Life on the Border*, 340.

20 This story was told by George Mayes of Oklahoma City. It is cited in "Journal of John Lowery Brown of the Cherokee National enRoute to California in 1850," *Chronicles of Oklahoma*, XII, 2 (June 1934), 202n.

21 Marcy, *Thirty Years of Army Life*, 343.

22 This story is an excerpt from Othniel C. Marsh's uncompleted autobiography, an unpublished manuscript in the Marsh Papers at Yale University Library. It was brought to my attention by James Penick, Jr. of Loyola University, Chicago and is used with his and the library's permission. A longer excerpt, focusing around this same incident, has been published by Prof. Penick as "A Ride for Life in a Buffalo Herd," *American Heritage* (June 1970), 46ff.

23 Henry Inman, *Buffalo Jones' Forty Years of Adventure*, 260.

24 Martin S. Garretson, *The American Bison*, 57.

25 E. N. Andrews, "A Buffalo Hunt By Rail," *The Kansas Magazine*, III (May 1873), 455.

26 Earl Drummond manuscript.

Chapter XII   Trails and Travel

1 [Nicholas Biddle (ed.),] *The Lewis and Clark Expedition*, III, 743.

2 John Bradbury, *Westward with the Astorians under Wilson P. Hunt*, Vol. V of *Early Western Travels 1748-1846*, edited by Reuben G. Thwaites, 99.

3 Charles Scott, "The Old Road and Pike's Pawnee Villages," *Kansas Historical Collections*, XVII (1926-1928), 311-17.

4 Carl Coke Rister and Bryan W. Lovelace (eds.), "A Diary Account of a Creek Boundary Survey—J. R. Smith," *Chronicles of Oklahoma*, XXVII, 3 (1949), 268.

5 Richard I. Dodge, *The Plains of the Great West*, 122.

6 Henry Inman, *Buffalo Jones' Forty Years of Adventure*, 122.

7   J. A. Allen, *The American Bisons, Living and Extinct,* 465.
8   Frank Gilbert Roe, *The North American Buffalo,* 77.
9   *The Journals of Alexander Henry,* for example, provide numerous accounts of buffalo wintering in Montana, the Dakotas, southern Canada, Wyoming, etc., as do many other early narratives from the northern plains.
10 Frank H. Mayer and Charles B. Roth, *The Buffalo Harvest,* 21.
11 Roe, *North American Buffalo,* 7-8.
12 Ernest Thompson Seton, *Lives of Game Animals,* III, Pt. II, 654.
13 G. E. Lemmon, *Developing the West,* Section 11, 19.
14 Roe, *North American Buffalo,* 674.

Chapter XIII   Cows and Calves

1   Ernest Thompson Seton, *Lives of Game Animals,* III, Pt. II, 695.
2   Penelope Jane Marjoribanks Edgerton, "The Cow-Calf Relationship and Rutting Behavior in the American Bison," M.A. thesis, University of Alberta, Edmonton, 1962.
3   Conversations between the author and Elmer Parker at the Wichita Mountains Wildlife Refuge, Cache, Oklahoma, March 1969.
4   Letter to author from L. Roy Houck, April 25, 1973.
5   See Chapter XX and Appendix A for details of author's 1969 North American Buffalo Survey.
6   Howard Louis Conrad, *"Uncle Dick" Wootton,* 85-98.
7   Author's North American Buffalo Survey.
8   Henry Inman, *Buffalo Jones' Forty Years of Adventure,* 234.
9   Conversations between the author and Julian Howard at the Wichita Mountains Wildlife Refuge, Cache, Oklahoma, March 1969.
10 Conversations between the author and C. J. Henry at the National Bison Range, Moiese, Montana, June 1969, plus earlier correspondence.
11 Earl Drummond manuscript.
12 Author's North American Buffalo Survey.
13 Letter to author from Mrs. Toots Marquiss, Little Buffalo Ranch, Gillette, Wyoming, April 8, 1969.
14 *Forest and Stream,* July 25, 1908, as cited in Seton, *Lives of Game Animals,* III, Pt. II, 695-96.
15 Edgerton, "The Cow-Calf Relationship," op cit.
16 George Bird Grinnell, "The Last of the Buffalo," *Scribner's Magazine,* XII, 3 (September 1892), 274.
17 Conversations between the author and Elmer Parker at the Wichita Mountains Wildlife Refuge, Cache, Oklahoma, March 1969. The incident mentioned took place about 1948.
18 Richard I. Dodge, *The Plains of the Great West,* 125.
19 *The Classmate,* December 28, 1940.

20 Seton, *Lives of Game Animals,* III, Pt. II, 685.
21 Conversations between the author and Elmer Parker, op. cit.
22 Conversations between the author and C. J. Henry at the National Bison Range, June 1969.
23 Conversations between the author and Victor "Babe" May at the National Bison Range, June 1969.
24 Conversations between the author and Elmer Parker, op. cit.
25 Earl Drummond manuscript.

Chapter XIV   Heads and Horns

1 Fritz A. Toepperwein (ed.), *Footnotes of the Buckhorn,* 11.
2 Frank A. Root, *The Overland Stage to California,* 37.
3 *Prose and Poetry of the Live Stock Industry,* 334.
4 Letter to author from Joe P. Jonas, Sr., Denver, Colorado, June 19, 1969; letter to author from Henry Zietz, Denver, June 30, 1969.
5 William T. Hornaday, *The Extermination of the American Bison,* 529-48.
6 Letter to author from Ronald H. Pine, Associate Curator, Division of Mammals, Smithsonian Institution, April 28, 1969.
7 Conversations between the author and Boone McClure, February 1969.
8 Earl Drummond manuscript.
9 Letters to author from Russ Greenwood, Sundre, Alberta, during 1968 and 1969.
10 Carroll D. Murphy, "What the Buffalo Offers Us," *Texas Farm and Ranch,* XXIX, 49 (December 3, 1910).
11 Conversations between the author and Edwin Drummond at the Wichita Mountains Wildlife Refuge, Cache, Oklahoma, March 1969.

Chapter XV   White Buffalo

The facts, legends, and tales concerning white buffalo and the Plains Indians in the early part of this chapter were for the most part gathered into a notebook over a period of many years from published journals, early Plains narratives, old newspapers, early government publications; unfortunately, at the time each item was entered, its source was not always noted.
1 *Topeka State Journal,* August 4, 1897.
2 J. A. Allen, *The American Bisons, Living and Extinct,* 447.
3 Ibid.
4 Edwin James, *Account of an Expedition from Pittsburgh to the Rocky Mountains . . . ,* Vol. XV *of Early Western Travels 1748-1846,* edited by Reuben G. Thwaites, 244-45.
5 George Bird Grinnell, "The Bison," in *Musk-Ox, Bison, Sheep and Goat,* by Casper Whitney, George Bird Grinnell, and Owen Wister, 126.

6 James Willard Schultz, *Blackfeet and Buffalo*, 45-46.

7 Lawrence Burpee (ed.), "The Journal of Antony Hendry," *Proccedings and Transactions of the Royal Society of Canada*, 3rd Ser., I, Sec. II (1907), 337, as cited in Frank Gilbert Roe, *The North American Buffalo*, 720.

8 Elliott Coues (ed.), *The Henry-Thompson Journals*, I, 242.

9 *Weekly Missoulian*, May 2, 1873.

10 Theodore R. Davis, "The Buffalo Range," *Harper's New Monthly Magazine*, XXXVIII (January 1869), 158-59.

11 *Topeka Capital-Commonwealth*, December 18, 1888; see also *Topeka State Journal*, November 19, 1903.

12 *Topeka State Journal*, August 4, 1897.

13 Sir Cecil E. Denny, manuscript of reminiscences, 40, Provincial Legislative Library, Edmonton, Alberta, as cited in Frank Gilbert Roe, *The North American Buffalo*, 79.

14 Amy Lathrop, *Tales of Western Kansas*, 34.

15 W. E. Webb, *Buffalo Land*, 193. See note #40, Chapter XVII (infra), for reference to a Hollywood-whitened buffalo.

16 Henry Inman, *Buffalo Jones' Forty Years of Adventure*, 80-81. See Chapter XVIII (infra).

17 Robert M. Wright, *Dodge City, The Cowboy Capital*, 197.

18 *Holland's Magazine*, LII, 5 (May 1933), 11-12, fifth in a series of eight articles entitled "Buffalo Days," reminiscences of J. Wright Mooar, as told to James Winford Hunt. See also J. Wright Mooar interview with Brockman Horne, April 12, 1936. Transcript in Vandale Collection, University of Texas Library, Austin; microfilm copy at Kansas State Historical Society, Topeka.

19 Conversations with Cy Young at the National Bison Range, June 1969. Also, conversations with Victor "Babe" May, National Bison Range, and letter from May to author, December 20, 1966.

20 As far as I know, this white buffalo never had a name. The staff at the National Bison Range said he was not named there, and it is against the policy of the National Zoo to name animals, except in rare instances— mainly for publicity purposes.

21 Letter to author from Victor "Babe" May, National Bison Range, December 20, 1966. Also, conversations with May at the National Bison Range, June 1969.

22 Material in the files of the Montana Historical Society Library, Helena.

23 Correspondence between the author and William H. Griffin, Fairbanks, and Julius L. Reynolds, of the Alaska Department of Fish and Game.

24 *Lawrence Journal World*, October 26, 1973; conversation between the author and Al Johnson, Anchorage, November 1973.

25 Bon V. Davis, "Get off the Earth: The Life of Vern Bookwalter," unpublished manuscript, 152.

Chapter XVI **Those Who Saved the Buffalo**

Unless otherwise noted, the material in this chapter which deals with the American Bison Society is based on the organization's papers, now housed in the Conservation Library Center, Denver Public Library.

1 [Tom Jones,] *The Last of the Buffalo, A History of the Buffalo Herd of the Flathead Reservation, and An Account of the Great Round Up;* this is a souvenir booklet, with many photographs, containing the history of the Allard-Pablo herd and the roundup of the Pablo herd. See also John Kidder, "Montana Miracle: It Saved the Buffalo," *Montana, the Magazine of Western History,* XV, 2 (April 1965), 52-67, a detailed account of the Walking Coyote story and the Allard-Pablo herd; Larry Barsness, "Superbeast and the Supernatural: The Buffalo in American Folklore," *The American West,* IX, 4 (July 1972), 12, a debunking article about buffalo fakelore; Martin Garretson, *The American Bison,* 215-16, or Garretson, *A Short History of the American Bison,* 52, both of which give a brief history of the Pablo-Allard herd, though a few of Garretson's facts seem incorrect. Barsness labels the Walking Coyote story variously as myth, yarn, and folktale, but I am strongly inclined toward believing Kidder's summary (see especially Kidder, 58n11). An interesting addendum to the Pablo roundup story is provided by Lincoln Ellsworth. In October and November of 1911 he went up into the Bitter Root Mountains of Montana for "the last wild buffalo hunt" (as he entitled the booklet he wrote about it). Actually, the "wild buffalo" were, as Ellsworth said, "outlaws of the [Pablo] herd, unconquerable old veterans who live high up in these timbered fastnesses" (15); these were animals that had escaped the roundups since 1906. Appended to Ellsworth's account of his almost unsuccessful hunt is a curious pair of letters deploring Pablo's alleged practice of permitting hunters, for a fee, to hunt his strays: a letter from W. T. Hornaday to Ellsworth, in which Hornaday indicates that he has written the Montana state game warden, Henry Avare, and received a telegram in reply; and a letter from Avare to Hornaday.

2 William T. Hornaday, *The Extermination of the American Bison,* 458-59; see also Garretson, *The American Bison,* 216-17.

3 Letter to Military Secretary, Department of Dakota, from Major John Pitcher, November 28, 1906, Yellowstone Archives, Letters Received, VII, 25-28.

4 Letter to Major John Pitcher from C. J. "Buffalo" Jones, July 5, 1903, Yellowstone Archives, Letters Received.

5 Letter from E. A. Hitchcock, Secretary of Interior, July 6, 1905, ibid., VI, 58-59.

6 J. Frank Dobie, *Guide to Life and Literature of the Southwest,* 160.

7 J. Evetts Haley, *Charles Goodnight,* 453-54.

8 Larry Barsness, "Superbeast and the Supernatural," op. cit., 12-13. Goodnight was, of course, not the first to have visions of turning buffalo wool

to profitable commercial advantage. We recall the Buffalo Wool Company of 1821-22, described above in Chapter VI. Further, well over a century earlier North American explorers had had similar high hopes. For example, in 1699 Pierre Le Moyne d'Iberville made two voyages to the mouth of the Mississippi and established a French colony, and buffalo played a significant role in getting the king's support. As Justin Winsor put it:

> If they [the French] could establish themselves at its entrance, and were able to control its navigation, they could hold the whole valley. Associated with these thoughts were hopes of mines in the distant regions of the upper Mississippi which might contribute to France wealth equal to that which Spain had drawn from Mexico. Visions of pearl-fisheries in the Gulf, and wild notions as to the value of buffalo-wool, aided Iberville in his task of convincing the Court of the advantages to be derived from his proposed voyage. [*Narrative and Critical History of America*, V, 15-16]

Iberville carried royal instructions to bring back samples of buffalo wool and to establish a herd of buffalo at the colony. Though the undersupplied and discouraged settlement struggled along for its first few years, "the sages across the water still pressed upon their attention the possibility of developing the trade in buffalo-wool, on which they built their hopes of the future of the colony." (Ibid., 21) Even without the buffalo and its wool, New Orleans and Louisiana eventually grew and prospered.

9 The best source of information on Goodnight is Haley, *Charles Goodnight;* see especially 453-54. See also Hornaday, *Extermination of the Bison,* 460-61; Carroll D. Murphy, "What the Buffalo Offers Us," *Texas Farm and Ranch,* December 3, 1910; Carrie J. Crouch, "Buffalo in the Panhandle," *National Republic,* XX, 4 (August 1932), 14; *Panhandle-Plains Historical Review,* III (May 1932), 7-17; and a pamphlet, *Goodnight's American Buffalo.*

10 A basic source of information on Pete Dupree and Scotty Philip is George Philip's article, "James (Scotty) Philip," *South Dakota Historical Collections,* XX (1940), 357-406. See also *The Country Gentleman,* March 20, 1920; *Ellis County News,* November 5, 1925; Wayne C. Lee, *Scotty Philip: The Man Who Saved the Buffalo.*

11 This account of the McCoy brothers is from "The Return of the Southern Herd," an unpublished manuscript by Harry E. Chrisman, Denver, who found Allen M. McCoy's reminiscences in the museum at Beaver, Oklahoma, and then followed the trail that led to Keokuk, Iowa and Adrian, Michigan.

12 *Boston Evening Transcript,* June 25, July 2, August 6, and August 27, 1904.

13 *Congressional Record,* 59 Cong., 1 Sess., Pt. I, 103; see also *Annual Report of the American Bison Society, 1905-1907,* 2.

14 Ibid., 3.

15 Ibid., 4-5.

16 Ibid., 55-69.

17 Quoted by William T. Hornaday in a letter to R. J. Carrigan, dated December 10, 1908, American Bison Society Papers, Conservation Library Center, Denver Public Library.

Chapter XVII   In Captivity

1 This story of Dick Wootton and his buffalo comes from several sources: Howard Louis Conrad, *"Uncle Dick" Wootton*, 85-98; Glenn D. Bradley, *Winning the Southwest, A Story of Conquest*, 225; LeRoy Hafen, *The Mountain Men and the Fur Trade of the Far West*, III, 397-411; *Denver Republican* August 23, 1893.

2 *Missouri Democrat*, March 16 and May 6, 1844.

3 *Niles' National Register*, LXVI (March 30 and August 10, 1844), 160, 394.

4 Hubert H. Bancroft, *History of Arizona and New Mexico*, XVII, 138.

5 Jean Louis Berlandier, unpublished manuscript written (in Spanish) in 1828, Library of Congress, Washington, D.C.

6 David Lavender, *Bent's Fort*, 399.

7 William T. Hornaday, *The Extermination of the American Bison*, 451.

8 Peter Kalm, *Travels in North America*, I, 162; see also two-page letter from Martin S. Garretson, secretary of the American Bison Society, to W. E. Connelley, secretary of the Kansas State Historical Society, Topeka, dated February 26, 1918, in files of the Kansas State Historical Society, Topeka.

9 John J. Audubon and Rev. John Bachman, *Quadrupeds of North America*, II, 52-54.

10 Bethel Coopwood, "Route of Cabeza de Vaca," *Quarterly of the Texas State Historical Association*, III, 4 (April 1900), 234.

11 George Catlin, *Letters and Notes on the Manners, Customs, and Condition of the North American Indians*, I, 255-56.

12 Frank A. Root, *The Overland Stage to California*, 27.

13 Joseph G. McCoy, *Historic Sketches of the Cattle Trade of the West and Southwest*, 180-81.

14 Louis Charles Laurent, "Reminiscences by the Son of a French Pioneer," *Kansas Historical Collections*, XIII (1913-1914), 369.

15 William Ansel Mitchell, *Linn County, Kansas; A History*, 352.

16 *The Norton Champion*, March 5, 1914.

17 Henry Inman, *Buffalo Jones' Forty Years of Adventure*, 250.

18 Ibid., 253-54.

19 *Prose and Poetry of the Live Stock Industry*, 324.

20 *Kansas City Star*, January 24, 1905.

21 Ibid.

22 *Leslie's Weekly*, March 10, 1910.

23 Numerous letters to author from Russ Greenwood, Sundre, Alberta, Canada, during 1968 and 1969.

24 "Man and Beast with Same Hirsute Motif," *The Plainsman,* VIII (Summer 1962), 100.

25 J. Evetts Haley, *Charles Goodnight,* 441-42.

26 J. A. Allen, *The American Bisons, Living and Extinct,* 471-72.

27 Robert M. Wright, *Dodge City, The Cowboy Capital,* 204-05.

28 Conversations between the author and Victor "Babe" May at the National Bison Range, Moiese, Montana, June 1969, plus earlier and later correspondence.

29 Earl Drummond manuscript.

30 Conversations between the author and Julian Howard at the Wichita Mountains Wildlife Refuge, Cache, Oklahoma, March 1969.

31 Letter to author from Owen McEwen, April 24, 1969.

32 Letter to author from Erwin H. Weder, March 5, 1969.

33 *Amarillo Globe-Times,* October 3, 1966.

34 *Foard County News,* May 11, 1972; *El Paso Times,* May 13, 1972; letter from Sheriff Emmett R. Howard, May 23, 1972.

35 Letter to author from Adrian W. Reynolds, March 15, 1969. In addition to private herd strays, sheer false rumor is another source of the lost herd legends. Even in the late 1880s William T. Hornaday was reporting such hoaxes or rumors and was warning:
    > From this time it is probable that many rumors of the sudden appearance of herds of buffaloes will become current . . . [but] in these days of railroads and numberless hunting parties, there is not the remotest possibility of there being anywhere in the United States a herd of a hundred, or even fifty, buffaloes which has escaped observation. [*Extermination of the Bison,* 525]

36 Letter to author from Jerry Marr, Curator, March 2, 1969.

37 Letter to author from Freddy Rice, April 22, 1969.

38 The early history of the Catalina Island buffalo herd can be found in *The Catalina Islander,* December 24, 1924 and January 21, 1925; additional material furnished in letters to author from Allen "Doug" Propst, superintendent of ranches, Catalina Rock and Ranch Co., March 11 and May 9, 1969.

39 *Center Report, A Center Occasional Paper,* June 1971, 20-21.

40 Ibid. Another Hollywood film of the 1970s employing buffalo did not use the Catalina Island herd. The Cinema Center Films production "A Man Called Horse" used buffalo from the Don Hight herd in Murdo, South Dakota. At one point in the movie, the protagonist, Lord John Morgan (Richard Harris), has a white buffalo vision; the buffalo used in this scene was one of Hight's big brown buffalo bulls, named Ol' Bill, who was turned white for the filming by a special make-up crew. ("The Making of 'A Man Called Horse'," unpublished manuscript by Clyde Dollar, Chapter VI)

41 *New York Times,* January 2, 1972; see also New Mexico State Game Commission press release (mimeo), December 21, 1971, 3.
42 Letter from George Merrill, assistant chief of game management, January 11, 1972.
43 *Arizona Big Game Bulletin,* n.d.
44 Letter to author from Robert A. Jantzen, August 28, 1973.
45 Conversation between the author and Bill Sizer, August 28, 1973.
46 Bill Richards, "Buffalo—Area's Next Meat Source?" *Washington Post,* July 22, 1973. The agility of the buffalo to jump high obstacles was noted earlier in this chapter when we mentioned Charles Goodnight's bull, Old Sikes, disappearing over a six-foot corral fence. A similar feat is attested to by James Hertel. He and Bruce Poel, both of Fremont, Michigan, own five buffalo enclosed in a six-acre area surrounded by a six-foot-high barbed wire fence. Hertel told how he drove out to the pasture one day during the first winter (1971-72) they owned the buffalo and found the bull standing on the *out*side of the fence. Hertel concluded that the animal had *walked* over the fence the day before on a large, crusted snow bank that had by now melted down enough to be of no help to the bull in getting back over the fence. But Hertel was curious; he thought he would doublecheck—besides, he had to get the animal back inside somehow. So, he left some grain inside the fenced area and drove away. That evening when he returned the bull was *in*side. He had to have jumped that fence. "And he must have cleared it with height to spare," Hertel marveled, "because I couldn't find a tuft of hair on any of those barbs." (Conversation with James Hertel, August 30, 1973)

Chapter XVIII   Cattalo

1 Henry Inman, *Buffalo Jones' Forty Years of Adventure,* 243.
2 Ibid., 242.
3 See note #8, Chapter XVII.
4 Albert Gallatin, "A Synopsis of the Indian Tribes of North America," American Antiquarian Society *Transactions and Collections,* II (1836), 139.
5 John J. Audubon and Rev. John Bachman, *The Quadrupeds of North America,* II, 52-54.
6 Inman, *Buffalo Jones,* 244.
7 J. Evetts Haley, *Charles Goodnight,* 343-44.
8 *Goodnight's American Buffalo,* 3. This twenty-page booklet, published by ·H. A. Fleming & Co., Dallas, was part of a publicity campaign to attract stockholders to the Goodnight American Buffalo Ranch Company, for which Fleming & Co. were fiscal agents.
9 Alvin Howard Sanders, "The Taurine World," *National Geographic,* XLVIII (December 1925), 644-45.

10 Letter to author from H. F. Peters, September 10, 1968. See also on Canadian experiments: V. S. Logan and P. E. Sylvestre, *Hybridization of Domestic Beef Cattle and Buffalo;* H. F. Peters and S. B. Slen, *Brahman-British Beef Cattle Crosses in Canada;* H. F. Peters, *Experimental Hybridization of Domestic Cattle and American Bison;* H. F. Peters, *Experience With Yak Crosses in Canada.*

11 Conversations between the author and D. C. "Bud" Basolo, August 1973; see also "Have a Slice of Roast Beefalo," *Time,* CII, 2 (July 9, 1973), 55.

Chapter XIX   The American Buffalo as a Symbol

1 Information provided author by the public affairs office, U.S. Treasury Department, Washington, D.C.

2 Letter to author from Frederic G. Renner, Washington, D.C., August 9, 1971. For detailed list of Russell's buffalo art see Karl Yost and Frederic G. Renner, *A Bibliography of the Published Works of Charles M. Russell.*

3 Letter to author from Don Russell, August 17, 1971. See also *New York Posse Brand Book,* XVIII, 1 (1971); *Buckskin Bulletin,* V, 2 and 3 (Winter and Spring 1971), publication of Westerners International.

4 "Secretary Hickel Orders Return to Buffalo Seal," U.S. Department of the Interior news release, October 12, 1969; see also *The American Buffalo,* Department of Interior Conservation Note 12, p. 8.

5 Black Diamond in the Bronx Park Zoo was the model used by Fraser for the nickel. The National Cowboy Hall of Fame in Oklahoma City has an excellent exhibit detailing the design of the buffalo nickel.

6 Letter to author from Gordon C. Morison, manager, Philatelic Affairs Division, Post Office Department, November 12, 1971.

7 Letter to Mr. Bryant from Charles R. Knight, New York City, November 3, 1939; copies in files of Smithsonian Institution and Bureau of Engraving and Printing, U.S. Treasury Department, Washington, D.C. The Bureau furnished additional details:

A wash drawing by Charles R. Knight was the original subject material used in preparing the vignette for the series 1901 $10 bill. The note was engraved by G. F. C. Smillie and M. W. Baldwin, who were Bureau employees. However, there is no record of who is credited as being the designer of the note [and] nothing can be found concerning the identity of C. R. Knight. [Letter to author from Michael L. Plant, October 20, 1971]

Poor Charles Knight. In 1915 Frank Baker, superintendent of the National Zoo, wrote that Knight had done the $10 bill drawing from a buffalo in his zoo, but a quarter of a century later Knight had to clarify the matter again. In 1915 Baker could easily refer to Knight as "the well-known illustrator" (Smithsonian *Annual Report* for 1914, 447), but in 1971 the Bureau of Engraving and Printing did not even know who Knight was. For the record, Charles Robert Knight (1874-1953) was an American painter, illustrator, sculptor, and author, whose biographical facts can be found in many standard encyclopedias.

8 The question of the artist and the model for the 30-cent buffalo stamp was clarified in the same letter Knight wrote in 1939 about the $10 bill. He said: "Although I have forgotten the Buffalo on the 30¢ stamp, I believe it was adapted from my drawing which I made originally for the so-called Buffalo-bill." (op. cit.) Catherine L. Manning, Smithsonian philatelist, commented:

> With information contained in this [Knight's] letter and the fact that Mr. E. M. Weeks (recently retired from the Bureau) informed me that the original subject material was a sketch made from a Zoo specimen, I am convinced that the group of bison in the New National Museum (National History Building) was not the original subject of the 30-cent bison. Previous to the discovery of the wash drawing by Mr. C. R. Knight, stamp collectors were aware of the fact that the bison on the $10.00 bill was identical with the bison on the stamp, but the artist was unknown. [Quoted in letter to Alvin W. Hall, director, Bureau of Engraving and Printing, from J. E. Graf, associate director, U.S. National Museum, Smithsonian Institution, October 25, 1940]

On the basis of the information from Knight and the Smithsonian, it was agreed that henceforth the Post Office and Bureau of Engraving and Printing records would indicate the source of illustration for the stamp and the treasury note as: "a drawing by Charles R. Knight done at the Zoological Park in Washington from a Zoo specimen." (Letter to Post Office Department from Alvin W. Hall, November 6, 1940)

9 Modeler for the stamp was Robert J. Jones and engravers were Edward P. Archer (vignette) and Robert G. Culin (lettering), all of the Bureau of Engraving and Printing. Information relating to this 6-cent stamp provided in letter to author from Gordon C. Morison, manager, Philatelic Affairs Division, Post Office Department, November 12, 1971.

10 Richard F. Ferguson, "Who's Selling the Brooklyn Bridge Now?" *True West,* January-February 1966, 52.

Chapter XX   The Future of the American Buffalo

1 In 1933, for example, there were only 200 buffalo outside North America. Such figures are the basis for the statement that "over 99 percent" of the North American buffalo are in North America; outside North America all the buffalo are in zoos (there may be one or two privately owned).

2 The first seven listings in this table come from Ernest Thompson Seton's *Lives of Game Animals* (Vol. III, Pt. II, 670-71; a summary can also be found in Seton, "The American Bison or Buffalo," *Scribner's,* XL, 4 [October 1906], 402), and detailed bibliographical information about

them can be found there; see also William T. Hornaday's *The Extermination of the American Bison* for his own 1889 survey (525) and for Christy's 1888 *London Field* survey (523-25). The American Bison Society figures appear in their various annual reports; see also Seton, *Lives* (op. cit.) and Martin S. Garretson, *The American Bison* (212-13).

3 Henry H. Collins, Jr., *1951 Census of Living American Bison;* this survey was taken by mail.

4 Memorandum, dated November 10, 1964, with attached list of buffalo herds in the United States, from E. A. Schilf, chief staff veterinarian, Brucellosis Eradication, Cattle Diseases, U.S. Department of Agriculture, Hyattsville, Maryland, to Animal Disease Eradication Division Veterinarians in Charge. In a letter to the author dated March 14, 1969, E. A. Schilf said that the 1964 buffalo herd list was the last listing published by the U.S.D.A. "The publication of this listing was discontinued because of the probability that it would be inaccurate. It was called to our attention that there were some herds in existence that had not been listed in the '64 publication, and the owners of these herds felt that they were done an injustice. It is extremely difficult to maintain an up-to-date listing of this type, since herds are established and depopulated rather frequently. Since then, we have asked that each of the States maintain a listing of herds within their own State, which is simpler to maintain on a current basis."

5 Letters to author from N. S. Novakowski, Canadian Wildlife Service, Ottawa, May 1, 1968, October 18, 1968, and March 21, 1969.

6 Letter to author from Jack T. Errington, manager, Durham Meat Company, San Jose, California, July 22, 1969; conversation between the author and Armando Flocchini, San Jose, California, August 27, 1973.

7 Ibid.; see also *Wall Street Journal,* December 4, 1968.

8 *Washington Star,* October 17, 1966.

9 *Wall Street Journal,* December 4, 1968.

10 Ibid.

11 Howard Pankratz, "Beef Squeeze Gives Rosy Tint to Buffalo Ranchers' Dreams," *Denver Post,* August 14, 1973.

12 *Denver Post,* August 1973.

13 Letter to author from G. E. Speers, April 30, 1969.

14 By-Laws of the National Buffalo Association.

15 Letter to author from L. Roy Houck, March 31, 1969.

16 Dana Close Jennings, *Where the Buffalo Roam Again,* 43.

17 Ibid.

18 Rufus B. Sage, *Rocky Mountain Life; or, Startling Scenes and Perilous Adventures in the Far West, During an Expedition of Three Years* [in the 1840s], 69.

19 Ibid.

20 *Capper's Weekly,* March 16, 1971.

21 John Palliser, *Solitary Rambles and Adventures of a Hunter in the Prairies,* 113-14.
22 The information concerning brucellosis was compiled from numerous sources including the National Buffalo Association and their newsletter "Buffalo Chips"; Dr. C. B. Favour, M.D., Denver, Colorado, chief, Experimental Epidemiology at the National Jewish Hospital; and many interviews with buffalo men, cattlemen, and veterinarians.

Indicative of increased business interest in buffalo was this page-4 ad in the August 29, 1973 *Wall Street Journal.*

Bibliography

Bibliography is a necessary nuisance and a horrible drudgery that no mere drudge could perform. It takes a sort of inspired idiot to be a good bibliographer and his inspiration is as dangerous a gift as the appetite of the gambler or dipsomaniac—it grows with what it feeds upon and finally possesses its victim like any other invincible vice.

Elliott Coues 1892

BOOKS

Allen, Joel A. *The American Bisons, Living and Extinct.* Cambridge, 1876.

Allen, William A. *Adventures With Indians and Game or Twenty Years in the Rocky Mountains.* Chicago, 1903.

American Bison Society. *Annual Report.* 13 vols. New York, 1907-1931.

Ashe, Thomas. *Travels In America.* London, 1808.

Audubon, John J., and Bachman, Rev. John. *The Quadrupeds of North America.* 2 vols. New York, 1854.

Bancroft, Hubert H. *History of Arizona and New Mexico.* 7 vols. San Francisco, 1889.

Barsness, Larry. *Heads, Hides & Horns. The Complete Buffalo Book.* Fort Worth, 1985.

Baynes, Ernest H. *War Whoop and Tomahawk.* New York, 1929.

Berkeley, Grantley F. *The English Sportsman in the Western Prairies.* London, 1861.

[Biddle, Nicolas (ed.).] *The Lewis and Clark Expedition.* 3-vol. edition of the 2-vol. 1814 publication, *History of the Expedition Under the Command of Captains Lewis and Clark . . . 1804-5-6.* Philadelphia, 1961.

Bliss, Frank E. *The Life of Hon. William F. Cody Known as Buffalo Bill the Famous Hunter, Scout, and Guide.* Hartford, 1879.

Boller, Henry A. *Among the Indians, Eight Years in the Far West: 1858-1866.* Philadelphia, 1868.

Bonney, Orrin H. and Lorraine. *Battle Drums and Geysers: The Life and Journals of Lt. Gustavus Cheyney Doane, Soldier and Explorer of the Yellowstone and Snake River Regions.* Chicago, 1970.

Bradley, Glenn D. *Winning the Southwest, A Story of Conquest.* Chicago, 1912.

Branch, E. Douglas. *The Hunting of the Buffalo.* New York, 1929.

Brewer, W. H. *Animal Life in the Rocky Mountains of Colorado.* Denver, 1871.

Brewerton, George D. *Overland With Kit Carson.* New York, 1930.

Burdick, Usher L. *Tales From Buffalo Land: The Story of Fort Buford.* Cheltenham, Maryland, 1940.

Burton, Sir Richard F. *The City of the Saints.* New York, 1860.

Butler, Sir. W. F. *Wild North Land.* London, 1878.

Cartwright, David W., and Bailey, Mary F. *Natural History of Western Wild Animals and Guide For Hunters, Trappers, and Sportsmen.* Toledo, 1875.

Casement, Dan D. *Random Recollections.* Kansas City, 1955.

Catlin, George. *Letters and Notes on the Manners, Customs, and Condition of the North American Indians,* 2 vols. London, 1842.

Chittenden, Hiram M. *The History of the American Fur Trade of the Far West,* 3 vols. New York, 1902.

————. *The Yellowstone National Park.* Cincinnati, 1904.

Cloudsley-Thompson, J. L. *Animal Behaviour.* New York, 1961.

Cody, William F. *Life and Adventures of Buffalo Bill.* Chicago, 1917.

Coman, Katharine. *Economic Beginnings of the Far West.* 2 vols. New York, 1912.

Connelley, William E. (ed.). *A Standard History of Kansas and Kansans.* 5 vols. New York, 1918.

Conrad, Howard L. *"Uncle Dick" Wootton.* Chicago, 1890.

Cook, James H. *Fifty Years on the Old Frontier As Cowboy, Hunter, Guide, Scout, and Ranchman.* New Haven, 1923.

Cook, John R. *The Border and the Buffalo.* Topeka, 1907.

Cooke, P. St. George. *Scenes and Adventures in the Army.* Philadelphia, 1857.

Coues, Elliott (ed.). *The Expeditions of Zebulon Montgomery Pike.* 3 vols. New York, 1895.

————. *The Henry-Thompson Journals, 1799-1814.* 3 vols. New York, 1897.

Davidson, Charles. *The North West Company.* Berkeley, 1918.

Davies, Henry E. *Ten Days on the Plains.* New York, 1871. This rare work was reprinted by Southern Methodist University Press, Dallas, 1985, edited by Paul Andrew Hutton.

Davis, W. W. *The Spanish Conquest of New Mexico.* New York, 1869.

DeVoto, Bernard. *Across the Wide Missouri.* Boston, 1947.

Dixon, Olive K. *Life and Adventures of Billy Dixon of Adobe Walls.* Guthrie, Oklahoma, 1914.

Dobie, J. Frank. *Life and Literature of the Southwest.* Dallas, 1958.

————. *On the Open Range.* Dallas, 1931.

Dodge, Richard I. *Our Wild Indians.* Hartford, 1882.

————. *The Plains of the Great West.* New York, 1877.

Duncan, Bob. *Buffalo Country.* New York, 1959.

Ellsworth, Lincoln. *The Last Wild Buffalo Hunt.* New York, 1916.

Ford, Alice (ed.). *Audubon's Animals: The Quadrupeds of North America.* New York, 1951.

[Freeman, James W. (ed.)] *Prose and Poetry of the Live Stock Industry.* Kansas City, 1905.

Frémont, John C. *Report of the Exploring Expedition to the Rocky Mountains, Oregon and California.* Buffalo, 1850.

Gard, Wayne. *The Great Buffalo Hunt.* New York, 1959.

Garretson, Martin. *The American Bison.* New York, 1938.

George, Wilma. *Animals and Maps.* Berkeley, 1969.

Gerstaecker, Frederick. *Wild Sports in the Far West.* Philadelphia, 1884.

Greeley, Horace. *An Overland Journey from New York to San Francisco in the Summer of 1859.* New York, 1860.

Grey, Zane. *The Last of the Plainsmen.* New York, 1908.

Grinnell, George Bird. *The Cheyenne Indians.* New Haven, 1923.

———. *The Fighting Cheyennes.* New York, 1915.

———. *When Buffalo Ran.* New Haven, 1920.

Hafen, LeRoy (ed.). *The Mountain Men and the Fur Trade of the Far West.* 8 vols. Glendale, 1965-1970.

Haines, Francis. *The Buffalo.* New York, 1970.

Haley, J. Evetts. *Charles Goodnight.* New York, 1936.

Hampton, H. Duane. *How the U.S. Cavalry Saved Our National Parks.* Bloomington, 1971.

Hart, John A., et al. *Pioneer Days in the Southwest From 1850-1879.* Guthrie, Oklahoma, 1909.

Hartley, Cecil B. *Hunting Sports In The West.* Philadelphia, 1859.

Hearne, Samuel. *A Journey from Prince of Wale's Fort in 1770, 1771, and 1772.* Toronto, 1911.

Hennepin, Louis. *A New Discovery of a Vast Country in America,* edited by Reuben G. Thwaites. Chicago, 1903.

Herbert, Henry William. *Frank Forester's Field Sports of the United States and British Provinces of North America.* 2 vols. New York, 1866.

Hewitt, C. Gordon. *The Conservation of the Wild Life of Canada.* New York, 1921.

Hind, Henry. *Report on the Assiniboine and Saskatchewan Exploring Expedition of 1858.* Toronto, 1859.

Hodge, F. W. (ed.). *Handbook of the American Indians North of Mexico.* 2 vols. Washington, D.C., 1910.

Hornaday, William T. *The Minds and Manners of Wild Animals.* New York, 1922.

———. *Our Vanishing Wild Life.* New York, 1913.

———. *Thirty Years War for Wild Life.* Stamford, 1931.

———. *A Wild Animal Round-Up.* New York, 1925.

Hunter, J. Marvin (ed.). *The Trail Drivers of Texas.* 2 vols. San Antonio, 1920-1923.

Hunter, John D. *Manners and Customs of Indian Tribes*. Philadelphia, 1823.

Inman, Henry. *Buffalo Jones' Forty Years of Adventure*. Topeka, 1899.

————. *The Great Salt Lake Trail*. New York, 1898.

————. *The Old Santa Fe Trail*. New York, 1897.

————. *Stories of the Old Santa Fe Trail*. Kansas City, 1881.

Innis, Harold A. *The Fur Trade in Canada*. New Haven, 1930.

Irving, Washington. *The Adventures of Captain Bonneville*. Philadelphia, 1849.

————. *A Tour of the Prairies*. London, 1835.

Jennings, Dana C. and Judi Hebbring. *Buffalo Management & Marketing*. Custer, S.D., 1983.

Kalm, Peter. *Travels in North America*. 3 vols. London, 1771.

Kennerly, William Clark: *Persimmon Hill: A Narrative of the Far West*. Norman, 1948.

Lathrop, Amy. *Tales of Western Kansas*. Norton, Kansas, 1948.

Lavender, David. *Bent's Fort*. New York, 1954.

Lee, Wayne C. *Scotty Philip: The Man Who Saved the Buffalo*. Caldwell, Idaho, 1974.

Lemmon, G. E. *Developing the West* [reprints from *Belle Fourche Bee*]. N.p., n.d.

Long Lance, Buffalo Child. *Chief Buffalo Child Long Lance*. New York, 1928.

Mails, Thomas. *The Mystic Warriors of the Plains*. Garden City, 1972.

Marcy, Randolph B. *Thirty Years of Army Life on the Border*. New York, 1866.

Martin, Cy. *The Saga of the Buffalo*. New York, 1973.

Mayer, Frank H., and Roth, Charles B. *The Buffalo Harvest*. Denver, 1958.

McCoy, Joseph G. *Historic Sketches of the Cattle Trade of the West and Southwest*. Kansas City, 1874.

McCreight, Major I. *Buffalo Bone Days*. DuBois, Pennsylvania, 1950.

McDonald, Jerry N. *North American Bison, Their Classification and Evolution*. Berkeley, 1981.

McGowan, Dan. *Animals of the Canadian Rockies*. New York, 1936.

McHugh, Tom. *The Time of the Buffalo*. New York, 1972.

Mitchell, William A. *Linn County, Kansas: A History*. Kansas City, 1928.

Mumey, Nolie. *Rocky Mountain Dick*. Denver, 1953.

Murray, Joan. *The Last Buffalo: The Story of Frederick Arthur Verner, Painter of the Canadian West*. Toronto, 1984.

Myres, S. D. (ed.). *Pioneer Surveyor, Frontier Lawyer: The Personal Narrative of O. W. Williams, 1877-1902*. El Paso, 1966.

Paine, Bayard H. *Pioneers, Indians and Buffaloes.* Curtis, Nebraska, 1935.

Palliser, John. *Journals, Detailed Reports, and Observations Relative to Palliser's Exploration of British North America.* London, 1863.

————. *Solitary Rambles and Adventures of a Hunter in the Prairies.* London, 1853; Rutland, Vermont, 1969 reprint.

Parkman, Francis. *The Oregon Trail.* Boston, 1892.

Powell, Peter J. *Sweet Medicine: The Continuing Role of the Sacred Arrows, the Sun Dance, and the Sacred Buffalo Hat in Northern Cheyenne History.* 2 vols. Norman, 1969.

Powers, Alfred (ed.). *Buffalo Adventures on the Western Plains.* Portland, 1945.

Purchas, Rev. Samuel. *Hakluytus Posthumus, or Purchas his Pilgrimes: Contayning a History of the World in Sea Voyages and Lande Travells by Englishmen and Others.* 20 vols. Glasgow, 1905-1907.

Rath, Ida Ellen. *The Rath Trail.* Wichita, 1961.

Renner, Frederic G., and Yost, Karl. *A Bibliography of the Published Works of Charles M. Russell.* Lincoln, Nebraska, 1971.

Ridings, Sam P. *The Chisholm Trail.* Guthrie, Oklahoma, 1936.

Roe, Frank Gilbert. *The Indian and the Horse.* Norman, 1955.

————. *The North American Buffalo.* Toronto, 1951.

Roenigk, Adolph. *Pioneer History of Kansas.* Denver, 1933.

Roosevelt, Theodore. *The Wilderness Hunter.* New York, 1921.

————. *The Winning of the West.* 4 vols. New York, 1904.

Root, Frank. *The Overland Stage to California.* Topeka, 1901.

Rorabacher, J. Albert. *The American Buffalo in Transition.* Saint Cloud, Minnesota, 1970.

Ross, Alexander. *The Red River Settlement.* London, 1856.

Rush, W. M. *Wild Animals of the Rockies.* New York, 1942.

Russell, Don. *The Lives and Legends of Buffalo Bill.* Norman, 1960.

Ruxton, George F. *Adventures in Mexico and the Rocky Mountains.* New York, 1848.

Rye, Edgar. *The Quirt and the Spur.* Chicago, 1909.

Sage, Rufus B. *Rocky Mountain Life; or Startling Scenes and Perilous Adventures in the Far West, During an Expedition of Three Years.* Boston, 1859.

Salmon, T. *The Present State of Virginia.* London, 1737.

Sandoz, Mari. *The Buffalo Hunters.* New York, 1954.

Schoolcraft, Henry R. *Information Respecting the History, Conditions and Prospects of the Indian Tribes of the United States.* 4 vols. Philadelphia, 1854.

Schultz, James Willard. *Blackfeet and Buffalo: Memories of Life among the Indians.* Norman, 1962.

Seton, Ernest Thompson. *Life-Histories of Northern Animals.* 2 vols. New York, 1910.

―――. *Lives of Game Animals.* 4 vols. New York, 1929; Boston, 1953 reprint.

―――. *Studies in the Art Anatomy of Animals.* London, 1896.

Shields, G. O. *Hunting in the Great West.* Chicago, 1884.

Smith, Winston O. *The Sharps Rifle: Its History, Development and Operation.* New York, 1943.

Stoke, Will E. *Episodes of Early Days in Central and Western Kansas.* Great Bend, Kansas, 1926.

Streeter, Floyd B. *Prairie Trails and Cow Towns.* Boston, 1936.

Thwaites, Reuben G. (ed.). *Early Western Travels, 1748-1846.* 32 vols. Cleveland, 1904-1907.

Tibbles, Thomas H. *Buckskin and Blanket Days.* New York, 1957.

Tilden, Freeman. *Following the Frontier with F. Jay Haynes, Pioneer Photographer of the Old West.* New York, 1964.

Toepperwein, Fritz A. (ed). *Footnotes of the Buckhorn.* Boerne, Texas, 1960.

Townsend, Thomas. *The History of the Conquest of Mexico by the Spaniards* (translated from original Spanish of Don Antonio De Solis). London, 1753.

Townshend, F. Trench. *Ten Thousand Miles of Travel, Sport, and Adventure.* London, 1869.

Vasey, George. *The Natural History of Bulls, Bisons, and Buffaloes.* London, 1851.

VegLahn, Nancy. *The Buffalo King.* New York, 1971.

Vestal, Stanley. *Queen of the Cowtowns: Dodge City.* New York, 1952.

Webb, W. E.: *Buffalo Land.* Philadelphia, 1872.

Webb, Walter P. *The Great Plains.* New York, 1931.

Weekes, Mary. *The Last Buffalo Hunter.* New York, 1939.

Westerners, The. *New York Posse Brand Book.* New York, 1971.

Wheeler, Homer W. *Buffalo Days.* Indianapolis, 1923.

Whitney, Casper; Grinnell, George Bird; and Wister, Owen. *Musk-Ox, Bison, Sheep and Goat.* New York, 1904.

Winsor, Justin. *Narrative and Critical History of America.* 8 vols. Boston, 1887.

Wissler, Clark. *Indians of the United States.* New York, 1940.

Wright, Robert M. *Dodge City, the Cowboy Capital.* Wichita, 1913.

ARTICLES

"About Buffalo." *Forest and Stream,* XLV, November 1895.

Andrews, E. N. "A Buffalo Hunt by Rail." *Kansas Magazine,* May 1873.

Badger, George A. "Recollections of George Andrew Gordon." *Kansas Historical Collections,* XVI, 1923-1925.

Barnett, LeRoy. "The Buffalo Bone Commerce on the Northern Plains." *North Dakota History,* XXXIX, Winter 1972.

Beeson, Chalkley M. "A Royal Buffalo Hunt." *Kansas Historical Collections,* X, 1907-1908.

Biggers, Don H. "Buffalo Butchery in Texas was National Calamity." *Texas Farm and Ranch,* November 14, 1925.

"The Buffalo Hunt." *Harper's Weekly,* XI, July 6, 1867.

Burlingame, Merrill G. "The Buffalo in Trade and Commerce." *North Dakota Historical Quarterly,* III, July 1929.

Cauley, T. J. "What the Passing of the Buffalo Meant to Texas." *Cattleman,* XIV, December 1929.

Clark, J. S. "A Pawnee Buffalo Hunt." *Chronicles of Oklahoma,* XX, December 1942.

Connelley, William E. "Life and Adventures of George W. Brown." *Kansas Historical Collections,* XVII, 1926-1928.

Davis, Theodore R. "The Buffalo Range." *Harper's Monthly,* XXXVII, January 1869.

Day, James M. "A Preliminary Guide to the Study of Buffalo Trails in Texas." *West Texas Historical Association Year Book,* 1960.

Draper, Benjamin. "Where the Buffalo Roamed." *Pacific Discovery,* March-April 1950.

Dumont, Ella. "A Woman on the Buffalo Range: The Journal of Ella Dumont." *West Texas Historical Association Year Book,* 1964.

Dunbar, Gary S. "Some Notes on Bison in Early Virginia." Virginia Place-Name Society *Occasional Papers.* January 30, 1962.

Dunbar, John B. "Journal of John Dunbar." *Kansas Historical Collections,* XIV, 1915-1918.

———. "The White Man's Foot in Kansas." *Kansas Historical Collections,* X, 1907-1908.

Egan, Gail N., and Agogino, George A. "Man's Use of Bison and Cattle on the High Plains." *Great Plains Journal,* Fall 1965.

Forbis, Richard G. "Early Man and Fossil Bison." *Science,* CXXIII, February 24, 1956.

Fouquet, L. C. "Buffalo Days." *Kansas Historical Collections,* XVI, 1923-1924.

Fryxtell, F. M. "The Former Range of the Bison in the Rocky Mountains." *Journal of Mammology,* IX, May 1928.

Gallatin, Albert. "A Synopsis of the Indian Tribes of North America." American Antiquarian Society *Transactions and Collections,* II, 1836.

Gibson, Gail M. "Historical Evidence of the Buffalo in Pennsylvania." *The Pennsylvania Magazine,* April 1969.

"Goodnight Ranch." *Texas Farm and Ranch,* XXII, April 4, 1903.

Graham, I. D. "The Kansas State Board of Agriculture: Some Highlights of History." *Kansas Historical Collections,* XVII, 1926-1928.

Grinnell, George Bird. "Bent's Old Fort and Its Builders." *Kansas Historical Collections,* XV, 1919-1922.

————. "The Last of the Buffalo." *Scribner's Magazine,* XII, September 1892.

Henderson, John G. "The Former Range of the Buffalo." *The American Naturalist,* VII, 1872.

Holden, W. C. "The Buffalo of the Plains Area." *West Texas Historical Association Year Book,* II, June 1926.

Hough, Emerson. "Forest and Stream's Yellowstone Park Game Exploration." *Forest and Stream,* XLIII, May 5 to August 25, 1894.

Hunt, James W. "Buffalo Days, The Chronicle of An Old Buffalo Hunter." *Holland's Magazine,* five-part series, January, February, March, April and May 1933.

"Journal of John Lowery Brown . . . enRoute to California in 1850." *Chronicles of Oklahoma,* XII, June 1934.

Kidder, John. "Montana Miracle: It Saved the Buffalo." *Montana, the Magazine of Western History,* XV, Spring 1965.

McClernand, Edward J. "With the Indians and Buffalo in Montana." *Cavalry Journal,* No. 36, Jan.–April 1927.

McKinley, Daniel. "The American Bison in Pioneer Missouri." *The Bluebird,* XXXI, 1, 2, 1964.

McLaren, Allie, "A Texas Buffalo Herd." *Outdoor Life,* XXVI.

Malouf, Carling, and Conner, Stuart (eds.). "Symposium on Buffalo Jumps." Montana Archaeological Society *Memoir,* No. 1, May 1962.

"Man and Beast with Same Hirsute Motif." *The Plainsman,* VIII, Summer 1962.

Martin, Handel T. "On the Occurence of *Biston latifrons* in Comanche County, Kansas." University of Kansas *Science Bulletin,* XVII, September 1927.

Mayer, Frank H. "The Rifles of Buffalo Days." *American Rifleman,* September and October 1934.

Merriam, C. Hart. "The Buffalo in Northeastern California." *Journal of Mammalogy,* VII, August 1926.

Mills, John. "A Brush With The Bison." *Harper's Monthly,* July 1851.

"Monarch of the Old West." *The Humble Way,* April 1962.

Mooar, J. Wright. "The First Buffalo Hunting in the Panhandle." *West Texas Historical Association Year Book,* 1930.

Moore, Ely. "A Buffalo Hunt with the Miamis in 1854." *Kansas Historical Collections,* X, 1907-1908.

Moorehead, Warren K. "The Passing of Red Cloud." *Kansas Historical Collections,* X, 1907-1908.

Murphy, Carrol D. "What the Buffalo Offers Us." *Texas Farm and Ranch,* XXIX, December 30, 1910.

"Narrative of Sir George Gore's Expedition." *Contributions,* Montana Historical Society, I, 1876.

"Notes On A Buffalo Hunt: The Diary of Mordecai Bartram." *Nebraska Historical Society Collection,* XXXV, 1954.

Nye, Wilbur S. "The Annual Sun Dance of the Kiowa Indians." *Chronicles of Oklahoma,* XII, September 1934.

Patten, James I. "Buffalo Hunting with the Shoshone Indians in 1874." *Annals of Wyoming,* IV, October 1926.

Philip, George. "South Dakota Buffaloes Versus Mexican Bulls." *South Dakota Historical Review,* II, January 1937.

Power, James B. "Bits of History Connected with the Early Days of the Northern Pacific Railroad." North Dakota State Historical Society *Collections,* III, 1910.

Rees, D. S. "An Indian Fight on the Solomon." *Kansas Historical Collections,* VII, 1901-1902.

Riggs, Thomas L. "The Last Buffalo Hunt." *Independent,* LXIII, July 4, 1907. Reprinted as a booklet in 1935.

Rister, C. C. "The Significance of the Destruction of the Buffalo in the Southwest." *Southwestern Historical Quarterly,* XXXIII, July 1929.

Russell, Hamlin. "The Story of the Buffalo." *Harper's New Monthly,* April 1893.

Sanders, Alvin H. "The Taurine World." *National Geographic,* December 1925.

Satterthwaite, Franklin. "The Western Outlook For Sportsmen." *Harper's New Monthly,* May 1889.

Scott, Charles. "The Old Road and Pike's Pawnee Villages." *Kansas Historical Collections,* XVII, 1926-1928.

Seton, Ernest T. "The American Buffalo or Bison." *Scribner's Magazine,* October 1906.

Skinner, Morris F., and Kaiscn, Ove C. "The Fossil Bison of Alaska and Preliminary Revision of the Genus." American Museum of Natural History *Bulletin,* LXXXIX, Art. 3, 1947.

Smith, J. R. "A Diary Account of a Creek Boundary Survey." *Chronicles of Oklahoma,* XXVII, 1949.

Snell, Joseph (ed.). "Diary of Henry Raymond." *Kansas Historical Quarterly,* XXXI, Winter 1965.

Soper, J. Dewer. "History, Range, and Home Life of the Northern Bison." *Ecological Monographs,* XI, October 1941.

Spencer, Rev. Joab. "The Kaw or Kansas Indians: Their Customs, Manners, Folk-Lore." *Kansas Historical Collections,* X, 1907-1908.

Street, William D. "The Victory of the Plow." *Kansas Historical Collections,* IX, 1905-1906.

Strickland, Rex W. (ed.). "The Recollections of W. S. Glenn, Buffalo Hunter." *Panhandle-Plains Historical Review,* XX, 1949.

Thompson, W. F. "Peter Robidoux: A Real Kansas Pioneer." *Kansas Historical Collections,* XVII, 1926-1928.

Vanderhoof, V. L. "A Skull of *Bison latifrons* from the Pleistocene of Northern California." University of California *Publications in Geological Science,* 1942.

Wallace, Lew. "A Buffalo Hunt in Northern Mexico." *Scribner's Monthly,* XVII, March 1879.

Weichselbaum, Theodore. "Statement of Theodore Weichselbaum, of Ogden, Riley County, July 17, 1908." *Kansas Historical Collections,* XI, 1909-1910.

Wheat, Joe Ben. "The Olsen-Chubbuck Site: A Paleo-Indian Bison Kill." *American Antiquity,* XXXVII, Part 2, January 1972.

————. "A Paleo-Indian Bison Kill." *Scientific American,* January 1967.

Wright, Robert M. "Personal Reminiscences of Frontier Life in Southwest Kansas." *Kansas Historical Collections,* VII, 1901-1902.

GOVERNMENT DOCUMENTS

Allen, Joel A. "History of the American Bison." *Ninth Annual Report* of the U.S. Geological and Geographical Survey for 1875, Washington, D.C., 1877.

"The American Buffalo." U.S. Department of the Interior, *Conservation Note,* No. 12 (revised), Washington, D.C.

Bison Facts in Relation to the Preservation of American Bison in the U.S. and Canada. Senate Document 445, 57th Congress, 1st session, vol. 30, Washington, D.C., July 1, 1902.

Congressional Globe, 1871, 1872.

Congressional Record, 1874, 1876, 1881, 1905, 1906, 1908, 1912.

Ewers, John C. "The Horse in Blackfoot Indian Culture." Smithsonian Institution *Bureau of American Ethnology Bulletin,* No. 159, Washington, D.C., 1955. The Ewers section of the *Bulletin* (pp. 1-374) was subsequently published as a separate book (with original pagination), Washington, D.C., 1969.

Fuller, W. A. "The Biology and Management of the Bison of Wood Buffalo National Park." *Wildlife Management Bulletin,* Series 1, No. 16, Canadian Wildlife Service, Ottawa, 1966.

Hay, Oliver P. *The Extinct Bisons of North America with Descriptions of One New Species, Bison Requis,* Washington, D.C., 1913.

Hornaday, William T. "The Extermination of the American Bison." Smithsonian Institution *Annual Report* for 1887, Washington, D.C. 1889. The Hornaday section of the *Report* (pp. 367-584) was subsequently bound as a separate book (with original pagination).

————. "How to Collect Mammal Skins for Purposes of Study and for Mounting." Smithsonian Institution *Annual Report* for 1886, Part II, Washington, D.C., 1889.

Logan, V. S., and Sylvestre, P. E. *Hybridization of Domestic Beef Cattle and Buffalo.* Canada Department of Agriculture, Ottawa, 1950.

National Bison Range. U.S. Department of the Interior, Washington, D.C., 1967.

Palmer, T. S. "Chronology and Index of the More Important Events in American Game Protection, 1776-1911." U.S. Department of Agriculture, *Biological Survey Bulletin,* No. 41. Washington, D.C., 1912.

Peters, H. F. *Experimental Hybridization of Domestic Cattle and American Bison.* Canada Department of Agriculture, Manyberries, Alberta, 1964.

————. *Experience With Yak Crosses in Canada.* Canada Department of Agriculture, Ottawa, 1968.

Peters, H. F., and Slen, S. B. *Brahman-British Beef Cattle Crosses in Canada.* Canada Department of Agriculture, Lethbridge, Alberta, 1966.

Report of the Acting Superintendent of the Yellowstone National Park to the Secretary of the Interior, 1894. Washington, D.C., 1895.

PAMPHLETS

Armstrong, T. R. *My First Buffalo Hunt.* N.p., 1918.

Burdick, Usher L. *Tales From Buffalo Land: The Story of Frederick Zahl.* N.p., 1938.

Collins, Henry H., Jr. *1951 Census of Living American Bison.* Bronxville, New York, 1952.

————. *The Unvanquished Buffalo.* Bronxville, 1952.

Fleming, H. A., and Company. *Goodnight's American Buffalo.* Dallas, 1910.

Garretson, Martin S. *A Short History of the American Bison.* New York, 1927, 1934 (revised).

Hill, J. L. *The Passing of the Indian and Buffalo.* Long Beach, California, n.d.

[Jones, Tom.] *The Last of the Buffalo, A History of the Buffalo Herd of the Flathead Reservation, and An Account of the Great Round Up.* Cincinnati, 1909.

Log Cabin Days. Manhattan, Kansas, 1929.

Loring, J. Alden. *The Wichita Buffalo Range.* New York Zoological Society, New York, 1906.

McCreight, Major I. *Buffalo Bone Days.* Sykesville, Pennsylvania, 1939.

Shoemaker, Henry W. *A Pennsylvania Bison Hunt.* Middleburg, Pennsylvania, 1915.
Smythe, Bernard Bryan. *The Heart of the New Kansas.* Great Bend, Kansas, 1880.
Thomson, Frank. *Last Buffalo of the Black Hills.* N.p., 1962.

THESES

Burlingame, Merrill G. "The Economic Importance of the Buffalo in the Northern Plains Region, 1880-1890." State University of Iowa, 1928.
Crockett, Glenn. "The Life of William Mathewson, the Original Buffalo Bill." Wichita University, 1932.
Dolph, James A. "The American Bison Society: Preserver of the American Buffalo and Pioneer in Wildlife Conservation." University of Denver, 1965.
Edgerton, Penelope Jane Marjoribanks. "The Cow-Calf Relationship and Rutting Behavior in the American Bison." University of Alberta, Edmonton, 1962.
Grieder, Theodore G. "The Influence of the American Bison or Buffalo on Westward Expansion." State University of Iowa, 1928.

MANUSCRIPTS

American Bison Society. Papers covering the period 1905-1949, Conservation Center, Denver Public Library, Denver, Colorado.
Berlandier, Jean Louis. Narrative, Library of Congress, Washington, D.C.
Bryant, Peter. Letters, Kansas State Historical Society, Topeka.
Davis, Bon V. "Get Off the Earth: The Life of Vern Bookwalter," College, Alaska.
Dixon, William. Papers, Panhandle-Plains Historical Museum, Canyon, Texas.
Dollar, Clyde. "The Making of 'A Man Called Horse'," Vermillion, South Dakota.
Drummond, Earl. In possession of Edwin Drummond, Cache, Oklahoma.
Garretson, Martin S. Letter, Kansas State Historical Society, Topeka.
Gauthier, Carl P. "The Bone Trail," Santa Ana, California.
Gilman, Musetta. "Pump on the Prairie: Story of the Gilman Road Ranch," Lincoln, Nebraska.
Graf, J. E. Letter, Bureau of Engraving and Printing, U.S. Treasury Department, Washington, D.C.
Knight, Charles R. Letter, Bureau of Engraving and Printing, U.S. Treasury Department, Washington, D.C.

McCombs, Joe S. Narrative, University of Texas Library, Austin.
McKinley, J. W. Narrative, Panhandle-Plains Historical Museum, Canyon, Texas.
Mooar, J. Wright. Interviews. Transcript in Vandale Collection, University of Texas Library, Austin; microfilm copy at Kansas State Historical Society, Topeka.
Pitcher, Major John. Letter, archives, Yellowstone National Park.
Raymond, H. H. Diary, 1872-1873, Kansas State Historical Society, Topeka.
Trautwine, John Hannibal. Diary of a buffalo hunt in Kansas, 1873, Kansas State Historical Society, Topeka.
Wilson, Wenton. Account of a buffalo hunt in 1876, Kansas State Historical Society, Topeka.

NEWSPAPERS

Alma Signal, Kansas
Beatrice Express, Nebraska
Boston Evening Transcript, Massachusetts
Brown County World, Hiawatha, Kansas
Capper's Weekly, Topeka, Kansas
Catalina Islander, Avalon, California
Commercial Gazette, Wyandotte, Kansas
Denver News, Colorado
Denver Republican, Colorado
Denver Tribune, Colorado
Dodge City Times, Kansas
El Dorado News, Kansas
Ellis County News, Hays, Kansas
Forest and Stream, New York
Hutchinson Herald, Kansas
Junction City Statesman, Kansas
Kansas City Enterprise, Missouri
Kansas City Star, Missouri
Kansas Daily Tribune, Lawrence, Kansas
Kansas National Democrat, LeCompton, Kansas
Kansas Press, Cottonwood Falls, Kansas
Kansas Press, Council Grove, Kansas
Kansas State Record, Topeka, Kansas
Kansas Tribune, Topeka, Kansas
Lawrence Journal World, Kansas
Leavenworth Daily Commercial, Kansas
Leavenworth Daily Conservative, Kansas
Leavenworth Times, Kansas
McPherson Daily Republican, Kansas
Missouri Democrat, St. Louis, Missouri

Missouri Republican, St. Louis, Missouri
New York Times, New York, New York
Newton Kansan, Kansas
Niles' National Register, Baltimore, Maryland
Norton Champion, Kansas
Railway Advance, Hays City, Kansas
Salt Lake Tribune, Utah
Sioux City Journal, Iowa
Topeka Capital-Commonwealth, Kansas
Topeka Daily Capital, Kansas
Topeka Daily Commonwealth, Kansas
Topeka Herald, Kansas
Topeka State Journal, Kansas
Turf, Field and Farm, New York, New York
Wakeeney Weekly World, Kansas
Wall Street Journal, New York, New York
Washington Star, Washington, D.C.
Weekly Missoulian, Missoula, Montana
Western Journal of Commerce, Kansas City, Missouri
Wichita Eagle, Kansas
Yellowstone Journal, Miles City, Montana

Acknowledgments

The Notes show my indebtedness to many persons, writers and non-writers. For their aid, I thank them. To certain others who more than fifteen years ago made the first edition possible, I wish to express a special note of gratitude: to Joseph Snell, friend, writer, and historian and retired executive secretary of the Kansas State Historical Society in Topeka, who went out of his way to help me and to provide many answers that otherwise might have been lost along the trail; to the late Frank Gilbert Roe of Cadboro Bay, Victoria, British Columbia, a leading authority on the wild North American buffalo, who offered helpful suggestions and advice; to the late Wayne Gard of Dallas, Texas, who traveled many of these same trails a decade or more before me and who retraced many to assist me; and to N. S. Novakowski then of the Canadian Wildlife Service, Ottawa, whose assistance was above and beyond the call.

In addition there was the late Nyle Miller, long-time executive secretary of the Kansas State Historical Society; the staff of the Oklahoma Historical Society; Jack Haley, now retired, of the Western History Collections, University of Oklahoma; Chester V. Kielman and Miss Llerena Friend of the University of Texas; and Mrs. Willie Belle Coker, long-time curator of the J. Frank Dobie Collection at the University of Texas, who assisted me in using the late Texas author's outstanding library in my research.

Appreciation goes also to many people with whom I have lost contact in the intervening years including C. Boone McClure, formerly of the Panhandle-Plains Historical Society, Canyon, Texas; James E. Hartmann of the Colorado State Historical Society; Mrs. Roberta Winn and Miss Kay Collins, Conservation Library Center, Denver Public Library; John Aubrey and Robert W. Karrow, Jr., Newberry Library, Chicago; and the late Mary K. Dempsey, long-time librarian at the Montana State Historical Society, Helena.

But not all assistance came from the hallowed halls. In buffalo country there was C. J. Henry, Coot Haven, Charlo, Montana, retired manager of the National Bison Range in Montana, who helped greatly, as did Marvin R.

Kaschke and Victor "Babe" May of the National Bison Range. Of tremendous help were Melvin Evans, Elmer Parker, D. W. Jackson, Edwin Drummond, and Julian Howard all once associated with the Wichita Mountains Wildlife Refuge in southwestern Oklahoma. Together these men had more than eighty years of experience in handling buffalo. And there were other buffalo men and a few buffalo women in forty states and several Canadian provinces who either owned buffalo or managed herds for state or national governments. For their help, thanks.

I must not forget western bookmen; Bob Marsh, Topeka, Kansas; Eric Lunberg, Ashton, Maryland; and Fred Rosenstock, Denver, Colorado, all of whom have since crossed the great divide. And there is Jeff Dykes, College Park, Maryland, who still helps the cause by uncovering numerous old books, articles, and pamphlets dealing with the buffalo.

Thanks also to Lurton Blassingame, who has now retired to his native South, and the late John B. Bremner of the University of Kansas. The same appreciation is felt for the efforts of Durrett Wagner of Chicago whose advice and guidance was greatly appreciated.

Finally, a warm "thanks" to my wife, Carolyn Sue Dary, who carefully read each word of the first edition and the revisions for this new edition, offered helpful suggestions, and still puts up with the buffalo as almost another member of our family.

David Dary

Index

On Thanksgiving Day 1925 Charles Russell wrote a letter to Ralph Budd, president of the Great Northern Railway:

> turkey is the emblem of this day and it should be in the east but the west owes nothing to that bird but it owes much to the humped backed beef in the sketch above
> the Rocky mountains would have been hard to reach with out him. . . . he was one of natures bigest gift and this country owes him thanks. . . . [*Good Medicine,* 32]